T0339901

INTELLIGENT VEHICLES

INTELLIGENT VEHICLES

Enabling Technologies and Future Developments

Edited by

FELIPE JIMÉNEZ

Universidad Politécnica de Madrid, Madrid, Spain

Butterworth-Heinemann
An imprint of Elsevier

Butterworth-Heinemann is an imprint of Elsevier
The Boulevard, Langford Lane, Kidlington, Oxford OX5 1GB, United Kingdom
50 Hampshire Street, 5th Floor, Cambridge, MA 02139, United States

Notices
Knowledge and best practice in this field are constantly changing. As new research and experience broaden
our understanding, changes in research methods, professional practices, or medical treatment may become
necessary.

Practitioners and researchers must always rely on their own experience and knowledge in evaluating and
using any information, methods, compounds, or experiments described herein. In using such information or
methods they should be mindful of their own safety and the safety of others, including parties for whom they
have a professional responsibility.

To the fullest extent of the law, neither the Publisher nor the authors, contributors, or editors, assume any
liability for any injury and/or damage to persons or property as a matter of products liability, negligence or
otherwise, or from any use or operation of any methods, products, instructions, or ideas contained in the
material herein.

British Library Cataloguing-in-Publication Data
A catalogue record for this book is available from the British Library

Library of Congress Cataloging-in-Publication Data
A catalog record for this book is available from the Library of Congress

ISBN: 978-0-12-812800-8

For Information on all Butterworth-Heinemann publications
visit our website at https://www.elsevier.com/books-and-journals

Working together
to grow libraries in
developing countries

ELSEVIER Book Aid International

www.elsevier.com • www.bookaid.org

Publisher: Matthew Deans
Acquisition Editor: Carrie Bolger
Editorial Project Manager: Carrie Bolger
Production Project Manager: Kiruthika Govindaraju
Cover Designer: Mark Rogers

Typeset by MPS Limited, Chennai, India

CONTENTS

5.   Big Data in Road Transport and Mobility Research   **175**

Sergio Campos-Cordobés, Javier del Ser, Ibai Laña, Ignacio (Iñaki) Olabarrieta,
Javier Sánchez-Cubillo, Javier J. Sánchez-Medina and Ana I. Torre-Bastida

Part II Applications   **207**

8.   Automated Driving   275

Jorge Villagra, Leopoldo Acosta, Antonio Artuñedo, Rosa Blanco,
Miguel Clavijo, Carlos Fernández, Jorge Godoy, Rodolfo Haber,
Felipe Jiménez, Carlos Martínez, José E. Naranjo, Pedro J. Navarro,
Ana Paúl and Francisco Sánchez

Part III Additional Aspects   343

9.   Human Factors   345

Subchapter 9.1 Human Driver Behaviors   346

Luis M. Bergasa, Enrique Cabello, Roberto Arroyo, Eduardo Romera
and Ángel Serrano

Subchapter 9.2 User Interface   384

Alfonso Brazález, Olatz Iparraguirre and Joshua Puerta

10. Simulation Tools   395

Subchapter 10.1 Driving Simulators   396

Alfonso Brazález, Luis Matey, Borja Núñez, and Ana Paúl

LIST OF CONTRIBUTORS

Leopoldo Acosta
Universidad de La Laguna, Santa Cruz de Tenerife, Spain

Jorge Alfonso
Universidad Politécnica de Madrid, Madrid, Spain

Nourdine Aliane
Universidad Europea de Madrid, Villaviciosa de Odón, Spain

Arrate Alonso
Mondragon Unibertsitatea, Mondragón, Spain

José M. Armingol
UC3M, Leganés, Spain

Rafael Arnay
Universidad de La Laguna, Santa Cruz de Tenerife, Spain

Roberto Arroyo
Universidad de Alcalá, Alcalá de Henares, Spain

Antonio Artuñedo
CSIC, Madrid, Spain

Luis M. Bergasa
Universidad de Alcalá, Alcalá de Henares, Spain

Rosa Blanco
CTAG - Centro Tecnológico de Automoción de Galicia, Porriño, Spain

Alfonso Brazález
Ceit-IK4, San Sebastián, Spain

Enrique Cabello
Universidad Rey Juan Carlos, Móstoles, Spain

Sergio Campos-Cordobés
TECNALIA, Bizkaia, Spain

Miguel Clavijo
Universidad Politécnica de Madrid, Madrid, Spain

Arturo de la Escalera
UC3M, Leganés, Spain

Javier del Ser
TECNALIA, University of the Basque Country (UPV/EHU) and Basque Center for Applied Mathematics (BCAM), Bizkaia, Spain

Carlos Fernández
Universidad Politécnica de Cartagena, Cartagena, Spain

Javier Fernández
Universidad Europea de Madrid, Villaviciosa de Odón, Spain

José A. Fernández
CTAG – Centro Tecnológico de Automoción de Galicia, Porriño, Spain

Fernando García
UC3M, Leganés, Spain

Jorge Godoy
CSIC, Madrid, Spain

Rodolfo Haber
CSIC, Madrid, Spain

Olatz Iparraguirre
Ceit-IK4, San Sebastián, Spain

Felipe Jiménez
Universidad Politécnica de Madrid, Madrid, Spain

Ibai Laña
TECNALIA, Bizkaia, Spain

Antonio M. López
CVC-UAB, Barcelona, Spain

Maria E. López-Lambas
Universidad Politécnica de Madrid, Madrid, Spain

David Martín
UC3M, Leganés, Spain

Carlos Martínez
Universidad Politécnica de Madrid, Madrid, Spain

Mario Mata
Universidad Europea de Madrid, Villaviciosa de Odón, Spain

Luis Matey
Ceit-IK4, San Sebastián, Spain

José M. Menéndez
Universidad Politécnica de Madrid, Madrid, Spain

Basam Musleh
ABALIA, Madrid, Spain

José E. Naranjo
Universidad Politécnica de Madrid, Madrid, Spain

Pedro J. Navarro
Universidad Politécnica de Cartagena, Cartagena, Spain

Borja Núñez
Ceit-IK4, San Sebastián, Spain

Ignacio (Iñaki) Olabarrieta
TECNALIA, Bizkaia, Spain

Ana Paúl
CTAG – Centro Tecnológico de Automoción de Galicia, Porriño, Spain

Joshua Puerta
Ceit-IK4, San Sebastián, Spain

Eduardo Romera
Universidad de Alcalá, Alcalá de Henares, Spain

Francisco Sánchez
CTAG – Centro Tecnológico de Automoción de Galicia, Porriño, Spain

Javier Sánchez-Cubillo
TECNALIA, Bizkaia, Spain

Javier J. Sánchez-Medina
Universidad Las Palmas de Gran Canaria, Las Palmas de Gran Canaria, Spain

Ángel Serrano
Universidad Rey Juan Carlos, Móstoles, Spain

Rafael Toledo-Moreo
Universidad Politécnica de Cartagena, Cartagena, Spain

Ana I. Torre-Bastida
TECNALIA, Bizkaia, Spain

David Vázquez
CVC-UAB, Barcelona, Spain

Jorge Villagra
CSIC, Madrid, Spain

Gabriel Villalonga
CVC-UAB, Barcelona, Spain

PREFACE

The development of road vehicles is unstoppable and disciplines such as Electronics, Control, and Communications have increasingly important roles in this advance. This has opened up numerous fields of research and development that, although they end up being interconnected, correspond to very diverse areas. Therefore, there is often a lack of a comprehensive and integrated vision of the intelligent vehicle.

With the intention of bringing together research groups in Spain dealing with aspects of the intelligent vehicles from different perspectives, the Thematic Network on Intelligent Vehicles (RETEVI) has been launched, funded by the Ministry of Economy and Competitiveness of Spain (TRA2015-69002-REDT). This national network, led by the University Institute for Automobile Research (INSIA) of the Technical University of Madrid (UPM), brings together a large set of research groups active in transport systems and intelligent vehicles: Carlos III University of Madrid, University of Alcalá, Spanish National Research Council (CSIC), Technical University of Cartagena, Computer Vision Center & Autonomous University of Barcelona, TECNALIA Research & Innovation Foundation, Ceit-IK4 Technology Center, Galician Center for Automotive Technology (CTAG), University of La Laguna, University of Las Palmas de Gran Canaria, University Rey Juan Carlos, Complutense University of Madrid, European University of Madrid, and other UPM groups. This plurality of partners allows taking into consideration the main methodologies and technological tools relevant to ITS research (such as autonomous vehicles, intervehicle communications, cooperative services, and vehicle sensing) in the workings of the Thematic Network. The Network's mission is to create a framework for joint action, to promote the organization of technical and informative events, to write scientific papers, and to coordinate a shared participation in national or international fora.

This book appears as one of the initiatives of the Thematic Network, although these iniatives are not restricted to the work of the Thematic Network alone, since having opened the possibility of collaboration with other groups that, due to different circumstances, were not officially integrated in the Network, nevertheless, these other groups can contribute with their extensive knowledge in some of the research areas involved.

Firstly, it should be noted that this book focuses on intelligent vehicles, not on intelligent transport systems in general, which would be a much broader field. On the other hand, although we will focus especially on technical aspects, mention will be made of other aspects of a less technical nature such as economic, political, social, etc., which have a clear influence on the development and implementation of all systems that end up constituting what we call intelligent vehicles. We believe that an overall view of these should not neglect these collateral aspects and that, even from a technical perspective, they should be considered.

As noted above, the development of intelligent vehicles has implied that new work fields are involved. Thus, there are many ramifications that can escape the attention of a researcher of a specific area, mainly when initiating the research activities or when focusing in a more closed scope. However, it is considered essential to have a vision that is as complete as possible, although then specialization is inevitable. This broad vision allows us to better perceive the shortcomings to be overcome and where each development is framed with respect to the intelligent vehicle.

This book is mainly intended for PhD students, postgraduate or masters students, and for researchers in their early stages, in order to obtain in a single work a global vision of intelligent vehicles. It has been designed in such a way that it could be easily adapted for the organization of a complete academic year or a group of subjects. But it is also targeted at more experienced researchers who want to get up-to-date knowledge on other areas related to intelligent vehicles. In this sense, each chapter has been written by relevant representatives of that area, since they are the most qualified to select the essential contents of each research field.

It is also necessary to explain at this point a note about the scope of the book. The book has a clear generalist approach to offer a global vision, since the specific partial visions of each area can be consulted in other specific works, and the most recent advances are more dynamically focused in journals and congresses. However, there has been an attempt to avoid excessively superficial treatment, and preliminary technical details have been provided for the nonspecialist in a particular subject matter but involved in one way or another in the world of intelligent vehicles. In fact, we hope that the details offered will be useful to senior researchers who, focusing on a specific area, want to start in other areas or want to know how some areas link with others. Thus, the concepts presented here should constitute a departure point from which to continue inquiring.

Finally, as the coordinator of this book, I would like to thank all the coauthors and especially those who have led the chapters, as I am aware of the task this has entailed, especially with the clear intention of providing a broad and integrative vision, transmitted by the main experts in each area but, in turn, organized, unified and coherent for the reader. We trust that this book will be of interest and usefulness, at least in the area we had in mind, and will provide a clear global vision of intelligent vehicles.

Felipe Jiménez

CHAPTER 1

Introduction

Felipe Jiménez
Universidad Politécnica de Madrid, Madrid, Spain

Contents

1.1 INTELLIGENT TRANSPORT SYSTEMS (ITS)

Transport has a clear link with economic development. However, the great growth of mobility in recent years, and above all, the preponderance of the mode of road transport over the others, has some associated negative effects in different areas: congestion, pollution, and accidents, mainly. In particular, in the European area:

- Approximately 10% of the road network is affected by daily congestion.
- Transport is the sector that has grown most in energy demand in all sectors, 83% of which corresponds to road transport, which in turn has a direct relationship with CO_2 emissions.
- Vehicles are the main source of pollution in cities and 20% of cities suffer unacceptable levels of noise.

Intelligent Vehicles
DOI: http://dx.doi.org/10.1016/B978-0-12-812800-8.00001-1

- Every year, in the European Union, there are around 4000 deaths and 1.7 million injured in traffic accidents.
- In economic terms, the effects of congestion represent 0.5% of GDP, the environmental impact represents 0.6%, and accidents represent 1.5%.

On the other hand, these negative effects can be expected to continue to rise because of the growth in demand associated with development itself. Given the magnitude of the problems, Governments have set specific targets, such as the reduction of road deaths or the reduction of pollutant emissions, through increasingly restrictive regulations. With regard to congestion, the construction of new infrastructures provides a temporary response, but not enough to achieve transport sustainability. Instead, a policy based on better capacity management offers greater long-term benefits.

The technological and social evolution has led to demands for transport, safety, comfort, reliability, performance, efficiency, etc., that have grown exponentially in recent years. In addition to this pressure, regulatory requirements must be added, as well as the requirements for competitiveness in a global market, where there is a need for constant improvement and renewal, continually adding value to the products and services offered.

Intelligent Transport Systems (ITS), understood as the systems that apply the technologies of Electronics, Automatic Control, Computer, and Communications to the field of transport, are aimed at the reduction of accidents, energy savings, pollution reduction, and the increased efficiency of the transport system as a whole, and can be studied in a very broad way. Thus, it can be extended to all modes of transport and all their elements can be considered: vehicle, infrastructure, and user.

Although ITS may be associated with recent developments in the 21st century, it is not entirely true. Specifically, their beginnings came from the 1980s, with more or less simultaneous initiatives in Europe, the United States, and Japan. Already in those years strategic visions were presented as to how road transport should be taking into account the major problems detected in this mode of transport. This vision already considered the flows of information between all the actors present on the roads, their handling, and their diffusion. However, in those years it was not possible to have a clear vision on the specific technologies that could provide a viable and effective practical deployment. These were later developments both within the field of transport and, above all, outside it, which marked the route to deploying the vision raised years ago.

The concept behind ITS is the use of new technologies in the world of road transport infrastructures and vehicles, in order to reduce the major problems of this mode of transport (accidents, pollution, congestion, etc.) through different solutions to the conventional ones. The acquisition, classification, processing, and use of information are the key elements in intelligent systems and this flow of information, with its positive aspects and its difficulties, gives them a competitive advantage over conventional solutions.

Thus, from a technical point of view, intelligent transport systems can be understood as a chain of information, which includes the acquisition, communication, processing, exchange, distribution, and finally the use of this information. Thus, the following elements related to information flows can be defined, among others:

- Information captured by onboard sensors
- Intravehicular communications (communication buses)
- Vehicles positioning, based primarily on satellite location, but also beacon and other technologies have been developed
- Communication between vehicles (V2V)
- Infrastructure bidirectional communications (V2I)
- Centralized units of information: traffic management, fleet management, emergency systems, etc.
- Communications with such units that receive infrastructure or "floating vehicles" information
- Information provided to transport users (users of public transport, goods tracking, pre-trip information, etc.), fixed or variable, offered at specific points (variable panels on the roads or bus stops, for example) or in the vehicle itself or to the user directly.

It is worth mentioning the aspect of the generic name that these systems are called. The character of "intelligent" must be used with caution for a system that has been devised and materialized by human beings and that, at least at the moment, has little capacity to replace them. Moreover, as the years pass, systems that originally were considered highly "intelligent," are treated in a much more conventional way now we are used to them, while other new systems and services are the ones that now make up the group of intelligent systems.

Finally, it should be noted that the implementation of ITS services, in addition to the technological problems involved, presents difficulties of a social (acceptance), political, and legal nature. Its introduction implies profound cultural changes, among which we can mention the creation of multidisciplinary groups, higher levels of cooperation between companies and administrations, integration of vehicles and infrastructure, etc. In addition, cooperation between the public and private sectors in transport is identified as essential. Thus, it is essential that, from the Governments' (and, sometimes, broader) point of view, the appropriate framework for the development of these services be established.

1.2 EARLY INITIATIVES

Vehicles have undergone a spectacular evolution. It has not been simply an esthetic change, but has been a continuous technological progression. New control systems are increasingly being introduced to improve comfort and safety, firstly in expensive vehicles and, later, by generalizing to the most economical ranges. This flow of information in the control systems motivated the introduction of internal communication buses in vehicles, but the demands of the speed of communication and the volumes of information have led to the incorporation of more than one bus in vehicles at present.

The same situation happens for infrastructures. Several differences can be found between roads of decades ago and the current high capacity roads equipped with sensors, variable information panels, cameras, etc. The advances are not so evident on conventional roads (which has its reflection in the high number accidents of these roads). The progress in urban areas has also been remarkable (although often not as visible as in the case of vehicles or high capacity roads), but it must be taken into account that traffic management (in its most basic versions) has been present for many years, even before ITS systems were a reality or even a project.

As indicated above, it was in the 1980s that the need for major changes in road transport began to be raised by large administrations in Europe, the United States, and Japan, in order to face problems that were becoming increasingly unsustainable, and it was recognized that there was a need to address problems from different perspectives than usual, including the ITS approach. The ideas of the first initiatives are briefly presented below.

1.2.1 Europe

This concern was materialized in Europe in the Prometheus project (Program for a European Traffic System with Highest Efficiency and Unprecedented Safety) between 1987 and 1995. It emerged from the vision of "Creating Intelligent Vehicles as part of an Intelligent Road Traffic System." It was a large project, involving autonomous vehicles and communications, and many other areas. Its objective was to create new concepts and solutions aimed at improving traffic in terms of safety, efficiency, and environmental impact. It was intended to develop new information management systems with the incorporation of electronics and artificial intelligence. Led by major European manufacturers, it involved electronics companies, component manufacturers, authorities, traffic engineers, etc. It was composed of four subprojects of basic research and three of applied research. Among the former, PRO-ART aimed at the use of artificial intelligence, PRO-CHIP focused on the development of hardware for intelligent processing in vehicles, PRO-COM established communication methods and standards, and PRO-GEN focused its activity on the creation of scenarios for the evaluation of systems. On the other hand, the applied research subprojects included PRO-CAR, which dealt with vehicle sensing, actuators, user interface, and general architecture, PRO-NET, which focused on communication between vehicles, and PRO-ROAD, which was oriented to the vehicle's communications with the environment, although the interaction between the last two subprojects was contemplated. In parallel, the equivalent approach was implemented from the infrastructure through the DRIVE project (Dedicated Road Infrastructure for Vehicle Safety in Europe).

It is important to note that, several years after its kick-off, it was argued that it was difficult to see new challenges in the road transport sector that exceeded Prometheus' vision, considering that its objectives were still fully valid. However, there were identified aspects on which to continue working to complete the objective. These areas included feasibility studies, establishment of marketing plans, implementation and monitoring plans, and evaluation of results against the original plan.

Nowadays, many of the European projects that are part of the large consortia have only materialized the Prometheus vision using new technologies that have been emerging and maturing over many years.

1.2.2 United States

In the United States, the approach to addressing road transport problems through the application of new technologies began in 1986 in an informal discussion group called Mobility 2000 (Sussman, 2005). The motivation of this group came from the congestion levels and the traffic accident figures, as well as the incipient environmental awareness. In 1990 the need for a permanent organization was clear, and the Intelligent Vehicle Highway Society (IVHS) was created as a committee of the Department of Transportation. In 1992, a Strategic Plan for the IVHS in the United States was disseminated that set a 20-year horizon for research, development, field testing, and ITS deployment. The plan focused on three pillars: development and implementation of technologies that allow the deployment of ITS; the integration of technologies to support systems; and ways to deal with the essential public—private institutional relationships within the framework of ITS. In 1995, ITS were classified into six groups in the USA:

- Advanced Traffic Management Systems (ATMS), which are intended to predict traffic and provide alternative routes, with real-time information management for dynamic traffic control.
- Advanced Traveler Information Systems (ATIS), which are intended to inform the driver and the passengers, on or off the vehicle, about the state of the roads, optimal routes, incidents, etc. (Austin et al., 2001).
- Advanced Vehicle Control Systems (AVCS) for safer and more efficient driving. Different temporal scenarios arise. In the nearest, systems are considered to warn the driver of collision hazards and can act automatically on the controls of the vehicle. In a second stage, a greater interaction with the infrastructure in considered, including driving in platoons.
- Commercial Vehicle Operations (CVO) improve the productivity of transport companies.
- Advanced Public Transportation Systems (APTS) improve information provided to users and improve fleet management (McQueen et al, 2002).
- Advanced Rural Transportation Systems (ARTS) are where actions are not clearly defined as there are economic restrictions on these low density routes.

This classification is not very different to those posed later by other authors (Comisión de Transportes, 2003; Miles and Chen, 2004; Aparicio et al, 2008), except in small details such as the incorporation of a specific area for electronic payment and another one for emergency management, and the integration of measures for any type of road in a single group without distinguishing between urban scope, high capacity roads, and secondary roads.

At this point, it should be noted that the objectives established almost 20 years ago, largely coincide with the current ones, although, in some of them, significant levels of implementation have already been achieved, such as the information to drivers about traffic or information to public transport travelers, although several issues remain to be achieved in the dynamic routing of vehicles and problems have been encountered regarding automated driving, for example. On the other hand, the problems on conventional roads are still present with few specific solutions for them.

1.2.3 Japan

In Japan, the first initiatives in the ITS field date back to the 1970s with the Comprehensive Automobile Control System project (CACS). Given that the ultimate objectives could not be achieved, new projects such as the Automobile Traffic Information and Control System (ATICS) were proposed between 1978 and 1985. They were more conceptual than practical projects, essentially related to traffic management. But two other projects were the ones that actually launched ITS activities in the country. One of them is the RACS Project (Road/Automobile Communication System), which was the basis of current navigation systems and started in 1984. The other was called Advanced Mobile Traffic Information and Communication System (AMTICS), which sought to present traffic information to drivers on an onboard equipment and the use of communication terminals for the provision of such information.

1.3 SERVICES

The ITS services have undergone a very strong development, which has given rise to an extensive number of systems and services, which can fall into different classifications. One of the most widespread classifications is formed by the following seven areas:
- Provision of information to the user.
- Traffic management.
- Operation of commercial vehicles.
- Public transport operation.
- Electronic payment.
- Emergencies.
- Control of vehicles and safety systems.

In addition, it is common to find synergies and overlapping areas that face the same problem. For example, the passenger information and public transport operation areas can share information, as can vehicle safety systems, driver information systems (information such as driving assistance), etc.

1.3.1 Provision of Information to the User

This area has a strong link with other ITS areas. Under the heading of information to the user, several systems can be found. For example, information points to public transport users, information to the driver before and during the trip, etc., could be mentioned.

There are two types of information that have relevant implications in the technology needed for its acquisition and treatment:

- Static information: works, events, tolls, public transport schedules;
- Dynamic information: congestion, weather conditions, parking.

For the private transport user, navigation and route selection systems have become increasingly important in recent years. The evolution of these systems has been similar to the growth of demand so that more and more functionalities and more information are offered. In this sense, the trend is to provide these systems with greater intelligence, so that they perform a guide in terms of variables they receive, such as congested points of the route, free parking lots, etc.

As a preliminary step, the traffic information service Radio Data System-Traffic Message Channel (RDS-TMC) and the informative variable panels (VMS) are highlighted. The first one is an effective means to provide a large amount of information, with immediate access by the user. The second allows the remote control of traffic, reaches all drivers of the road, and can alert or simply inform. This information in real time of the traffic and other conditions, like the meteorological ones, allows the drivers to choose the route and the most suitable conditions of circulation in each moment. These systems involve, in addition to the information terminals, information collection and sensing systems and management centers. The technologies used to obtain this information are very diverse and include onboard sensors, which transmit what they capture, video cameras, static sensors in the infrastructure, etc.

In terms of information to users of public transport, information on multimodal options, arrival times, en route information, etc., can be provided in order to be able to plan the trip before starting it or to receive information during the trip or to know real-time issues. Information to the traveler is a consequence of all the information gathered within a fleet management system, although its provision to users is not completely immediate since preprocessing is required and especially when the interrelation with other modes of transport, frequently operated by third parties, is sought.

Finally, it should be noted that, although these kinds of services have had a great development, it is still necessary to move forward in the following aspects:

- Greater precision and reliability in the information provided.
- Better information on multimodal options.
- More efficient methods to provide incident information quickly.
- Better transmission of information, so that it is easier to understand and it will be available through multiple channels.

All this should result in a reduction in travel times, a better multimodal choice, less passenger stress, and, in short, an improvement in the public service.

1.3.2 Traffic Management

Traffic management means achieving an efficient balance between travelers' needs and network capacity, bearing in mind that many users of diverse means of transport share the limited space of available infrastructure.

A prerequisite for traffic management is monitoring the traffic. This monitoring, which is not an end in itself, supports other services, in addition to traffic management itself, such as information to the user, emergency management, support for public and commercial transport, etc. This field has made significant progress, with solutions such as road-embedded detectors, television circuits, and "floating" sensors and information transmitters, as well as the centralization of data dynamically.

Urban traffic management can be considered as the origin of telematic systems applied to road transport. This is due to the fact that the urban environment revealed very soon the problems caused by the mass use of vehicles and the potential of solutions based on communications to address them.

Other relevant services in the urban environment are parking lots management and priority management. In the former case, there are solutions to reserve parking spaces based on the hours of arrival or the suggestion of alternative locations in case all the available ones are full. In the latter case, the deployment of green-wave systems for emergency vehicles or public transport vehicles is noteworthy. These corridors are materialized in systems of communication with the infrastructure in order that the traffic management is adapted. The simplest solution is that in which only the interrelation with a single intersection regulated semaphorically arises. This solution is quite simple from a technological point of view, although it does not end up being totally efficient since the incidence of this type of actions can have negative effects on the traffic of other vehicles. To solve this limitation, the interrelation with a more extensive area of intersections that operate in a coordinated way is considered.

Although part of the problem of urban traffic is also extrapolatable to the interurban one, other ITS systems are specific to these routes. Thus, we can highlight:

- Access control of vehicles that are incorporated into a certain flow of traffic, in order to avoid congestion.
- Road warnings of incidents, accidents, or breakdowns, to accelerate the response and offer the most convenient one in each case.
- Payment of automatic tolls that facilitate operations to the user and increase the capacity by avoiding the retentions, by the forced stop in manual toll areas.

Tunnels represent a particular situation of road section with very specific characteristics, which require different management and, for this reason, they are expressly separated from urban and interurban traffic management. This has resulted in the centers of integral control of tunnels that centralize all the information and have the capacity for action. Thus, various aspects can be controlled, such as tolls (if necessary), incidents, lighting, ventilation, control of dangerous goods, information to the user, etc.

Finally, other applications such as services to support infrastructure maintenance, detection and management of infractions, and incident management can be cited in this area.

1.3.3 Freight Transportation Operation

Fleet management systems are applicable to both freight transport and passenger transport, presenting numerous common points. This management is based, firstly, on the positioning of vehicles at all times and on the continuous exchange of information with a control center. This allows the adjustment of the service, intervals, and compensates for mismatches. Thus, vehicles, infrastructure points, and control centers are constantly exchanging information that supports high-level services, both for the data exploitation by the company and for the improvement of the service to the user.

The subsystems within transport fleet management systems (freight or public transport) are as follows:

1. An embedded subsystem, consisting of an embedded computer, a driver terminal, a positioning system, a communications system, and other elements installed in vehicles. Given the current architecture of modern vehicles and in order to take full advantage of the potential of these systems, the best solution is to access information from the vehicle's internal communications bus and send it wirelessly to external servers.

2. Central subsystem, consisting of system servers and different "customer" elements. This subsystem is responsible for the management strategies.

3. The subsystem in infrastructure composed of, among other elements, information panels connected to the central subsystem in the case of public transport.

In addition, the fact of using a central processing unit in each vehicle enables the possibility of connecting other peripherals for additional advantages.

The fundamental idea of this kind of management is to create an integrated environment that allows acting in a more agile way, in real time, to adjust to the demand and to solve incidences in the service, occasioned by its own causes or those external to the fleet (breakdown, retentions, demand changes, for example). The complete management of the fleet allows a better understanding of the demand, which can better adjust the service, with short and medium-term actions.

In the case of freight transport in Europe, there is a serious problem that, for the most part, appears at short distances trips of less than 200 km, where intermodal transport is not viable. As a result, and given the negative impact that road transport currently has, the introduction of ITS in the sector has been seen as a clear opportunity to alleviate the problem and promote sustainability.

The freight operation is closely linked to the location of freights and vehicles. The continuous communication between the vehicle and a central unit allows the control of the operations, being able to make changes in real time if needed. This communication is also expected to have a positive impact on intermodal transport, by speeding up the processes to reach logistics centers. Access to information such as the one that presented in previous sections allows for more accurate route choices and tighter forecasting of arrival times, which are some more advantages that could be obtained.

Another fundamental aspect, especially in fragile or valuable goods, is the continuous monitoring of its state, in addition to its location, acquiring data such as temperature, humidity, etc.

The new technologies allow communication between the driver and the central office that allows exchanging messages and automating administrative processes, such as the delivery confirmation. This is a service that gives added value and allows the planning of other service challenges. A better allocation sequence calculation can also be achieved, thereby saving time and fuel, in addition to more convenient vehicle service assignment, and vehicle redirection.

On the other hand, incident management is closely linked to previous services. In this regard, the transport of dangerous goods receives a specific treatment, oriented towards the management of possible incidents in the most efficient way. This is achieved through proper planning and obtaining detailed information on the transport.

However, there are clear problems for implementation such as the low technological level of many small companies, the high initial economic effort, and the low inter-company coordination.

1.3.4 Public Transport Operation

The main objectives of public transport management systems are to improve the quality of service and regularity, improve the information provided to the user, adapt to the demand, reduce operating costs and investments, and achieve greater flexibility and better control of the fleet. These systems are based on the positioning of vehicles, the processing of real-time information, and historical information to adapt the operation of the vehicles. Thus, it is a question of answering the demand with the available means, analyzing both and the conditions of operation in each moment.

Among the main functions that a public transport fleet management system can develop are the following:

- Information to the user, both inside the vehicle and at stops (through variable message panels or apps in smartphones). In this regard, it is sought that the information is more and more accurate and updated, so that the confidence of the users grow and can more reliably program their trips.

- Real-time operation management, oriented to the management of incidents in a fast and efficient way, in vehicles, infrastructure, traffic, etc. In this way, with precise information of the operation of the vehicles, any incidence in the service can be solved in a shorter time. Likewise, it is possible to provide more accurate and up-to-date information to travelers. Within this management is included the interrelation with traffic and intersections regulated semaphorically.

- Emergency communications, as well as accidents, damage, or criminal actions, for a faster intervention.

- Stops and transport intermodal areas management. According to the new concept of cities of the future, transport intermodal areas should become increasingly important, enabling changes from one transport mode to another quickly. Likewise, they can be considered for the change from private vehicle to public transport, which implies the provision of parking areas.

- Vehicle demand information. To do this, the onboard integration with a vending machine is required, as well as the passengers who leave the vehicle at each stop (for that, different solutions based on computer vision, pressure sensors, etc. have been developed) in such a way that a more accurate view of the demand is obtained that allows modifications in the future planning, not so much in real time.

- Web services with data in real time, planning data, and multimodal information. Beyond the provision of information in a single way, the challenges include the faster updating of information and the integration of all information in a consistent format, allowing the evaluation of alternatives that affect different modes operated, more generally, by different companies.

- Analysis of the deployment using the data collected. Beyond the use of real-time data, information collected historically can be used for future planning, adjusting schedules to reality, for example, in terms of travel times, as well as to try to adapt the offer to demand as much as possible. This analysis is based on the continuous monitoring of the vehicles, their position, and operating variables and being able to do an exhaustive study of vehicles, trips, drivers, etc.

- Remote diagnosis of vehicles. Collection of operation data of the vehicles allows a diagnosis and an approach to the maintenance of the vehicles in a more agile and effective way, mainly in those vehicles that require intervention at each moment. In the simplest state, diagnosis can be reduced to the detection of errors in the vehicle's control units, although the desired objective comprises a predictive maintenance based on operating data, so that the faults can be anticipated before they appear and maintenance tasks can be scheduled more efficiently.

- Drivers monitoring, with two fundamental aspects: promote safe and efficient behavior. This monitoring can be done online, providing warnings to the driver to improve their behavior, or offline, for example, focused on personalized training courses.

1.3.5 Electronic Payment

There has been a clear trend in recent years to eliminate or minimize economic transactions with cash to purchase goods and services, and to use electronic payment, for which various physical solutions have been proposed, such as payment by credit/debit cards, smart cards, or by mobile phones, due to their widespread use.

The electronic payment of services must be understood from the broader view since this is how the greatest effectiveness is achieved. In this way, it pursues a unified payment of public services, parking, tolls, etc. Flexibility should be its primary feature, although there are difficulties in involving different sectors and direct competitors. In this way, in an ideal scenario, you could pay in any country, receive the rate more adjusted to the use you make, and benefit from loyalty offers if it is billed at the end of travel or at the end of a certain period of time.

In the case of public transport, the following points differentiate these payments from those of other goods and services (McDonald et al, 2006):

- The traveler must have a document that verifies that he/she has made the payment.
- Delays in boarding are not allowed due to the payment.
- The payment must comply with the local rate structure (by number of stops, distance, etc.).
- Discounts must be considered for frequent travelers.
- The protection of personal data should be ensured in order to avoid tracking the journeys made by an individual person.

On the other hand, the payment of tolls should be understood in a broader way than is done today. Thus, in addition to the evolution of payment in fixed stations automatically, free-flow models should evolve that eliminate the formation of queues, or pay-as-you-drive systems where the vehicle is located at all times and the distances traveled by each type of road are known, proceeding to the calculation of the resulting toll.

Finally, although it is something that is far from the strict scope of electronic payment, some insurance companies are implementing the pay-how-you-drive models so that the insurance premium is recalculated according to the type of driving that the user is doing and it is registered and transmitted through a black box.

1.3.6 Emergencies

The application of ITS to emergencies involves emergency warnings and information, unified management of emergency vehicles, management of incidents with hazardous materials, and integrated emergency and traffic management. In this regard, the aim is to reduce the response time, ensure the use of adequate vehicles, and thus reduce the consequences of the incidents for both the people involved and the negative effects on traffic.

1.4 INTELLIGENT VEHICLES

Within intelligent transport systems, intelligent vehicles are one of the key elements, involving tasks of sensors, processors, and transmitters of information. Accordingly, driving in the future is conceived as an automated and cooperative driving.

It should be noted that the task of driving a vehicle is, for the most part, a low-risk activity, which although requiring a number of abilities, can be developed by most drivers in a short period of time. However, at critical times, the information that the user must process exceeds their capabilities. According to studies carried out on accident data, more than 90% of them have the human factor as one of its causes (Hobbs, 1989). For this reason, intelligent vehicles are aimed at helping and assisting the driver, reducing their load, increasing and improving the information available to promote safe and efficient driving without errors, and achieving a better response.

Making a single definition of the concept of an intelligent vehicle is complex because it is necessary to identify from what point we can talk about intelligence in a certain system designed by the human being. In a generic way, you can define an intelligent vehicle as a vehicle that is capable of:

- Taking information about its state and/or the environment (more or less distant);
- Processing that information;
- Making decisions, providing information, and/or acting.
  The collection of information can come in several ways:
- Onboard sensors in the vehicle that provide information to the internal communications buses. This was the first way to obtain data about the status of the vehicle, although the progress of the vehicles has caused a notable increase in the number and complexity of onboard sensors.
- Positioning the vehicle on a digital map. Although this function has been carried out for many years, the requirements in terms of precision in the positioning and accuracy and detail of the maps have been growing with the specifications of the systems that are being incorporated into the vehicles. In this regard, the problems caused by deterioration in positioning in adverse situations or the pernicious effect of inaccuracies in critical safety applications or autonomous driving have been highlighted.
- Monitoring of the near environment using onboard sensors, for which there are different technologies for short and long range. The current trend proposes the coordinated use of several sensors and the application of sensor fusion techniques that allow overcoming the limitations of each system individually.
- Reception of distant information through wireless communications. These data have greater possibilities by extending the horizon of information beyond the physical vision of the vehicle. However, the great potential of this information is the integration of large volumes of data from diverse sources.
- Monitoring of the driver and the interior of the vehicle, for which some already used solutions for the mentoring of the exterior can be accepted. However, other algorithms are required since the variables that are searched for are completely different.

Based on the above data, the vehicle must make decisions. These can be materialized in information to the driver or in actions on the systems of the vehicle to modify its dynamic behavior. As a result, more and more assistance systems are being included in vehicles that improve visibility, provide a better understanding of the environment to the driver so that he can anticipate his actions, detect situations that may be dangerous, improve the dynamic response (longitudinal, lateral and vertical), etc.

The incorporation of driving assistance systems is leading to the concept of autonomous and connected vehicles and, in the future, autonomous and cooperative vehicles. In this case, vehicles must be able to perform the driving tasks in full automatic form and, in addition, collaborate with each other with the exchange of information for optimal decision making, especially in complex environments where the single action of a single vehicle may be ineffective.

This research area is still open and, in order to achieve progress, advances must be introduced simultaneously in various fields such as electronics, processing units, sensors, communications, etc., as well as the development of robust and reliable algorithms capable of facing any driving situation. However, the anticipation of positive effects justifies the efforts being made. For example, the introduction of the fully autonomous vehicle can lead to the following benefits according to Morgan Stanley in the United States (Gill et al, 2015):

- Savings from collision avoidance will be $488 billion.
- Productivity gains from driver time savings will be $507 billion.
- Fuel savings will be $158 billion.
- Productivity gains from congestion mitigation will be $138 billion.
- Fuel savings from congestion mitigation will be $11 billion.

This book focuses on the study of intelligent vehicles, although mention is made of their interrelation with infrastructure since it has become clear that in the future they will not be able to be treated independently. The purpose is to offer a vision of the main technologies that enable them, as well as the systems that can be placed finally on them.

1.5 BOOK STRUCTURE

The book has been divided into three major parts, differentiating the enabling technologies, the systems, and other relevant aspects related to these.

The first part is dedicated to the technologies that allow the deployment of the systems that are explained later. In this sense, the following topics are addressed:

- Chapter 2, Environmental Perception For Intelligent Vehicles: Perception of the environment, including the perception from the vehicle, as well as from the infrastructure itself. It analyzes the main technologies that exist and their advantages and limitations, as well as the need for using sensor fusion to complement them.

- Chapter 3, Vehicular Communications: Communications between vehicles or between vehicles and infrastructure, which support cooperative systems, distinguishing aspects of standardization from the technological ones.
- Chapter 4, Positioning and Digital Maps: This chapter describes the most commonly used method for positioning vehicles, satellite positioning, but mention is also made of other alternatives that are necessary to improve the location data or to compensate losses of the satellite positioning signal.
- Chapter 5, Big Data in Road Transport and Mobility Research: Massive information capture, an essential aspect to support numerous applications, including specific techniques for acquiring and processing information to have reliable data in real time.

It should be noted that a specific chapter on the electronic structure of vehicles, including internal communications buses, sensors, and actuators, is not included because it is outside the scope of this book focusing on the most characteristic applications that are in line with the definition of the previous section of intelligent vehicles. However, it should be noted that these aspects are also essential to implement those applications, and can be considered as a low-level layer on which the rest of the technologies and systems are based. The justification for this choice, which is motivated by reasons of scope and extent, lies in the fact that research teams dedicated to applications of intelligent vehicles are generally different to those who are in charge of the electronic structure of the vehicle, although the actual implementation in a vehicle that is present in the market implies the interrelation of both groups.

The second part of the book focuses on the description of intelligent vehicles applications. It should be noted that the fast evolution of the systems described in this part and the widespread introduction in the market in some cases, makes it a very dynamic field. Therefore, it is not intended to make a comprehensive description of all systems implemented or under development, but to establish a conceptual framework and provide an idea of the areas being treated, as well as to provide an overview of the main developments being undertaken. Specifically, the following main themes are addressed:

- Chapter 6, Driver Assistance Systems and Safety Systems: This chapter provides an overview of the integrated safety model, which integrates assistance systems, primary, secondary, and tertiary safety systems, as well as primary—secondary interaction systems.

- Chapter 7, Cooperative Systems: A review of the systems offered based on communications between vehicles or with the infrastructure, which are proposed to improve information, traffic safety, and efficiency.
- Chapter 8, Automated Driving: Although the previous chapters have already referred to some applications involving a certain degree of automation of vehicle functions (the lowest levels of automation), this one focuses specifically on autonomous driving vehicles, mainly referring to those at the highest levels of automation.

Finally, the third part deals with a set of other aspects related to intelligent vehicles. The topics covered are as follows:

- Chapter 9, Human Factors: This chapter is divided into two separated parts:
  - Subchapter 9.1, Human Driver Behaviors: Driver monitoring and driver models are analyzed in order to study the state of the driver (fatigue, drowsiness, etc.);
  - Subchapter 9.2, User Interface: User interface and its implications on the driver are discussed.
- Chapter 10, Simulation Tools: This chapter is divided into three separated parts:
  - Subchapter 10.1, Driving Simulators: Architecture and Applications of Driving simulators are described;
  - Subchapter 10.2, Traffic Simulation: Fundamentals of Traffic simulators are explained with some examples;
  - Subchapter 10.3, Data for Training Models, Domain Adaptation: Tools are presented for generation of training data by simulation for testing algorithms (of perception, decision, etc.) that can be implemented in intelligent vehicles.
- Chapter 11, The Socio-Economic Impact of the Intelligent Vehicles. Implementation Strategies: Analysis of social and economic impact and policies for implementation. This chapter deals with nontechnical aspects, such as economic, political, and social, based on a definition of barriers that need to be overcome for the implementation of systems in the field of intelligent vehicles.
- Chapter 12, Future Perspectives and Research Areas: Prospects for the future and areas of work. The book ends with a reference to areas that are still open for work in future years, focusing mainly on the technical issues addressed in the first two parts of the book.

REFERENCES

Aparicio, F., Arenas, B., Gómez, A., Jiménez, F., López, J.M., Martínez, L., et al., 2008. Ingeniería del Transporte. Ed: Dossat. Madrid (in Spanish).

Austin, J., Duff, A., Harman, R., Lyons, G., 2001. Traveller information systems research: a review and recommendations for Transport Direct. Department of Transport. Local Environment and the Regions, London.

Comisión de Transportes, 2003. Libro Verde de los sistemas inteligentes de transporte terrestre. Colegio de Ingenieros de Caminos Canales y Puertos, Madrid (in Spanish).

Gill, V., Kirk, B., Godsmark, P., Flemming, B., 2015. Automated vehicles: The coming of the next disruptive technology. The Conference Board of Canada, Ottawa.

Hobbs, F.D., 1989. Traffic Planning and Engineering. Pergamon Press, Oxford.

McDonald, M., Keller, H., Klijnhout, J., Mauro, V., Hall, R., Spence, A., et al., 2006. Intelligent transport systems in Europe. Opportunities for Future Research. World Scientific.

McQueen, B., Schuman, R., Chen, K., 2002. Advanced Traveller Information Systems. Artech House.

Miles, J.C., Chen, K. 2004. The Intelligent Transport Systems handbook. PIARC.

Sussman, J.M., 2005. Perspectives on Intelligent Transportation Systems. Springer.

FURTHER READING

Comisión de Transportes del Colegio de Ingenieros de Caminos, Canales y Puertos, 2007. Libro Verde de los sistemas inteligentes de transporte de mercancías. Colegio de Ingenieros de Caminos Canales y Puertos, Madrid (in Spanish).

Kala, R., 2016. On-road intelligent vehicles. Motion Planning for Intelligent Transportation Systems. Elsevier, Waltham, MA.

MacCubbin, R.P., Staples, B.L., Mercer, M.R., 2003. Intelligent Transportation Systems. Benefits and Costs. US Department of Transportation.

Meyer, G., Beiker, S., 2014. Road Vehicle Automation. Springer, Switzerland.

Organisation for Economic Co-operation and Development, 2003. Road safety. Impact of new technologies. OECD.

U.S. Department of Transportation, 2003a. Intelligent Transportation Systems Benefits and Costs US DOT Washington DC.

U.S. Department of Transportation, 2003b. Intelligent Vehicle Initiative. Annual Report. US DOT Washington DC.

Whelan, R., 1995. Smart Highways, Smart Cars. Artech House, Boston, MA.

PART I

Enabling Technologies

CHAPTER 2

Environmental Perception for Intelligent Vehicles

José M. Armingol[1], Jorge Alfonso[2], Nourdine Aliane[3],
Miguel Clavijo[2], Sergio Campos-Cordobés[4], Arturo de la Escalera[1],
Javier del Ser[5], Javier Fernández[3], Fernando García[1],
Felipe Jiménez[2], Antonio M. López[6], Mario Mata[3],
David Martín[1], José M. Menéndez[2], Javier Sánchez-Cubillo[4],
David Vázquez[6] and Gabriel Villalonga[6]

[1]UC3M, Leganés, Spain
[2]Universidad Politécnica de Madrid, Madrid, Spain
[3]Universidad Europea de Madrid, Villaviciosa de Odón, Spain
[4]TECNALIA, Bizkaia, Spain
[5]TECNALIA, University of the Basque Country (UPV/EHU) and Basque Center for Applied Mathematics (BCAM), Bizkaia, Spain
[6]CVC–UAB, Barcelona, Spain

Contents

Intelligent Vehicles
DOI: http://dx.doi.org/10.1016/B978-0-12-812800-8.00002-3

Environmental perception represents, because of its complexity, a challenge for Intelligent Transport Systems (ITS) due to the great variety of situations and different elements that can happen in road environments and that must be faced by these systems. In connection with this, so far there are a variety of solutions as regards sensors and methods, so the results of precision, complexity, cost, or computational load obtained by these works are different.

Road safety applications require the most reliable sensor systems. During recent years, the advances in information technology have led to more complex road safety applications, which are able to cope with a high variety of situations. But a single sensor is not enough to provide the reliable results necessary to fulfill the demanding requirements that these applications need. Here is where Data Fusion (DF) presents a key point in road safety applications. Recent research in the ITS research field tries to overcome the limitations of the sensors by combining them. Also contextual information has a key role for robust safety applications to provide reliable detection and complete situation assessment.

The work presented in this chapter develops a set of algorithms and methods in order to give support to the implementation of a great variety of advanced driver assistance systems or traffic monitoring systems in these environments.

2.1 VISION-BASED ROAD INFORMATION

Lane marking and traffic signs are elements used to provide information and give instructions to road users, and their appropriate usage may help to improve the traffic flow as well as increasing safety (Miller, 1993). Thus, integrating into vehicles systems the ability to detect lanes and recognize traffic signs may entail economic and social benefits. In fact, lane and traffic signs detection can be carried out using images captured by onboard cameras, and roads information can be extracted with the help of image processing techniques. Nowadays, vision-based algorithms are able to estimate lane curvature and a vehicle's relative position in a lane as well as to detect traffic signs and recognize their meaning. More importantly, the synthetized information can be applied in multiple applications for driver assistance systems, such as providing relevant road information on the dashboard, road scene understanding, adaptive cruise control, and driving at adequate speed, braking and steering assistance, lane keeping and lane departure warning, and ultimately their use in autonomous vehicle guidance. In fact, vision-based lane detection and traffic signs recognition have been an active research topic over the last 15 years, and various techniques and methods have been presented in the scientific literature. Thus, the present chapter provides the reader with a broad picture of the overall methodologies and approaches for lane detection as well as traffic signs recognition developed in recent years. The chapter is organized as follows: Section 2.1.1 presents some road scenarios and environmental variability; Section 2.1.2 gives an overview of the different techniques used in lane detection; Section 2.1.3 is dedicated to the different techniques and algorithms used for traffic signs recognition. Finally, Section 2.1.4 comments some commercial systems.

2.1.1 Environmental Variability

Lane detection and traffic sign recognition depend on a number of parameters, such as the rapidly changing road morphology and lighting conditions, making it a not trivial problem. Designing effective lane and traffic sign detection and recognition systems requires the development of complex and specific algorithms taking care of many parameters and issues, with in many cases limited computational resources.

Roads, in general, are not homogeneous and traffic signs are not completely visible due to nearby objects (vehicles, trees, buildings, etc.) creating occlusion or shadow. Regarding lane detection, roads are not uniform and may present a varying morphology changing from straight segments to curved ones, and they may present varying slopes, intersections, and roundabouts. Roads may present several lanes with merging, splitting, and even with ending lanes. Road markings are not uniform and lanes may present different marking patterns (continuous, dashed), colors (white, yellow), and width, and even show degraded markings, and may have a low intensity and contrast. To be acceptable, lane detection and traffic signs systems should work properly in various light conditions and weather changes (sunny, rainy, snowy, cloudy, foggy, etc.). Illumination may also change drastically when entering or leaving tunnels, or crossing a road with overhead lights or lights on the walls, and many different circumstances may lead to poor visibility conditions. Thus, perfect and foolproof vision-based lane detection and traffic sign recognition, is utopic under uncontrolled illumination conditions in an unconstrained environment, with traffic signs subject to degradation by damage and aging, and many other environmental circumstances. Camera limitations may also affect the overall performance in the initial image detection. Fig. 2.1 shows some situations making lane detection difficult to

Figure 2.1 Examples of some extreme situation to handle by lane detection systems. (A) Different lane marking; (B) Lane occlusion; (C) Change of pavement texture; (D) Lanes with shadow; (e) Rainy road; (F) Saturated image at tunnel exit.

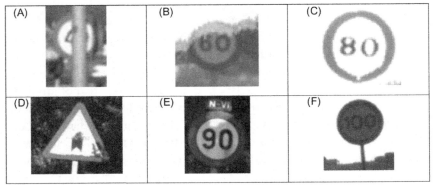

Figure 2.2 Examples of situations difficult to handle by traffic sign recognition systems. (A) Occlusion; (B) Smearing; (C) Saturation; (D) Degradation; (E) Shadows; (F) Backlight.

handle, while Fig. 2.2 shows some traffic signs recognition scenarios that are difficult to handle.

Many researchers have based their methods on some assumptions, meaning that it is a big challenge to make lane detection and traffic signs recognition systems work properly under all of the aforementioned conditions whilst ensuring an acceptable response time.

2.1.2 Lane Detection

Methods for lane detection can be classified into two large categories: sensor-based methods and vision-based methods (Hoang et al., 2016). Sensor-based methods mainly use some devices such as radar, laser, or global positioning systems to detect the vehicle lane, and this aspect is out of the scope of this study and readers can find valuable information in Li et al. (2014) and Lindner et al. (2009). Vision-based is the most important approach in lane detection because markings are made for human vision (Bar Hillel et al., 2012). This area has been an active research area during the last two decades, and significant progress has been made (Bokar et al., 2009; Satzoda et al., 2013; Bila et al., 2016). In their basic approach, lane detection algorithms for a short distance ahead are based on two different stages: namely, preprocessing and postprocessing. The preprocessing operation consists of applying suitable filtering techniques in order to reduce the noise present in the images, and have the images prepared for the subsequent processing. The postprocessing stage consists of thresholding the image intensity to detect potential lane edges followed by edge points grouping to detect lane markers.

2.1.2.1 Preprocessing

Initial images are hardly used for information extraction, and therefore, some preprocessing techniques are required. In this aspect, various tools have been used to reduce the noise present in the images. For example, median, Winner, or Gaussian filters are techniques used to blur images and to reduce salt and pepper noise (Srivastava et al., 2014; Aly, 2008). Thresholding is also a common way to reduce noise, but the use of a fixed threshold is unable to cope with variable illumination conditions, such as the weather changing or shadows. Thus, the use of adaptive threshold segmentation is appropriate to cope with these problems as well as to isolate bright objects like street lights and taillights of cars (Lu et al., 2008). Shadows are also a problem and are always present in lane detection applications, where their boundaries are confused with those of different markings. To remove strong shadows, the Finlayson-Hordley-Drew (FHD) algorithm proposed in Finlayson et al. (2006) is used in lane detection, for example in Assidiq et al. (2008). Inverse perspective mapping (IPM) algorithm (Mallot et al., 1991) is another preprocessing technique used for avoiding the perspective effect transforming an image into its corresponding bird's eye vision (Takahashi et al., 2002; McCall and Trivedi, 2006; Aly, 2008; Tuohy et al., 2010). Another important preprocessing operation is the pruning, which consists of removing unnecessary parts of the image, and regions of interest (ROI) are determined based on some preliminary knowledge. For example, the ROI is established in the bottom half of the image in Arshad et al. (2011), and in Zhang et al. (2013) the image is divided vertically to take account of only the parts where the main information appears, or in Han et al. (2015) where the vanishing point estimation technique is used. However, establishing ROIs may be somewhat problematic because of not having previous knowledge of the road geometry, or running the risk of eliminating useful parts of the image.

2.1.2.2 Postprocessing

As markings create strong edges on the road, which is one of the most significant features, and in order to distinguish the boundary between two dissimilar regions within an image, the first operation consists of performing edges detection. To this end are applied mainly the Sobel operator (Pratt, 2001) or the Canny edge detector (Canny, 1986). Afterwards, Hough Transform (Ballard, 1981) is used to link edges and detect boundary segments corresponding to the lane markings present on the road. In general, the lanes to be detected are assumed to be straight. This approach is used in many systems (Amemiya et al., 2004; Mori et al., 2004; Schreiber et al., 2005; Jung, 2016).

Even with its robustness to noise and missing data, the major inconvenience of the Hough Transform is the computational cost associated to its voting scheme for achieving acceptable time response. Thus, several methods referred to as modified Hough Transform have been proposed. For example, a modified Hough Transform for detecting both straight and curved lanes is proposed in Kaur and Kumar (2014), and a method operating on clusters of approximately collinear pixels, where for each cluster votes are cast using an oriented Elliptical Gaussian kernel modeling the uncertainty associated with the best-fitting line with respect to the corresponding cluster, is presented in Fernandes and Oliveira (2008).

Another method proposed in Kuk et al. (2010) is based on selecting a small set of line candidates passing through a circle centered at a previously detected vanishing point; this needs less memory than the conventional algorithm. Aly (2008) introduced a technique referred to as a simplified version of Hough Transform followed by RANSAC line fitting. More information regarding the RANSAC algorithm can be found in Fischler and Bolles (1981). In Jung and Kelber (2005) and Lee (2002) lane boundaries are detected using a combination of the edge distribution function (EDF) with a modified Hough Transform, where the EDF distribution is the histogram of edge magnitudes with respect to edge orientation angle, enabling the connection between edge-related information and the lane-related information. By using the orientation from the EDF peak, the Hough Transform is changed to a one-dimensional search problem.

In the literature are also some alternative methods to Hough Transform algorithms. For instance, a method combining a local line extraction method with dynamic programming as a search tool is presented in Burns (1986) and Kang (2003). The line extractor obtains an initial position of road lane boundaries, and then the dynamic programming approach improves the initial approximation to an accurate configuration of lane boundaries. To speed up the lane detection algorithm, McCall and Trivedi (2006) proposed steerable filters, which is a technique permitting to divide the search problem into their X and Y components, and then a finite number of rotation angles for a specific steerable filter are needed to form a basis set of all angles of a given steerable filter. This permits to see the response of a filter at a given angle, and therefore, tune the filter to specific lane angles. Additionally, steerable filters are used for detecting circular-reflector markings, solid-line markings, and segmented-line markings under varying lighting and road conditions. In Lin et al. (2009) markings are extracted by searching the lane model parameters in a special defined parameter space without thresholding, and the proposed method is based on the lateral inhibition property of human vision system. The approach does not use thresholding, and it uses 2D and 3D geometric information of lane markers instead. For lane detection, the positive and negative second difference map of the image is used, providing a strong contrast between the road and lane marker. In Kim and Lee (2014) a lane detection algorithm based on an artificial neural networks (ANN), followed by a RANSAC algorithm for validation is proposed, and the algorithm is able to handle challenging scenarios such as faded lane markers, lane curvatures, and splitting lanes.

Regarding lane models, even the straight-line is the commonly used geometric model in validating observed data, many works have created predefined geometrical models to fit the road in the image. For example, in Wang (2004), a B-snake model lane is proposed, where this model is able to describe a wide range of lane structures since B-Spline can be used to model any arbitrary shape by a set of control points. In Kim (2008), a lane-boundary hypothesis is represented by a constrained cubic-spline curve of two to four control points, which are then verified by a combined approach using a particle-filtering technique and a RANSAC algorithm. A similar approach is adopted in Aly (2008), where a RANSAC line fitting algorithm is used to give initial guesses and then apply a new RANSAC algorithm for fitting Bezier Splines. Li (2016) proposed an algorithm for multilane detection based on omnidirectional images with a feature extractor based on anisotropic steerable filter, and a parabola model is used to fit both straight as well as curved lanes. The approach in Fernandes and Oliveira (2008) is to classify the lines obtained by the Hough Transform into several line groups and calculate the central line for each group, and then a 3-degree B-Spline fitting algorithm is used to correct the straight lines to curved lanes.

After lane detection, many authors proposed some features extraction and classification strategies. For example, in Hoang et al. (2016) a lane detection discriminating between dashed and solid markings is proposed. In the work of Paula and Jung (2013) a set of features of lane markings are computed and a cascade of binary classifiers is adopted to distinguish five types of markings: dashed, dashed-solid, solid-dashed, single-solid, and double-solid. Maier (2011) presented a general geometric approach that uses prototype fitting for the recognition of predefined arrow models in real traffic scenes. Ozgunalp et al. (2016) proposed a detection of multiple curved lane markings algorithm based on vanishing point reference. And finally, Revilloud et al. (2016) proposed an algorithm for lane marking detection based on estimating the area of lane marking and using the profile of the lane estimation in a confidence map.

In the literature are some proposals for embedded system implementation of lane detection systems. For example, an experimental system for lane detection for a highways lane departure warning system is presented in Takahashi et al. (2002) using the extended Hough Transform and a bird's-eye view image transformation. This method was implemented using an embedded Hitachi SH2E microprocessor, reaching the execution of the algorithm in a period of time less than 66 ms. Lu et al. (2013) proposed a parallel Hough Transform implementation over FPGA architecture for real-time straight line detection in high definition videos. The solution consists of, firstly, applying an optimized Canny edge detection method with enhanced nonmaximum suppression to suppress most possible false edges and obtain more accurate candidate edge pixels, and then applying the accelerated computation of the parallel Hough Transform. Straight line parameters are calculated in less than 16 ms. Another FPGA-based real-time lane detection system is reported in Hwang and Lee (2016). An implementation using an embedded ARM-based real-time for lane departure warning system working during the day and nighttime is presented in Hsiao et al. (2009), where FPGA architecture are also used to implement part of the lane detection algorithms. Jiang et al. (2010) proposed a system to handle multiple lanes on a structured highway using an "estimate and detect" scheme. It detected the lane in which the vehicle is driving and estimates the position of two adjacent lanes. The vehicle recognizes if it is driving on a straight road or on a curve using additional information sources such as its GPS position and the OpenStreetMap digital map. Fan et al. (2016) reported the implementation of an embedded lane detection system using the TMS320C6678 DSP.

2.1.3 Traffic Signs Recognition

Automatic traffic sign recognition first appeared in computer vision literature in 1987 (Akatsuka and Imai, 1987), and since then it became a very active research field with relevant progress in the last two decades. This section is aimed at presenting the different approaches used in traffic signs recognition, providing readers with the state of the art from the research point of view as well as presenting some commercial applications integrated in todays vehicles. For a more in-depth review more detailed surveys such as Stallkamp et al. (2012) and Mathias et al. (2013) are recommended.

In their basic approach, traffic signs detection and recognition algorithms are based on two-step processes. The first stage consists of detection, where the presence and the location of traffic signs are obtained. The second step is about traffic signs classification according to their categories and meaning. Each of these two stages may be independently approached in many different ways, allowing a good degree of flexibility.

For instance, the detection stage may be common to a wide range of signs in different countries, while only the classification stage must be modified and fine-tuned to tackle different sign sets. In many different works, one can find extra steps referred to as tracking, consisting mainly in detecting a given traffic sign as the vehicle moves towards.

2.1.3.1 Sign Detection

The main clues that are exploited for traffic sign detection are color and shape, and in fact, many approaches use these features explicitly implementing ad hoc algorithms looking for a fast and efficient implementation. In other approaches, color and shape are implicitly used, since they are the basic relevance from which more complex features are derived, which are processed later by self-learning algorithms.

Color seems to be an obvious property in traffic sign detection since signs are designed using strong red or blue colors to be easily identified by humans. Restricting the regions of interest (ROI) in input images to those showing relevant colors is a straightforward approach. Several color spaces have been extensively used, mainly RBG, HSV, and HSL. They work properly under good illumination conditions, but the use of color has shown important limitations:

- Some signs of interest do not include red or blue backgrounds: overtaking and speed end of limits are white.
- Color is not constant under varying illumination conditions. The human brain implements what is known as color constancy: a red chair is perceived by humans as red under direct sunlight or inside a gloomy room. This fact does not happen within computer vision; although surface reflectance may be constant, illumination is not, so the color perceived by a camera is changing as lighting conditions do. Some attempts have been done to compensate for that using illumination models but sunlight, shadowing, clouds, day/night transitions, dawn and dusk conditions, or night illumination using the headlamps, constitute a formidable challenge.
- Sign materials reflectance is not really constant. It varies with material quality, and with degradation. Old signs should be replaced but often they are not. "Red signs" can be found in all grades from brown to orange to pink to bright red.

Although color may not be a decisive property, it helps to create harder transitions between the signs and their environment, helping in the definition of the signs' borders. So it is still a useful clue to improve the determination of the signs shape.

Sign shape is also designed to allow an easy discrimination for humans. Natural objects usually are not perfectly round, or triangular. Indeed "stop" and "give way" signs are designed to be detected by humans from behind without color clues, so a driver can understand that a given sign does not apply to him/her but it does for other drivers, so they have unique octagonal and inverted triangle shapes, respectively. Shape is a strong clue for traffic sign recognition since it is invariant to illumination and, although it is sensitive to partial occlusion, this is not a common situation in this kind of applications. Circle detection is often carried out using some variant of the classical circular Hough Transform, the Chamfer matching (Gavrila, 1999), or the fast radial symmetry transform (Barnes and Zelinsky, 2004). Triangles and octagonal shapes are extracted using border orientations, border analysis (Aliane et al., 2014), or oriented gradients (Mathias et al., 2013). Most successful approaches for exploiting shape start by obtaining a series of shapes descriptors, followed by a classification algorithm highlighting those that may correspond to the desired objects, i.e., traffic signs. Today's most popular shape description techniques are based on histograms of oriented gradients (HOG) (Dalal and Triggs, 2005). Shape descriptors are then classified with algorithms based on AdaBoost, presented by Viola and Jones (2001), such as integral channel features classifier (Dollar et al., 2009).

2.1.3.2 Sign Classification

The classification of the regions of the input image, which are candidates to be a traffic sign, can also be performed in many ways. Traffic sign recognition can be seen as a difficult classification problem mainly for two reasons:

- False positives from the detection stage will introduce random subimages to the classifier that are not real signs. Including all predictable cases in the training sets, for example, all standard signs from traffic regulations, even if not considered as possible outputs of the traffic signs recognitions algorithms (see Fig. 2.3B) and signs' back-sides (see Fig. 2.3C), will reduce the risk of wrong classifications. But, random subimages, such as circles, as shown in Fig. 2.3A, will eventually reach the classifier, and may lead to a wrong classification sign. The detection stage should be biased to minimize the occurrence of false positives, even at the cost of increased false negatives.
- Traffic signs are designed for human drivers, so the numbers and symbols are designed to be easily readable by humans but are not intended

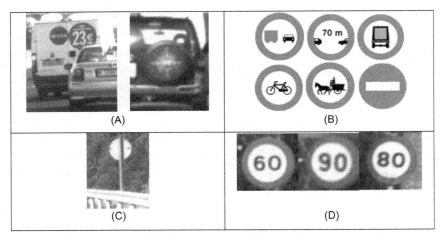

Figure 2.3 Examples of challenging situations for the classifier stage. (A) Random nonsign elements; (B) Signs not intended as outputs; (C) Back side; (D) Similar signs.

to maximize the differences between all patterns that would make machine classification easier. For example, the difference between some speed limit signs are limited to a small number of pixels in the tens digit, as depicted in Fig. 2.3D. Thus, classifiers should be extremely sensitive to these small differences. Reducing the set of possible signs that are considered is a usual approach to improve the classifier performance.

The most commonly used classification techniques are normalized correlation (Aliane et al., 2014), K-nearest neighbors and radial basis functions (RBF) (Gavrila, 1998), and polynomial classifiers (Kressel et al., 1999). More recently, artificial intelligence and deep learning techniques are proving to have superior performance, such as the use of the support vector machines (SVMs) (Maldonado et al., 2007), or convolutional neural networks (CNN) (Zhu et al., 2016), or extreme learning machines (ELM) (Huang et al., 2016), to cite few new techniques.

2.1.4 Commercial Systems

The efforts of many researchers in both lane detection and traffic sign recognition have led to commercial implementations, which became available as extra equipment for high-end vehicles around 2010. Nowadays, most car manufacturers include this feature not only in high-end vehicles, but it is also appearing integrated in the production of mid-range ones. Some of these implementations are:

- Bendix CVS launched the AutoVue system (Bendix web page, 2017), a lane departure warning system due to fatigue, distractions, and unfavorable weather conditions. It uses a proprietary image recognition software package. Data generated by this system are gathered to a central database to manage safety information and gives fleets the ability to develop targeted training programs to address the real issues occurring on the road.
- Bosch launched a hardware and software working on a CMOS color camera and a scalable multiprocessor architecture, launched as Multi-Purpose Camera (MPC2). Its lane departure warning (Bosch-1 web page, 2017) uses a video camera to detect lane markings ahead of the vehicle and to monitor the vehicle's position in its lane, and when it detects that the vehicle is about to unintentionally move out of the lane, it warns the driver by means of visual, audible, and/or haptic signal, such as steering wheel vibration. The lane detection algorithm records and classifies all common lane markings up to a distance of approximately 60 meters ahead, whether the road markings are continuous, dashed, white, yellow, red, or blue. Its traffic signs recognition, referred to as myDriveAssist (Bosh-2 web page, 2017), detects speed limits and cancellation signs together with their constraint signs by simply using the smartphone-camera and issues warnings in case of driving too fast. The myDriveAssist app is also country-specific and speed limit detection can offer you an optimal support in different countries.
- Delphi launched its intelligent forward view camera (IFV250), using a CMOS sensing device (Delphi web page, 2017), offering vehicle manufacturers a scalable architecture for their forward looking safety system needs. The IFV250 integrates several features, such as lane departure warning, alerting drivers when the vehicle approaches or crosses the lane markings, and traffic sign recognition facility providing traffic sign information including speed limits and curve speed alerts. The system is able to detect painted lane markers up to 25 m ahead.
- Continental has its own integrated system called as Multi-Functional Camera (MFC) (Continental web page, 2017). It integrates several facilities. The first one is the lane departure warning driver assistance system that alerts the driver with acoustical or haptic warnings before his vehicle is about to leave the lane. An additional facility is a lane keeping system that is initiated through a gentle intervention in the

steering. It includes also a traffic signs recognition allowing comfortable driving.

However, these systems are working under certain circumstances and with the support of additional information, such as satnav cartography, or any other background database, and such systems perform fairly well enough to be useful to drivers. Regarding traffic sign recognition, commercial systems only support circular signs, limiting their application to speed limits and overtaking signs. Other circular signs, such as "do not enter," "circulation restrictions," "forbidden turns," and many others, are not included. Their performance is intentionally biased for reducing false positives, at the cost of increasing false negatives. That is, missing some signals is desirable to recognizing a wrong or an inexistent speed limit. Undetected speed signals are compensated with information taken from the satnav cartography source. If no speed sign is detected after some time, the default speed sign taken from the satnav is displayed instead, which usually masks missed signs to drivers. After a "not allowed overtaking" sign, undetected "allowed overtaking" signs are compensated by removing the overtaking prohibition after some time. The satnav system is also used to help determine which of the detected signals are relevant to the driver and which ones are not. Other backup mechanisms are used to compensate the unavoidable detection system limitations. For instance, Bosch provides a Cloud database feed by detected signs and their respective locations by a free smartphone app, myDriveAssist (Bosch-2 web page, 2017). Thus, the system uses that database to feed the app and compensate for undetected signs. If previously stored signs are no longer detected by the apps, they are automatically deleted from the database.

2.2 VISION-BASED PERCEPTION

Robotic systems need to perceive their surrounding environment in order to perform their assigned tasks. When focusing on road vehicles, we can consider advanced driver assistance systems (ADAS), which have the task of acting as codrivers in the sense of warning human drivers when dangerous situations are detected or even taking control of the vehicle if necessary (e.g., performing an emergency brake to avoid an imminent collision). In fact, due to the many advantages in terms of safety, energy, and pollution efficiency, as well as enhanced mobility, industry and academia have moved their goals even further than traditional ADAS. In particular, the ultimate objective nowadays is to reach fully autonomous driving (AD); i.e., the aim is to fully replace human drivers.

In practice, this means that a full understanding of the driving scenario is required; or at least an understanding far beyond traditional ADAS. For instance, only detecting pedestrians can be used to develop an ADAS focused on avoiding vehicle-to-pedestrian fatalities. Another ADAS can focus on avoiding vehicle-to-vehicle accidents, and so on. However, for performing autonomous driving, we cannot rely on such partial percepts; a holistic understanding of the scenario is needed since there is not a human driver to decide about vehicle maneuvers in a continuous fashion. Accordingly, we can foresee the need of a really advanced artificial agent (or interconnected agents) for autonomous driving.

Developing such an agent is a multidisciplinary task, where different scientific and engineering disciplines converge. Appropriate sensors are needed to capture the relevant raw data of the world, transforming the raw data into situation knowledge is mandatory, and deciding and executing the maneuvers according to the situation too; all this in real-time.

At this point, and referring to the title of this section, the reader may wonder about the role of vision-based perception in this context. The answer is very simple, vision is the main sense used by human drivers to perceive the environment and we want to emulate and improve upon them. Therefore, using cameras to capture the world as images and computer vision techniques to raise their semantic content is crucial.

Certainly, by reviewing the different ADAS and autonomous prototypes developed in the last decade, we can see that they rely on different types of sensors to perform their tasks. This is to be expected because sensor fusion enhances system reliability in general terms. However, another reason is that computer vision state-of-the-art seemed to be still very far from providing a standalone environment perception for autonomous driving, and even for some ADAS (e.g., for pedestrian protection systems computer vision jointly operates with LIDAR/Radar sensors). In spite of this, computer vision has been essential in many ADAS, see Geronimo et al. (2017), and vehicles with a large capability for driving autonomously, see Franke (2017).

What we really want to highlight in this section is the fact that the potential contribution that computer vision can offer to autonomous driving has developed extensively since 2012. We can see that the state-of-the-art offered by this technology has exploded. The community has moved from the feeling of "we are years apart of robustly understanding general outdoor road scenes in detail, it is too challenging by using only images" towards a feeling of "the required vision-based scene understanding for autonomous driving is arriving, we have to research and develop more, but we are close."

Why has this change of feeling been produced? The answer is "deep learning," in this case applied to solve computer vision tasks. We must clarify, that autonomous vehicles will still rely on fusion sensors to provide a high reliability, for instance under adversarial weather conditions. However, the usefulness of computer vision will be quite high for the intelligent driving agent. In fact, deep learning can be used to raise semantics from any sensor raw data, not only from images. Thus, the computer vision view described here can be shared for these other sensors (LIDAR, Radar, Ultrasound).

Obviously, a detailed description of the deep learning field falls out of the scope of this book and the reader is referred to any of the many sources available today, for instance in Goodfellow et al. (2016). Here, we will introduce deep learning in contrast to the approaches used before its eruption into the field of computer vision; in particular, when computer vision is applied to ADAS and autonomous driving. With this purpose

Figure 2.4 (Top) Vehicle detection task: vehicle instances are framed by bounding boxes (pure 2D in this case; it could be projected 3D bounding boxes). (Bottom) Semantic segmentation considering road surface (pink), vehicles (dark blue), vegetation (green), poles and traffic signs (gray), fence (white), and sky (light blue).

we will focus on two very relevant tasks: vision-based object detection and semantic segmentation (see Fig. 2.4). First, we will review the traditional approaches used for addressing such tasks, and later we will introduce some of the promising methods used nowadays that rely on deep learning.

2.2.1 Vision-Based Object Detection and Semantic Segmentation

"Object detection" and "semantic segmentation" are terms borrowed from the computer vision community. Given an image, we may be interested in performing different tasks. For instance, we may be interested in knowing if the image contains (or not) an instance of a given class among a set of classes (e.g., planes, cats, cows, etc.), in this case we talk about "image classification" or "object recognition." Note that this does not mean that we know where the instance or instances are located within the image, not even how many instances are included in the image. When we search for such locations so that we highlight the instances with a bounding box, we are performing "object detection." Even more difficult, if we aim at delineating the silhouette of all such objects, eventually assigning a class to all image pixels, then we are doing "semantic segmentation." The reader is referred to the PASCAL VOC challenge by Everingham et al. (2010) as a classical example on different computer vision challenges.

In the context of ADAS and autonomous driving, the classes of interest are pedestrians, vehicles, traffic signs, free road surface, etc. Detecting their instances in images (assuming a monocular acquisition system) can be addressed by object detection techniques or semantic segmentation approaches. For instance, detecting pedestrians, vehicles, and traffic signs used to be addressed by object detection techniques, where the input is an image and the output a set of bounding boxes framing the instances of such classes. However, detecting the visible road surface cannot be tackled as an object detection problem; instead it is tackled as semantic segmentation problem since we need pixel-level classification. Of course, we can aim at detecting pedestrians, vehicles, and traffic signs with semantic segmentation methods, but it would remain the task of distinguishing among the different instances of the same class (e.g., within a region of connected pixels classified as "pedestrian" there can be different persons). We refer to Fig. 2.4 for a visual insight.

Comparing object detection and semantic segmentation in the context of ADAS and autonomous driving with respect to the usual settings followed in the computer vision community, we can see commonalities and differences. Let's call the former "onboard datasets" (e.g., Geiger et al., 2016; Dollar et al., 2012; Brostow et al., 2009; Enzweiler and Gavrila, 2009; González et al., 2016) and the latter as "CV-datasets" (e.g., Everingham et al., 2010; Lin et al., 2014). Let's take the detection of the "pedestrian/person" class as an example, and let's think in terms of differences introduced by intraclass appearance, environmental conditions, and acquisition conditions. In all cases, persons change appearance due to gender, clothes, and body size. However, in CV-datasets people tend to show a rather variable set of poses, while in onboard datasets pedestrians basically appear standing on side view or frontal/rear view. In CV-datasets we have indoor and outdoor environments, the latter mostly visualizing good environment conditions. In onboard datasets the images are acquired outdoors under different natural lighting and weather conditions. Finally, in CV-datasets persons tend to be in photographs, i.e., in focus and captured at high resolution, appearing in not too many instances; while in onboard datasets pedestrians appear at different distances of the camera (i.e., not everybody is in focus and the different instances are captured by a rather different number of pixels) and there can be many of them. In all cases, persons may appear as partially occluded. Yet another difference is that onboard processing must be real-time, while time processing requirements are not always as demanding for other computer vision applications.

Altogether, this implies that there are many synergies between general object detection in computer vision and onboard detection, but also many differences; and the same happens for semantic segmentation. In practice, onboard object detection and semantic segmentation must introduce its own constraints and algorithmic variants. Therefore, in the rest of this section we will focus on those.

2.2.2 Onboard Vision-Based Object Detection

Due to the previously mentioned phenomena, we need to detect class instances (objects) that are captured with varying appearance, pose, viewpoint, and scale. For the sake of simplicity, we assume the most widespread approach, which consists of framing the object instances by 2D bounding boxes (Fig. 2.4 top). Thus, given a rectangular image window, we must be able to determine if it is framing or not one object of interest.

Since appearance, pose, viewpoint, and scale are many sources of variability, we need to build an object detection process able to cope with them. The traditional approach involves three main steps: (1) generation of candidates; (2) classification of candidates; and (3) refinement. Since onboard computer vision images come as a continuous stream, applying temporal–coherence helps to obtain more robust methods as well as

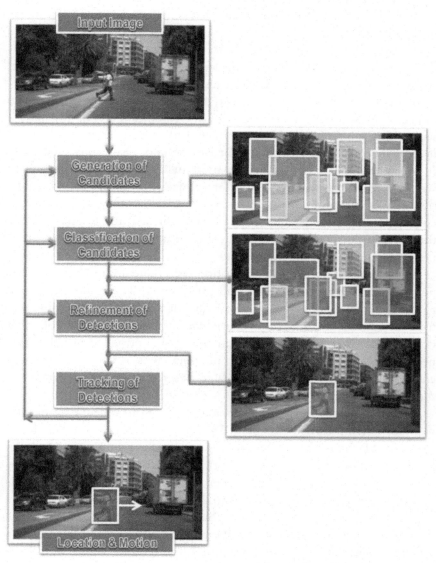

Figure 2.5 Conceptual processing pipeline in classical object detection exemplified by onboard pedestrian detection.

additional information (e.g., moving direction of pedestrians). In the case of object detection, this translates into a step (4) of tracking. Fig. 2.5 illustrates the idea for the case of pedestrian detection, more detailed information can be found in Geronimo and Lopez (2014a). In the following we review the goal of each of these steps, with special interest in candidate classification since it is the one with a larger impact in the final accuracy of the object detector.

Classification of candidates. This processing stage receives an image window of a canonical (fixed) size and decides which type of object frames the window (a pedestrian? a vehicle? just background? etc.); thus, we perform multiclass classification. In fact, the most widespread approach in onboard object detection has been binary classification. For instance, if we are performing just pedestrian detection then the goal is to determine if the window frames a pedestrian or background, i.e., a binary decision. It is important to mention that in most of the methods the processed windows reach this stage with a canonical size. This means that classification may or may not cope with intraclass variability due to scale; we will comment a bit more about this point later. But it is the responsibility of this classification stage to address the rest of intraclass variability sources, i.e., appearance, pose and viewpoint.

In order to perform this classification task, there are three main components: (1) features; (2) model; and (3) classifier. The reader is referred to Geronimo and Lopez (2014b) for in-depth details in the case of pedestrian detection, but that can be generalized to the detection of other objects of interest in onboard vision. Here, we introduce just the basic notions since, in fact, this separation of components is what deep learning changes dramatically as we will see later, i.e., removing such separation is the key of its success.

Let's start by the classifier. In practice, successful object classification methods rely on learned classifiers. Thus, the field of machine learning is the core. Machine learning algorithms such as Support Vector Machines (SVM), Adaptive Boosting (AdaBoost), or Random Forest (RF), are key for learning the desired object classifiers. This raises the question of needing training data with annotated ground truth, which is not a minor issue and will be treated in Section 10.3. No matter which of those machine learning algorithms we use, they output a classifier which can be seen as a function (linear or not) whose inputs are features and whose output is a score or probability that informs if the input features represent an object in which we are interested. The score or probability is then compared with a threshold value to take the classification decision. The optimum threshold is tuned during training by using a validation set (independent of the training set) and constitutes a trade-off between accepting false detections (i.e., determining that the window under classification contains an object of interest when it is false) and misdetections (i.e., determining that the window does not contain an object of interest when there is one).

The role of feature extraction is to take an input window of raw pixels and project it to a new space, the feature space, where windows containing the same class of objects are mapped closely, while windows containing different classes are mapped far apart. Accordingly, a classifier can be seen as a frontier in the feature space which must discriminate between the different classes of interest. Therefore, features are critical to make an effective classification. If the features are highly discriminative, simple frontiers can be built (e.g., linear hyperplanes), otherwise, more complex (nonlinear) frontiers are required. In general, simpler frontiers are desired since the corresponding classifiers tend to be more accurate.

But where do these features come from? In short, they are designed by researchers. By analyzing the addressed task (e.g., pedestrian detection), different researchers have proposed different transformations of the raw pixels aiming at reducing that intraclass variation in feature space while enhancing interclass differences. We could talk about the art of hand-crafting features. A popular and successful example is the so-called HOG (histogram of oriented gradients). Basically, HOG features are built by splitting a window in cells, computing the orientation histogram of each cell, and concatenating all the histograms as a feature vector that represents the window. Indeed, there are many more crucial details (blocks, normalizations, interpolations) that make HOG more complex than this intuition, but still this is the essence. What is relevant here is that HOG was designed by humans. This is the case of many more popular and useful feature types, for instance, local binary patterns (LBP) and Haar wavelets.

With the technologies mentioned so far we can already build object detectors with a reasonable accuracy for many applications. For instance, combining HOG features with a SVM classifier, see Dalal and Triggs (2005), was the state-of-the-art pedestrian detector in 2005, and posterior improvements keep including both HOG and SVM. This, in fact, drives us to the concept of model, one of the three components previously mentioned as part of the classification task. The original HOG/SVM classifier follows what we call a holistic model; i.e., an object is seen as a whole rigid entity without parts. This means that the features take the responsibility of coping with the object's variability due to appearance, pose, and viewpoint. Since this is really challenging, this holistic model was replaced by more sophisticated ones. In particular, one of the most widespread was the so-called deformable part-based model (DPM; see Felzenszwalb et al., 2010), which explicitly accounts for different poses (assumes that an object has mobile parts) and viewpoints (incorporated as different submodels that are further joint to form the overall model). The underlying reason for using such more elaborated models is that by explicitly taking into account the sources of variation, the models are more accurate, which was confirmed by different experimental results over the years. Fig. 2.6 illustrates the idea: when learning a classifier based on a holistic model all object instances are mixed as positive examples and it does not matter if the object has mobile parts that can appear at different relative locations. When learning a model like DPM (deformable part-based), different views are explicitly taken into account (e.g., in the middle: frontal/rear pedestrians; to the right: side-viewed pedestrians) as submodels so that the global model is a mixture of them. Moreover, each

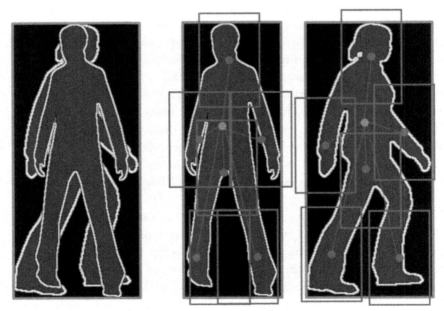

Figure 2.6 Holistic model (left) and a multiview multipart model (right).

submodel takes into account that the objects have mobile parts. As a result, for each submodel the features used are the concatenation of the features of each part plus the features of the object as a whole, plus the relative location of the parts.

Generation of candidates. This processing stage receives an image and must provide windows of a canonical size to be later classified by the processing stage previously seen. The simplest method consists of sliding the canonical window along the full image, given fixed row and column displacement steps. If these steps are large, we will need to classify few windows; however, we can miss objects that can be inbetween windows. If these steps are too close, we will process many windows (the extreme case is one per pixel), many of them redundantly because the classifiers themselves must already be robust to small displacements of the window's content. Yet, with this method we still would fail to account for variations of scale, a crucial issue in onboard computer vision. Therefore, rather than applying a sliding window only to the original image, in fact, a pyramid version of the image (from the original to progressively new versions of lower resolution) is constructed and the sliding procedure is applied at all the layers of the pyramid, using the same canonical window. At the bottom of the pyramid we have the original image, while at the top we have the smallest version of the original image where the canonical window still fits. In this way, the closets objects are detected at the top of the pyramid, the further away ones at the bottom. The overall

Figure 2.7 Pyramid of features to perform pyramidal sliding window.

procedure is known as pyramidal sliding window. Fig. 2.7 illustrates the idea. Note also that in practice a pyramid of the features is carried out to avoid redundant feature computations between neighboring windows.

Overall, we can see that scale variability is mainly treated at this processing stage. However, during the classification of candidates, it is possible to do more to explicitly treat scale. In particular, both during the training of the object classifiers and during operation time of the object detector, the original size of the windows is known; therefore, we can have scale-dependent classifiers. In this way, blurred examples due to distance (low resolution) are not mixed with well-focused examples close to the camera (high resolution). This consideration has been shown to improve the overall accuracy of the detectors (see Benenson et al., 2012); Park et al., 2010; Xu et al., 2016).

Despite the practical usefulness of this method, it has two main drawbacks. First, since it is a blind method with respect to the image content, a priori nothing prevents the search for "flying" pedestrians for instance. Thus, when we have a stereo rig we can use additional information to only generate windows at plausible locations; e.g., windows whose bottom matches a detected ground surface and only around relevant concentrations of 3D points, see Geronimo et al. (2010). In other words, we can place 3D-style windows in a depth or disparity image and project them into the 2D image. Usually, this reduces at least one order of magnitude the number of windows to be classified. Using a monocular system we can also perform ground plane estimation in order to try to reduce the number of windows in an analogous way, see Ponsa et al. (2005).

The second drawback is the single-class nature of the pyramidal sliding window and its variants. Note that, the generated candidate windows have a canonical size. This size is class-dependent. This means, for instance, that for vehicles it will be different than for pedestrians. Therefore, for each class of interest eventually we need to build a different pyramid; which is very time-consuming. Depending on the features, it is possible to share them between classes, but still it is difficult in general. Accordingly, an alternative approach for a multiclass classification setting consists of learning to generate windows of different sizes and aspect ratios so that they are later classified as belonging to one of the classes of interest in a single step. Fig. 2.8 shows alternative approaches to sliding window: (A) onboard image. In (B) we depict as yellow rectangular boxes the 0.1% of the windows we should process according to a pure

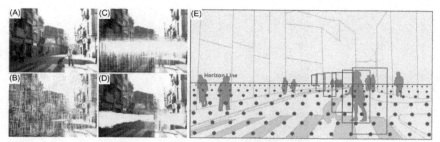

Figure 2.8 Different alternatives to pyramidal sliding window.

pyramidal sliding window approach. In (C) we see how the number of windows is reduced (5% of them shown) by taking into account only windows that touch the ground [see (D)]. In (D) we apply an additional reasoning based on 3D-point density according to stereo information [the image in (A) comes from the left camera of a stereo rig].

Refinement. For the sake of completeness, we note that after classifying the candidate windows there can be redundancies. Note that some windows are overlapping over the same object of interest. Therefore, they can all eventually being classified as containing such an object since object classifiers must show a certain degree of shift-invariance. Accordingly, a procedure to resume redundant detections in one is used. It takes into account the overlapping geometry and the classification scores/probabilities of the windows for removing them following a greedy strategy.

2.2.3 Onboard Vision-Based Semantic Segmentation

Semantic segmentation aims at assigning a semantic class to each image pixel. Fig. 2.9 shows its classical processing stages, see Ros et al. (2015). Since classifying image pixels individually can be very time-consuming and prone to errors, it is usual to do an image partition in regions called super-pixels. Such super-pixels are groups of connected pixels that are similar according to some low-level features, usually color or texture. Ideally, the contours of the interesting objects of the processed image match frontier segments of a subset of super-pixels. Once these super-pixels are created, each one is classified individually in an analogous way than windows in the object detection pipeline. Therefore, new features are involved (e.g., SIFT, HOG, Fisher Vectors) together with a trained classifier (SVM, AdaBoost, Random Forest). In the next step, spatial coherence is induced in the sense of reclassifying super-pixels by taking

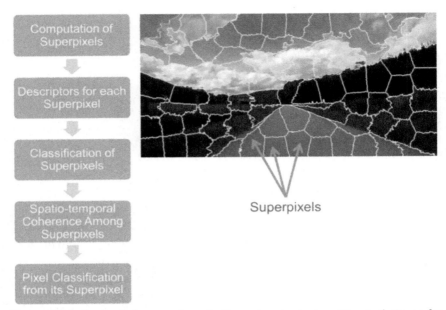

Figure 2.9 Left: classical processing pipeline using computer vision techniques for semantic segmentation. Right: example of mage partition based on super-pixels.

into account the classification of their neighbors, which can be done by using conditional random fields (CRFs) for instance. This step suppresses spurious misclassified super-pixels surrounded by properly classified ones. Moreover, it induces statistically dominant spatial relationships (e.g., the sky is on top of the rest of classes). Finally, all image pixels are classified with the class of the super-pixel to which they belong.

Semantic segmentation has been always a rather complex problem, very time-consuming both during training and operation. Many variants have been proposed to alleviate processing times as well as to improve the segmentation accuracy. Overall, when one looks at the different proposed pipelines they can be seen as a sequence of steps, where each steps tries to cope with problems remaining from the previous steps.

When processing images arriving in streaming, it is also possible to induce temporal coherence during super-pixel classification. This is done by projecting the super-pixels of the previous frame into the current one for instance with the help of optical flow. The previous classification of a super-pixel is taken into account to perform current classification.

Figure 2.10 Examples of road segmentation: red pixels have been classified as road, the rest as nonroad. These results correspond to the method in Alvarez and Lopez (2011).

For onboard vision, many times researchers have focused on the particular case of road segmentation. In other words, in only detecting the navigable space; thus, super-pixels are classified as road or background. Fig. 2.10 shows some examples or road detection.

2.2.4 Onboard Vision Based on Deep Learning

When talking about deep learning in the computer vision context, we mainly refer to deep Convolutional Neural Networks (CNNs); where the one known as LeNet is an early use case illustrating vision-based character recognition, and Krizhevsky et al. (2012) produced the disruptive paper on image classification that led the computer vision community to switch to address all kinds of visual tasks with deep CNNs. Such a breakthrough was due to the combination of three main factors: (1) the massive availability of images with labels (we talk more about this in Section 10.3); (2) the amazing performance of modern GPUs for training the millions of parameters of such CNNs; and (3) a set of best practices to make the training of CNNs affordable (e.g., SGD, drop-out, batch normalization, data augmentation, etc.). Fig. 2.11 illustrates a classical deep CNN architecture; of course, nowadays there are many variants: an input image is processed by a bank of K parallel convolutional layers, the output of the neurons is taken by a following layer of pooling with subsampling; this two-stage block can be concatenated. The output of a sequence of such blocks enters fully connected layers that in turn can be concatenated

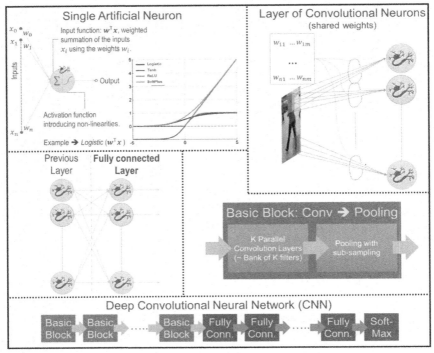

Figure 2.11 Classical deep CNN architecture.

again, until a last step of classification is performed (e.g., soft-max). There are many hyperparameters to decide: number of neurons per layer, number of blocks, number of convolutional filters in the corresponding layers, dimensions of such filters, nonlinearity type at the neuron level, etc. In addition, there can be millions of parameters to learn (weights of the convolutional filters and of the fully connected neurons); however, they are all jointly learned by optimizing a desired cost function (e.g., related to classification accuracy) using SGD (or any improvement to this standard) during backpropagation.

But why are deep CNNs so appealing? In short, thanks to the so-called end-to-end learning. In other words, learning a deep CNN implies to learn jointly both a hierarchical feature space as well as the desired classifier; i.e., there is no need for hand-crafting features. This approach has boosted accuracy in a variety of computer vision tasks.

The introduction of CNNs for tasks beyond image classification was progressive in the field of ADAS and autonomous driving and in computer vision in general; i.e., first attempts tried to combine preexisting image processing pipelines and CNNs.

For instance, in Razavian et al. (2014) CNN layers are used as off-the-shelf features for object detection and classification, fine-grained object recognition, and attribute detection. In Hariharan et al. (2015) hypercolumns of CNN layers are used for object segmentation and fine-grained localization.

Focusing on object detection, we can see how the first approaches followed the idea of Fig. 2.4, replacing the classifier by a CNN architecture and so avoiding hand-crafted features. For instance, in Sermanet et al. (2014) we can find the so-called OverFeat approach which relies on sliding windows for generating candidates and a CNN for classifying them. In Girshick et al. (2016) we can find the so-called R-CNN which uses the method in Uijlings et al. (2013) for generating candidate windows (in short, candidate windows based on learning to group super-pixels) and the CNN of Krizhevsky et al. (2012) for classifying them. R-CNN is too slow, and different subsequent modifications were introduced to gain speed. In particular, in the Fast R-CNN version of Girshick (2015), computations are shared at the classification level, while in the Faster R-CNN of Ren et al. (2015), the candidate generation step is also based on a CNN which shares computations with the CNN used in the classification step; thus, giving rise to an end-to-end trained object detector. Fast R-CNN is quicker by a factor of 25 approximately over R-CNN, and Faster R-CNN is quicker by a factor of 250 approximately. Typical PASCAL VOC images are processed at approximately 0.2 s by the Faster R-CNN. Further CNN-based approaches for object detection have been proposed, being significantly faster even than Faster R-CNN, for instance, the so-called YOLO9000 of Redmon and Farhadi (2016) and the SSD of Liu et al. (2016). It is worth mentioning, that these generic object detectors may need some adaptation to operate in the ADAS/AD context. For instance, Zhang et al. (2016) does an analysis and posterior improvement of Faster R-CNN that increases the accuracy on the task of pedestrian detection. Usually, adaptation of the new type of scenario (urban onboard), camera, and multiresolution is required when going from generic object detectors to detectors specialized in onboard driving.

Focusing on semantic segmentation, we can see how the very complex pipeline illustrated in Fig. 2.9, has been replaced by some type of deep CNN architecture. The benefits of this new approach are clear: (1) dramatic increase in accuracy results; and (2) simplification of the overall learning procedure. Note that each stage in the pipeline of Fig. 2.9 has been a research area on its own: methods to compute super-pixels; methods for describing the super-pixels (i.e., low-level features and descriptors based on them); classifiers to operate in the space of the descriptors; fast and accurate methods to work with conditional random fields, etc. Now, the responsibilities of all these stages are taken by the CNN architecture, which is trainable end-to-end. In other words, semantic segmentation is not the summation of a plethora of stages, each one being a problem in itself. The CNN architecture takes the image as input and outputs its semantic segmentation.

Popular CNNs for semantic segmentation are Fully Convolutional Networks (FCNs) by Long et al. (2015), Deconvolutional Networks by Noh et al. (2015), SegNet by Badrinarayanan et al. (2015), Dilated Convolutions by Yu and Koltun (2016), and ResNet by Wu et al. (2016). These networks have shown astonishing performance for onboard driving images. Thanks to this performance, even harder task are being taken, for instance, instance-level semantic segmentation as in Uhrig et al. (2016) where the aim is not only knowing the class to which each pixel belongs, but also to distinguish different instances of classes such as pedestrians, cars, etc. Moreover, CNN-based semantic segmentation is also being combined with other complementary world representations such as the so-called Stixels, see Schneider et al. (2016). In addition, CNN architectures for jointly performing semantic segmentation and object detection are under development for AD, see Teichmann et al. (2016).

It is also worth mentioning that there are lines of research which aim at avoiding the explicit computation of intermediate world representations such as the ones produced by object detectors and semantic segmentation. For instance, in Chen et al. (2015) a deep CNN is used to generate steering wheel angles directly from the "direct" processing of raw images, which is known as End-to-End Driving. More elaborated approaches can come from sensorimotor learning based on both deep learning and reinforcement learning, see Dosovitskiy and Koltun (2017).

At this point, it is clear that deep learning is bringing a revolution for ADAS/AD based on computer vision. Obviously, one can argue that the design of the CNNs and their training are an "art" at the level of hand-crafting features; the latter was many times used as an argument to criticize traditional computer vision approaches. However, the advantage of CNN is that "features and classifiers" are trained end-to-end in a way that we can obtain a hierarchical representation of the images, where deeper layers of the CNN can capture higher level concepts (e.g., a pedestrians) while lower layer capture more class agnostic and low-level concepts (e.g., edges). Moreover, from a practical point of view, we do not need to develop and fine-tune different compositional stages to perform a visual task, a CNN architecture is not only the representation of the knowledge to perform the addressed task, it is also the algorithm to execute it.

From the point of view of ADAS/AD, probably one of the main remaining issues with CNNs is their memory footprint, since they are eventually based on millions of parameters. Accordingly, distilling CNNs is also a key research topic, see Zhou et al. (2016).

2.3 LIDAR-BASED PERCEPTION

LIght Detection and Ranging (LIDAR) is a technology that measures distance to a target by illuminating that target with a pulsed laser light. The measurement system is based on the Time of Flight (ToF) method, which determines the time that a laser pulse needs to overcome a certain distance in a particular medium. The Lidar sensor emits a short laser pulse and measures the time that elapses whilst the laser pulse is reflected by an object and subsequently received by the sensor. Due to the constant propagation of light, the distance to the reflecting object can be calculated by measuring the time between the emission, the reflection, and the subsequent receipt by the sensor. In the automotive sector, laser pulses with a length of 3−20 ns are used for the ToF method whereas the shorter laser pulses provide a better accuracy. LIDAR sensors in the automotive industry can reliably detect objects within ranges of up to 300 m. By convention, Lidar outputs are generally specified in terms of wavelength rather than frequency. Lidar systems used in ITS typically operate in the near-IR region of the EM spectrum, between 750 and 1000 nm.

Current LIDAR systems use rotating hexagonal mirrors which split the laser beam, either to be used for different detections or for achieving 360-degree, however, innovative high resolution 3D flash Lidar systems are being developed which remove the moving parts of the Lidar system. This will be a great advantage in the future for obtaining more robust, reliable and smaller Lidar systems than the current ones, and will also allow a smoother integration in the vehicles and ITS.

LIDAR sensors for ITS systems have either a high angular resolution and a wide field of view, or a full 360-degree view for precise and reliable mid-range detection. The scanned data are processed and signal processing algorithms are employed to properly interpret the objects as well as adjust for limited visibility issues. With the help of software tools, the scanned point cloud and detected objects can be used to create a model of the vehicle's surroundings and to calculate several individual dynamic variables to support navigation.

LiDAR sensors are becoming one of main technologies that have found a new wide application field in autonomous vehicles. In this way, several developments for surroundings recognition are using this type of sensor. One of the major differences between LiDAR sensors and cameras is that the latter are very sensitive to light changes or challenging environment factors due to their intrinsic performance. Whereas LiDAR sensors avoid these drawbacks, illuminating objects with a singular laser pulse and measuring the time it takes to travel from the sensor to the object and to return. Moreover, cameras only display where they are placed, they are directional; instead, LiDAR sensors provide most of the time a 360 degree field of view, getting a complete point cloud of the vehicle's surroundings. However, the point density does decrease with the distance, due to the divergence between laser beams.

The point cloud data provided by a 3D laser scanner consists of a large number of points distributed vertical and horizontally, organized in layers. Each point provides the distance to the sensor in local coordinates and also can indicate the reflection intensity value. With these data, it is completely determined where the point is, its orientation, reflectivity, and even the exact moment of the impact. Using, different combinations of data available by the sensors, several algorithms for surroundings recognition have been developed. But, due to the high accuracy of data provided by scanner lasers, a great number of algorithms rely on LiDAR sensors to detect obstacles, identify reference elements for autonomous driving, calculate the passable or nonpassable path, etc.

2.3.1 Surroundings Recognition

Autonomous navigation for ground vehicles requires a very precise knowledge of the surroundings for safe driving. An essential requirement in surroundings understanding is to identify obstacles and path boundaries. Depending on the type of environment in which the vehicle is located it is necessary to tackle these tasks differently. There are mainly two types of environments, urbanized and nonurbanized areas. Urbanized areas comprise urban and interurban environments, where the surroundings are structured to a greater extent. Curbs, traffic signs, lanes lines, or buildings create a distinguishable pattern along the streets and roads where the vehicle circulates. On the other hand, off-road environments offer a greater complexity due to the lack of reference elements.

2.3.1.1 Obstacles Detection

While driving a car, we are always evaluating possible obstacles that may appear in our path and making the proper decision. In autonomous driving, the car's eyes are the sensors, and specifically LiDARs give a surrounding interpretation in 360 degree. Algorithms must analyze, identify, and track potential obstacles that involve risk to the vehicle's circulation.

A generalized form to tackle the obstacle detection problem with LiDARs is described as follows:

- *Selecting the region of interest (ROI)*:

    As mentioned before, LiDARs provide a large number of data; consequently, the computational cost could be a problem. In order to simplify the point cloud obtained, a ROI must be defined. Selecting only the area where the points' data are useful, it is possible to discard information without interest for the algorithm calculations. The ROI selected depends on the application and what is going to be identified. In this way, ROI can be defined as constant limits in the 3-axis, focusing the available data in a specific area; or also, it is defined as a function of the look-ahead distance, which depends on the vehicle speed—the higher the speed, the greater the distance; or, if the trajectory that the vehicle has to perform is known, the ROI can be fit to the oncoming path.

    After selecting the ROI, a denoise function is normally used to delete those points isolated in the space and if necessary, downgrading the point cloud in order to keep a constant density throughout the region due to the dispersion issue described above.

- *Features identification*:

    In urban areas, as the environment is more structured, geometric shapes are more easily recognizable. For this reason, looking for planes that fit in the point cloud or any other shape, like spheres or cylinders, is a commonly used method.

Another strategy employed in the identification stage is the study of normal vectors. With this study, it is possible to set thresholds limits that comprise the target element, or to calculate the angle evolution of the normal vector throughout the neighbor points. This normal vectors-based method can provide general features extraction. For that reason it is a preferable method for off-road situations.

In general, several filters are applied consecutively, taking into account different calculations. Those points that comply with all the proposed filters become part of the potential obstacles.

- *Data segmentation*:

After extracting all the possible obstacles, it is necessary to classify them in order to endorse that the potential obstacles are real obstacles.

Firstly, a clustering method is used. The clustering algorithm groups the points in classes depending on the features between points. In addition, if the number of clusters is known or not beforehand, algorithms that classify the point cloud into that number of cluster (e.g., k-means) or not (e.g., DBSCAN) can be applied.

A frequent issue in data segmentation is to split the same obstacle into two or more different clusters. Because of the shadow produced by a closer obstacle, another element placed behind is divided, and consequently, identified as various obstacles. For that reason and for improving the tracking performance, it is essential to take into account a data association step. A typical method to tackle data association is with a multiple hypothesis tracking (MHT) algorithm or that described in Petrovskaya and Thrun. (2009) called the "virtual scan" method.

- *Obstacles tracking*:

Once all the possible obstacles are classified, it proceeds to perform the obstacles tracking. The tracking stage consists of monitoring the obstacles previously identified in order to improve the reliability of the identification itself. Monitoring the obstacles permits noticing what elements are static and which ones are in motion by studying the transitory states.

It is common to use bounding boxes that comprise the obstacle and calculate different parameters such as its orientation, velocity, location, size, etc. Then, by variants of Kalman filter, matching between different instants of time can be made.

Fig. 2.12 shows an example of obstacles detection using LiDAR information after performing the different steps.

Obstacles identification

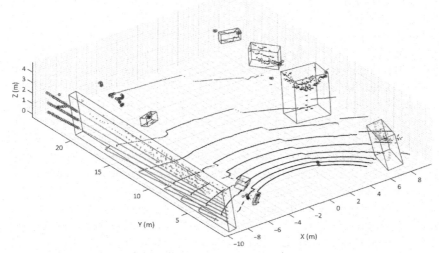

Figure 2.12 Example of onstacles detection using LiDAR.

The stages described above are the standard process for obstacle identification, but depending on the situation and the application to be developed, they can vary and adapt to another approach. All obstacles that may appear in the vehicle's trajectory can be categorized into two groups, positive and negative obstacles depending on its height relative to the ground.

1. Positive obstacles

In urban areas, the most common obstacles that may appear on the road are vehicles, pedestrians, cyclists, traffic signs, barriers, etc., while in off-road situations, most of the obstacles that may involve a dangerous situation are trees, large rocks, bushes, etc. In both cases, they have a positive height above the ground. There are several developments tackling this task.

As mentioned above, a simple approach would be a feature-based method (Gidel et al., 2010). This approach is only suitable when the environment offers a flat floor or there is barrier-separated road or, in general, when there are geometric features such as planes, lines, boxes, etc. Most of this geometric matching is based on variant of the RANSAC iterative algorithm. It is clear that when the obstacles present a complex shape, it is necessary to apply other strategies.

Another common approach is based on grid-occupancy calculation, as in Montemerlo et al. (2009), who employed it in their
self-driving car used in the DARPA Urban Challenge. They used a
grid-occupancy method in order to discard overhanging objects that
do not present any danger.

The next step to grid-occupancy methods is the features extraction
by voxels. By voxel, it means a 3D pixel. Thus, the evolution over the
time of points density in each voxel can be measured and, therefore, it
is possible to distinguish between static or dynamic obstacles, as in
Asvadi et al. (2016).

Then, more sophisticated methods for features extraction have
been proved in other works. That is the case described by Han et al.
(2012), where a line segment extraction in polar coordinates to detect
any type of obstacle on both, road and off-road situations is used.

2. Negative obstacles

On the other hand, obstacles will not always be positive, and different methods of those used for positive obstacles have to be studied.

For this purpose, LiDAR technology is being used for roads and
streets maintenance. In these areas, ruts, cracks, potholes, etc. may
result in a safety hazard. For that reason, using the accurate measurement and the intensity reflection data it is possible to identify these
types of elements on the road (Laurent et al., 2009; Li et al., 2009;
Wang et al., 2011).

When the autonomous vehicle is circulating along an off-road terrain, the traditional LiDAR-based obstacles detection approach is not
the optimal way to accomplish this task. Instead, as is developed by
Larson and Trivedi. (2011), comparing the ideal distance to the real
one when there is a negative obstacle, the latter can be located.
Focusing on the relationship between adjacent points or identifying
shadows produced by the negative obstacle are different strategies for
negative obstacles detection. Furthermore, there are developments
applying a novel setup method for the fusion of two 3D LiDARs in
order to get a better density in the ROI (Shang et al., 2016) to find
potholes on the path.

2.3.1.2 Path Boundaries Detection

In addition to recognizing what obstacles can lead to an imminently hazardous situation, knowing the limits of the road and distinguishing the
passable area from the nonpassable is fundamental for the system

integration. Decision-making by autonomous vehicles must have a perception of the environment that, besides avoiding any collision, knows the viable area through which they must circulate. The merging of the route-tracking system together with the estimation of the viable path, make possible a fully autonomous navigation. Moreover, being able to recognize the route by which the vehicle can circulate is fundamental for the strategies of path-planning.

LiDAR-based algorithms to estimate the passable area of the path work similarly to how they proceed in identifying obstacles. However, there are some distinguishable stages:

- *Selecting the region of interest (ROI)*:

    ROI can be selected following the same strategy as mentioned before: as constant limits, as a function of the velocity, etc. but this time, the ROI is focused at the floor level.

    In this case, the major difficulty lies in extracting those LiDAR points belonging to the floor. It is not a trivial task such as simply selecting points whose Z height is equal to zero. Oncoming slope changing or superelevation rates can misrepresent these measures.

- *Features identification*:

    After the floor-points identification, the features extraction stage of those points is applied. In these cases, calculations usually take into account different parameters such as the neighboring points' orientation, curvature, standard deviation, or another statistic measure.

    Moreover, LiDAR sensors offer the reflection intensity data. This intensity data depends mainly on four parameters of the element where the laser point is fired: what the material is made of; its color; its surface orientation with reference to the sensor placement; and the distance. Thus, when the element's material has high reflectivity (e.g., lane lines of the road), they can be easily recognized.

- *Data segmentation*:

    Data segmentation might seem like an easier task in these cases. Due to the fact that only two possible boundaries for the trajectory are searched, a k-means algorithm defining two clusters can be used. However, it is an error to think that this solution is valid only when circulating for example on a highway, identifying the lane line on the left and on the right. It is necessary to contemplate other more challenging situations. Intersections of streets, corners, incorporations and exits in roundabouts are some examples where the boundaries estimation becomes more complicated to identify. For this reason, it is important to understand the environment where the vehicle is navigating in order to have adaptive capacity.

As happened in the obstacles identification, shadows may appear due to an element placed between the sensor and the limit of the road. This time, data association follows a better-defined model, most of the time it will be a rectilinear pattern, a wide curvature turn, or a 90° corners.

- *Reference elements tracking*:

Once the boundaries are identified and classified, tracking these elements has the advantage of being continuous throughout the LiDAR's ROI most of the time. Therefore, estimating where the element must be in the following instant of time can be helpful for monitoring them. On the other hand, the disparity issue of LiDAR points, make tracking more laborious with large distances.

Some specific examples of how the boundaries limits of the path have been estimated are described below:

- *Lane lines position and curvature*

As was previously anticipated, the intensity value given by the LiDAR sensors is a measure of the reflectivity of the material. Thus, using this available data, lane lines are simply detectable due to their reflectivity and the contrast with the asphalt. The problem is not detecting this type of traffic element; the main issue is that the lines are drawn on the floor. Extracting the floor points can become a tough task. Different approaches have been done. In urban areas, using geometric features matching for a planar surface, may work, but oncoming slopes or different irregularities on the road/street may cause an error, for that reason, this geometric matching is made consecutively, as in Oniga and Nedevschi. (2010)

- *Curbs identification*

Additionally, curbs identification is also a common task. As described above, extracting points belonging to the floor requires specific calculations. In this manner, in Montemerlo et al. (2009) they calculated distances between the different layers of the 3D LiDAR when the pitch and roll angles of the vehicle frame varied. Specifically, to identify small curbs, they calculated the distance between rings and by comparing this "adjacency" with the expected distance, curbs were detected.

- *Path edges in off-road situations*

In a complex scene, like off-road environments, delimiting the path edges is not as clear as in urban situations. This time, the only reference elements that may help to limit the passable areas are usually trees, bushes, etc. Moreover, terrain often presents more irregularities than in a structured ground. This is why, an ICP algorithm is often used to increase the density of the point cloud. When there is more information about the terrain available, it is easier to identify reference elements on the ground such as the ruts. Furthermore, using the same calculation between the LiDAR layers in urban areas has no sense off-road as the irregularities that appear in the path distort this calculation too much.

2.4 SENSING FROM THE INFRASTRUCTURE

The "sensorization" of ITS has multiple applications, going from the detection of incidents, to the identification of vehicles or the measurement of the atmospheric conditions of the roadways as possible examples.

All the possible applications of the sensors included in ITS can be classified under these two groups: detection and vehicles monitoring, and the control of meteorological and environmental conditions.

Taking as starting point this first classification, the ITS sensors can be separated into two big categories:
• Traffic sensors; and
• Meteorological sensors.

In this section the evaluation of the traditional methods of information acquisition associated with mobility, security, and traffic conditions will be done. Considering the traffic sensors, it is possible to do a new classification based on the position of the sensors and on the requirement or not of devices in the vehicle. So this will get to the differentiation of two types:

Autonomous traffic sensors: those that do not require a device in the vehicle and for which the sensor element is situated in the infrastructure.

Dependant traffic sensors: those in which the sensor element is in the infrastructure and that require a device in the vehicle or in which the sensor element/s are in the vehicle.

In the following paragraphs the main "sensorization" technologies considered as included in the previous groups will be explained.

2.4.1 Autonomous Traffic Sensors

Two types of sensors for detection and vehicle monitoring can be specified (Klein et al., 2006): intrusive technologies that are installed on or along the pavement, and nonintrusive technologies in which the sensors are above or at the sides of the roadway creating as little as possible impact on the traffic flow.

2.4.1.1 Intrusive Sensors

In terms of intrusive sensors, it is possible to find a lot of types that will be described hereafter: the magnetic loop, pneumatic tubes, piezoelectric sensors, optical fiber sensors, and geomagnetic sensors.

Although the fundamental bases of all of them are similar—all of them detect the passage of a vehicle when going over the sensor—the first two are the most commonly used in roadways.

Intrusive sensors are *normally cheaper* in terms of installation costs, however they present a series of disadvantages:

- Traffic disruption during installation and repair processes.
- They can lead to errors in the measurements if pavement conditions are not adequate or if not a proper installation has been done.
- Re-asphalt of highways may require sensor reinstallation.

This type of sensor can provide information about *volume of vehicle traffic, detection and vehicle classification*, or even information about the *speed*.

2.4.1.1.1 Magnetic Loops

Magnetic loops are the most commonly used sensors in roads due to the fact that it is a cheap and mature technology, with a simple operation that is not influenced by environmental conditions.

Just like the other intrusive sensors, they have, as a handicap, the requirement for work on the infrastructure for their installation and

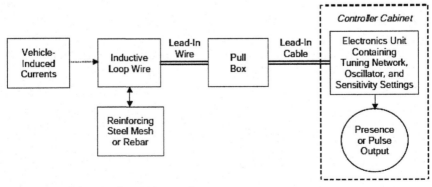

Figure 2.13 Scheme of an inductive loop.

repair. Additionally, they are liable to failure and malfunctioning, so there is a trend to look for new technologies to replace them (Fig. 2.13).

The functioning basis of a magnetic loop is the inductance variation that happens in the loop at vehicle passage. The loop is made up of four parts:

- *Loop* or *coil* made up by a cable with one or more turns, embedded in the highway pavement close to its surface.
- *Introduction wire* that connects the loop with a splice box.
- Another *join wire* between the splice box and the controller
- The *electronic cabin* situated in the control cabin. This unit includes an oscillator and an amplifier that feeds the inductive loop.

The physical explanation of this phenomenon is the following:

The electronic unit transmits current to the inductive loop.

The current that flows through the loop produces a magnetic field H around the cable (see Eq. 2.1). N is the number of turns of the cable loop, I is the current in the cable measured in Amperes and l is the length of the loop (or coil).

$$H = \frac{N \cdot I}{l} \tag{2.1}$$

The magnetic field produces a magnetic flux through the loop Φ represented by the Eq. (2.2), where B is the magnetic flux density, A is the area enclosed by the loop, μ_r is the relative magnetic permeability of the media and μ_o is a constant with a value of $4\pi \times 10^{-7}$ N/A^2.

$$\phi = B \cdot A = \mu_r \mu_o H \cdot A \tag{2.2}$$

The inductance L (expressed in Henrys) of the inductive loop varies as the magnetic flux that passes through it varies as well, following the expression (2.3). From this variation the presence of vehicles can be deduced.

$$L = \frac{N\phi}{I} = \frac{NBA}{I} \tag{2.3}$$

If a vehicle (or any other object that conducts electricity) enters into the mentioned magnetic field, or in a more precise manner, any component of the magnetic field is normal to the object area, a current is induced in the cable that generates another magnetic field opposite to the first magnetic field mentioned. This fact produces a decrease of the global

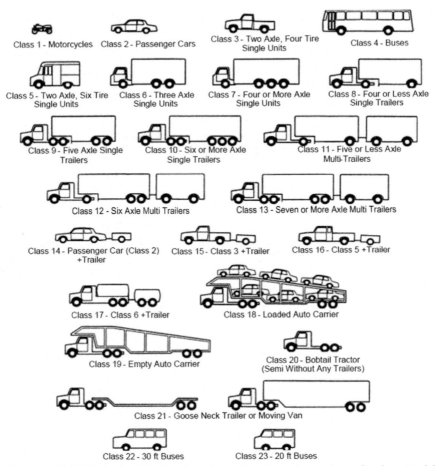

Figure 2.14 Vehicle classification based on length and number of axles. *Model IVS-2000 Installation and Operation Manual, Rev. 1.53, 1997.*

magnetic field. Taking into account that the inductance of the loop is proportional to the magnetic flux, this means that the inductance decreases as well.

This technology can detect variables such as *traffic volume*, *presence*, *speed*, or *vehicle class* (see Fig. 2.14) using a configuration of simple or double loop, and a prefixed distance separation between them.

Apart from the permanent installations of inductive loops (Fig. 2.15), there are temporal portable installations that do not require insertion in

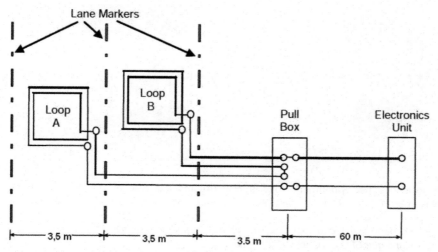

Figure 2.15 Installation scheme of loops for two lanes. *Klein, L.A., Mills, M.K., Gibson, D.R.P., 2006. Traffic detector handbook. McLean, VA, U.S. Dept. of Transportation, Federal Highway Administration, Research, Development and Technology, Turner-Fairbank Highway Research Center. <http://purl.access.gpo.gov/GPO/LPS91983>.*

the pavement but that instead are attached, one way or another, to the surface of the roadway, as shown in Fig. 2.16.

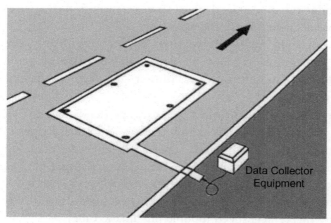

Figure 2.16 Temporal inductor loop fixed to the pavement. *Klein, L.A., Mills, M.K., Gibson, D.R.P., 2006. Traffic detector handbook. McLean, VA, U.S. Dept. of Transportation, Federal Highway Administration, Research, Development and Technology, Turner-Fairbank Highway Research Center. <http://purl.access.gpo.gov/GPO/LPS91983>.*

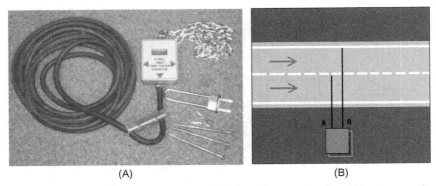

Figure 2.17 Real pneumatic tube (K-Hill) (A) and example of positioning on the pavement (B). *(A) Klein, L.A., Mills, M.K., Gibson, D.R.P., 2006. Traffic detector handbook. McLean, VA, U.S. Dept. of Transportation, Federal Highway Administration, Research, Development and Technology, Turner-Fairbank Highway Research Center. <http://purl. access.gpo.gov/GPO/LPS91983>, (B) Appendix F.*

2.4.1.1.2 Pneumatic Tubes

Pneumatic tubes are rubber tubes (see Fig. 2.17A) that are installed along the pavement over the roadway and that are capable of passage vehicle detection by the change of the pressure that the vehicle creates on the air that is inside the tube as the vehicle passes on it.

The burst of air created inside the tube feeds a membrane that closes a switch producing an electrical signal that is transmitted to a counter or to equipment adequate for its processing.

They normally include a register, that approximately every hour saves on a magnetic tape or on paper, the information collected in that time lapse, resetting the counter to zero to begin a new measurement period.

Pneumatic tubes are installed normally perpendicular to the driving direction (see Fig. 2.17B). They are a cheap and reliable sensor, but due to the fact that they are installed superficially they get damaged quickly and require regular monitoring. The electrical signal that they generate is very weak so it needs amplification to be able to work with it.

The installation of two tubes separated a prefixed short distance allows to compute the vehicle speed (double pneumatic tube).

These types of sensors are very useful in provisional installations or in those of short duration, as the rubber tubes are easily installed on the pavement and the register is fed by a battery. The most frequent failure that these sensors may suffer is the electric failure of the counter and the damage of the rubber tubes as they have to resist the impact of the vehicles' tyres.

With this technology it is possible to measure different variables such as *traffic volume, speed* (using the double tube mentioned before), and *classification* (based on the number and space between axles) of the vehicle.

2.4.1.1.3 Piezoelectric Sensors

The detection in these sensors is based on the pressure that the vehicle generates on them while passing over them. They detect the passage of the vehicle based on the electric charge that is generated by the piezoelectric material when it is pressed by a tire while distorted.

They are installed on the surface of the pavement, and this makes them appropriate for mobile measurements. Just as happens with the rubber tubes, these sensors require amplification of the signal they generate.

When compared with rubber tubes, these sensors are more robust and due to the fact that they are thinner, they have lesser impact on the driving experience. A generic scheme of these type of sensors can be seen in Fig. 2.18.

Regarding their functionality, it is similar to that of the rubber tubes, this means they are capable of measuring traffic volume, vehicle weight, and class (based on the number and space between axles).

2.4.1.1.4 Fiber Optic Sensors

Fiber optic sensors just like the already mentioned sensors, are based on the detection of the pressure caused by the vehicle when passing over them.

When the vehicle passes over the detector it causes a decrease of the optical transmittance. This variation is interpreted by an optoelectronic

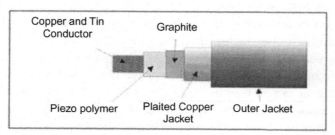

Figure 2.18 Scheme of a piezoelectric sensor.

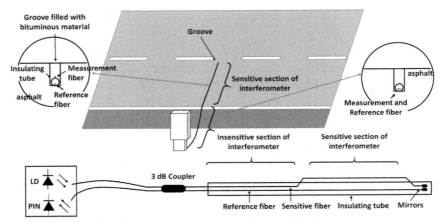

Figure 2.19 Positioning of fiber optic sensors and optical scheme.

interface that determines the presence of a vehicle or can even give the type of vehicle to be considered.

The main parameter of these sensors is the sensitivity. They are capable of detecting from objects of a few grams to vehicles of high tonnage. These sensors allow to measure variables such as number of vehicles, weight of vehicles, and, as mentioned before, their class (based on the number and space between axles).

In Fig. 2.19 the possible configuration for the positioning of these type of sensors on the highways is displayed.

2.4.1.1.5 Geomagnetic Sensors

The last intrusive technology that will be hereby described is based on the magnetic field variations caused by vehicles. This technology will be described more in detail in Chapter 3, where it will be approached as part of the new sensing technologies.

Differing from loops, magnetic sensors are passive elements that detect the presence of a ferromagnetic metal (so by extension the presence of a vehicle) through the change that it creates on the terrestrial magnetic field. This fluctuation is known as a magnetic anomaly.

The impact that a vehicle creates over the magnetic field when passing over the sensor can be seen in Fig. 2.20A—not only the variation created on the compass but also the variation of the signal that the sensor offers. In Fig. 2.20B an example of installation of these type of sensors on a highway can be seen.

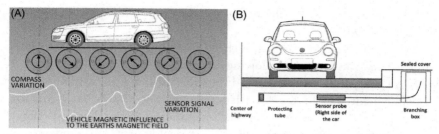

Figure 2.20 (A) Influence of the vehicle on the terrestrial magnetic field and (B) typical scheme of sensor positioning on the highways. *Klein, L.A., Mills, M.K., Gibson, D.R.P., 2006. Traffic detector handbook. McLean, VA, U.S. Dept. of Transportation, Federal Highway Administration, Research, Development and Technology, Turner-Fairbank Highway Research Center. <http://purl.access.gpo.gov/GPO/LPS91983>.*

2.4.1.1.6 Wireless Sensor Networks (Motes)

Wireless Sensor Networks are made up of a group of nodes with two main functionalities:

- Wireless communication capacity.
- Sensing capacity to measure parameters such as temperature, humidity, movement, sound, etc.

This technology has traditionally been used in areas such as industrial operation control or domotics, and only recently are being integrated into intelligent transport systems.

The different nodes that make up the wireless network are known as *motes*, and thanks to their wireless communication capability, they can transmit the collected information to a central control point that computes that information to register it or to take decisions based on them.

The communication protocol is normally *ZigBee*, as it allows a cheaper communication and is much simpler than protocols such as Bluetooth, and is very adequate for applications that require low transmission bit rate, security in the transmission, and long life for the batteries of the motes.

The motes that make up the Wireless network have a *self-organization* capacity, this means that from time to time, at predefined time intervals, they check the optimal communication routes available for them.

Apart from this, the power requirements for these networks are very low, using normally batteries for each mote. This feature allows its installation at sites with no electrical power supply, making possible very flexible topologies and network installations.

The nodes of the sensing network do not need to transmit continuously, it is only necessary for them to collect a sample at certain periods in time, staying at standby the rest of the time, optimizing in this way the power consumption.

To summarize and clarify the already mentioned till now, the motes wireless networks have the following advantages:

- *Easy deployment*: as mentioned before, they are networks of easy deployment due to the fact that they do not require configuration thanks to their capability of self-organization. Also the low energy consumption requirements allow their deployment independently of the availability of electrical power supply.
- *Scalability and flexibility*: thanks to the capability of self-organization, the integration of new motes in the network or a change in their position in the network do not require any type of configuration.
- *Endurance capability and high availability*: these types of networks are very robust as a failure in one or more motes does not influence the communication of the rest of the sensors of the network that are capable of finding new communication links. This implies a great advantage when compared with centralized designs, as they give a robust service of high availability.

However, just like any other technology, it has a series of disadvantages that make it of little use for its installation in certain scenarios:

- *Easiness*: although easiness can be shown as an advantage it can also be a disadvantage. This easiness can be excessive depending on the monitoring needs required and so can derive in a drastic utility decrease.
- *Environment*: the motes communicate using wireless protocols that have small coverage (it varies between 10 and 100 meters, depending on the model). This shortage can prevent the installation of motes in highways with a complex geography or that have elements that difficult wireless communication.
- *Maturity*: although there is a business tissue based on motes, it is a very recent technology. Still it does not have the acceptance, use, and maturity of the traditional ITS equipment that have been deployed for several decades, have.

As a summary, the use of motes for traffic monitoring opens a new world of possibilities at low cost, this is why it is a technology to be investigated as the cost−benefit ratio could be very high and positive.

2.4.1.2 Nonintrusive Sensors

Differing from intrusive sensors, these types of sensors do not disrupt traffic as they are installed over or at the side of the road. As a disadvantage, in general, they are more expensive, not only due to the cost of the sensor, but also due to their installation needs.

The nonintrusive detectors currently used are principally of two types:

- *Active sensors*: those that transmit a signal and receive the response from the vehicle. Included in this type it is possible to find the *microwave radars, laser radars*, and *ultrasonic sensors*.
- *Passive sensors*: this type of sensor captures variations of certain parameters caused by the vehicle passage. Passive sensors are for example *video cameras, infrared sensors,* and *acoustic sensors.*

2.4.1.2.1 Microwave Radars

Vehicle detection using RADAR (*RAdio Detection And Ranging*) is based on the transmission of a microwave signal (between 1 and 30 GHz, typically at 10,525 GHz). Microwave signals are reflected by objects so the radar computes the time lapse between signal transmission and the reception of the reflected signal.

This travel time is the basic variable from which it is possible to get the information about the vehicle, its position, speed, etc.

Its positioning depends on the desired function. It is possible to install it on the pavement, at the center of a lane, to measure the associated lane, or else it is possible to install it at the side of the lane to measure traffic parameters on all the lanes. Its functioning base can be seen in Fig. 2.21.

Apart from its positioning it is possible to distinguish two types of microwave radars:

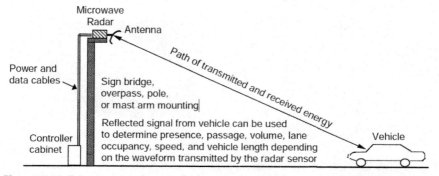

Figure 2.21 Functioning scheme of a microwave radar. *Klein, L.A., Mills, M.K., Gibson, D.R.P., 2006. Traffic detector handbook. McLean, VA, U.S. Dept. of Transportation, Federal Highway Administration, Research, Development and Technology, Turner-Fairbank Highway Research Center. <http://purl.access.gpo.gov/GPO/LPS91983>.*

- *CW radars (Continuous Wave)*: they transmit a continuous Doppler wave and can detect vehicle passage, allowing as well to measure its speed and to classify a vehicle based on its length.
- *Frequency Modulated Continuous Wave radars (FMCW)* detect not only vehicle passage and class, but also their presence, which means they detect stationary vehicles as well.

Fig. 2.22 shows frequency variations relative to time of the two types of microwave radars. The first type keeps frequency constant, while with the second type frequency changes with time.

2.4.1.2.2 Laser Sensors (Active Infrareds)

Vehicle detection with laser radars, also known as LIDAR, is based on the emission of an optical laser light in a pulse (in the near-infrared) to measure the distance to a specific object. The distance calculation is based on

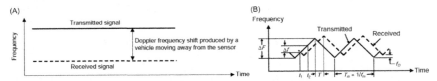

Figure 2.22 Frequency variations for the two types of radars, CW Doppler (A) and FMCW (B). *Klein, L.A., Mills, M.K., Gibson, D.R.P., 2006. Traffic detector handbook. McLean, VA, U.S. Dept. of Transportation, Federal Highway Administration, Research, Development and Technology, Turner-Fairbank Highway Research Center. <http://purl. access.gpo.gov/GPO/LPS91983>.*

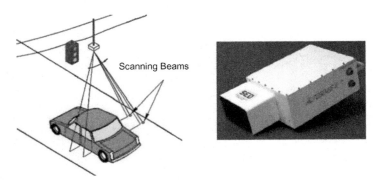

Figure 2.23 Example of laser radar application and photograph of a laser radar. *Klein, L.A., Mills, M.K., Gibson, D.R.P., 2006. Traffic detector handbook. McLean, VA, U.S. Dept. of Transportation, Federal Highway Administration, Research, Development and Technology, Turner-Fairbank Highway Research Center. <http://purl.access.gpo.gov/ GPO/LPS91983>.*

the time that the optical pulse takes to travel back and forth from the LIDAR to the vehicle (Fig. 2.23).

Knowing that this pulse travels at the speed of light, the distance is computed following Eq. (2.4), where D is the distance, c is the speed of light and t is the time lapse between emission and reception.

$$D = \frac{1}{2}c \cdot t \qquad (2.4)$$

These devices allow as well to compute a vehicle's speed. This is possible by transmitting at least two laser pulses that allow to compute the distance variation in a specified lapse of time.

Their traditional use has been vehicle classification based on their length and volume. Modern LIDAR sensors offer 2D, or even 3D images, increasing the reliability of vehicle classification.

New applications that have appeared in several investigation projects will be described in the part of the document that refers to new sensing technologies.

2.4.1.2.3 Ultrasonic Sensors

Ultrasonic sensors transmit acoustic waves of a frequency between 25 and 50 kHz, always above the human hearing range. A typical positioning scheme can be seen in the Fig. 2.25 with an example of installation above the pavement and at its side (Fig. 2.24).

Figure 2.24 Positioning scheme of ultrasonic sensors. *Klein, L.A., Mills, M.K., Gibson, D. R.P., 2006. Traffic detector handbook. McLean, VA, U.S. Dept. of Transportation, Federal Highway Administration, Research, Development and Technology, Turner-Fairbank Highway Research Center. <http://purl.access.gpo.gov/GPO/LPS91983>.*

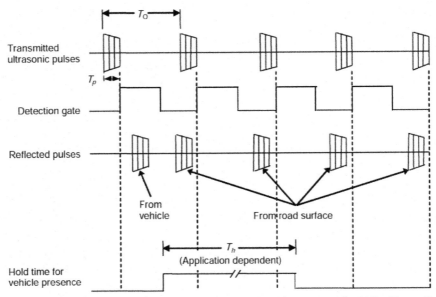

Figure 2.25 Emitted and reflected ultrasonic pulses. *Klein, L.A., Mills, M.K., Gibson, D. R.P., 2006. Traffic detector handbook. McLean, VA, U.S. Dept. of Transportation, Federal Highway Administration, Research, Development and Technology, Turner-Fairbank Highway Research Center. <http://purl.access.gpo.gov/GPO/LPS91983>.*

The system transmits ultrasonic pulses of a time length T_p with a spacing T_o bigger than the time lapse taken by the acoustic wave to go from the sensor to the pavement of the highway and backwards again. The sensor is "opened" at regular time intervals for vehicle detection. Emitted and reflected pulses are represented in Fig. 2.25. When a vehicle is present, the variations in the reception of reflected pulses are converted into an electrical signal, which is passed to a signal processing system for its interpretation.

These sensors are capable of detecting moving vehicles (or stationary) as well as being able to compute their speed.

2.4.1.2.4 Passive Infrared Sensors

Differing from the technologies described before, these are passive sensors. This type of sensor does not emit energy, instead it detects energy coming from two sources:

- Energy emitted by the vehicles, the surface of the pavement, and other objects inside the sensors' vision range.

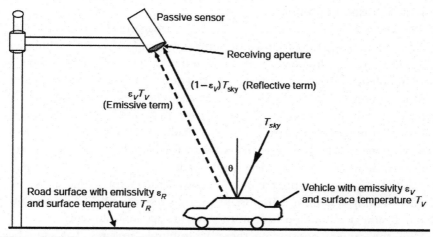

Figure 2.26 General functioning scheme of infrared passive sensors. *Klein, L.A., Mills, M.K., Gibson, D.R.P., 2006. Traffic detector handbook. McLean, VA, U.S. Dept. of Transportation, Federal Highway Administration, Research, Development and Technology, Turner-Fairbank Highway Research Center. <http://purl.access.gpo.gov/ GPO/LPS91983>.*

- Energy emitted by the atmosphere and reflected on the mentioned objects.

The received energy is redirected to a material sensitive to infrared. This material converts the infrared signal into an electrical signal. Then a processing phase analyzes the information included in the signal. The general scheme can be seen in Fig. 2.26.

It is possible to position it above the highway or at its side. It allows the measurement of speed and volume, as well as being able to do a classification of the vehicle.

2.4.1.2.5 Acoustic Sensors

The functioning of acoustic sensors is based on the detection of the acoustic energy (audible sounds) produced by vehicle traffic (vehicle noise and sound produced by the tyres of the vehicle on the road). When a vehicle crosses a detection area there is an increase of acoustic energy and this energy is converted into an electrical signal.

Acoustic sensors are usually made of a bidimensional array of microphones and they are capable of detection on one or several lanes. With them it is possible to perform the measurement of vehicle speed and its detection.

2.4.1.2.6 Video Cameras

Video cameras were introduced in highway management for remote surveillance by an operator. However, nowadays, *artificial vision* techniques are used to extract, in an automatic way, different types of information.

The application area of cameras has increased and they are used not only for surveillance, but also for *detection and vehicle counting*, as well as for vehicle identification based on *automatic plate recognition* between others.

A video processing system is typically made up of one or several cameras, equipment based on a microprocessor for digitalization and analysis, and software to analyze the images and obtain the required information.

Without going into the technical details of the elements that make up a video sensing system, it is important to stress the value that the positioning and camera calibration have in the system's accomplishments.

For the camera positioning there are multiple features that must be taken into account, such as the *specific position*, the height, the positioning over or at the side of the road, etc. A list of important and generic features, that enable to obtain useful images for the following phase of artificial vision for vehicle detection, is included hereafter.

- The positioning must minimize vibration and movement.
- The presence of obstacles in the vision field must be avoided. Also you should minimize occlusions between vehicles.
- When possible, avoid the presence of the horizon in the scene, so as to prevent focusing the sun.
- The camera position must minimize beam reception coming from the vehicles' lights, pavement reflections, etc.

As mentioned before, another important point is camera calibration. This process is necessary to *identify real dimensions* of the area captured by the camera.

In Table 2.1 a series of calibration parameters necessary for the use of video sensors in the localization and identification of vehicles that travel in the highway are included.

From the lens and camera position it is possible to compute the real dimensions (in meters) of the dimensions that appear in the image (in pixels), In Table 2.2 it is possible to see an example of dimension computation of the image of a camera with a lens with focal distance of 8 mm, positioned at a height of 12.2 m over the road. This position of the camera, at a great height over the highway is typical in traffic applications such as vehicle monitoring.

Table 2.1 Calibration information for video systems for vehicle detection

	Parameter	Description
Known	Focal distance	The focal distance of the lens of the camera (seen in cameras' specifications)
	Array dimension CCD/CMOS	CCD or CMOS Array dimensions used in the camera (seen in cameras' specifications)
Requires computation	Image area	A painted rectangle on the pavement's surface is fixed and shown in the monitor. It is necessary to know the dimension of one of the sides of that rectangle (or alternatively the distance that separates the lane markers of two adjacent lanes)
	Number of lanes	Number of lanes for which traffic data is collected
	Position of detection area	Position of the markers that determine two detection zones in each lane. Also the distance between them
	Traffic sense	For each lane determine if the traffic goes downwards or upwards or in both senses of the image
	Detection zone edition	Modification of the positioning and size of detection zones

Table 2.2 Example of real dimension computation based on the parameters of a camera and its positioning

Dimension	Symbol	How to compute	Value
Height of the camera over detection zone	h	Is a known input parameter	12.2 m
Half of the FOV vertical angle	α_V	Lens specification	16.7 degrees
Half of the FOV horizontal angle	α_H	Lens specification	21.8 degrees
Angle between the structure that holds the camera and its vision line	θ	$\theta = (90 - 5) - \alpha_V$; where the value 5 is the number of degrees that the vision field is below the horizon	68.3 degrees
Distance to the lower zone of the image	d_1	$d_1 = h\tan(\theta - \alpha_V)$	15.4 m
Width of the lower zone of the image	w_1	$w_1 = 2h\tan(A_H)/\cos(\theta - A_V)$	15.7 m
Distance to the higher zone of the image	d_2	$d_2 = h\tan(\theta + \alpha_V)$	139.4 m
Width of the higher zone of the image	w_2	$w_2 = 2h\tan(A_H)/\cos(\theta + A_V)$	112 m

Table 2.3 Summary table of intrusive sensors

Technology	Strengths	Weaknesses
Inductive loop	• Flexible design, capacity to adapt to multiple applications. • Mature and well-understood technology. • Provides basic traffic parameters (volume, presence, occupancy, speed, traffic direction, space) • High frequency models offer vehicle classification. • Give better results in vehicle counting that other technologies. • Not impacted by inclement weather	• Its installation impacts the pavement • Diminishes pavement endurance • Installation and maintenance require to disrupt traffic in the lane temporarily • They are impacted by temperature and excessive traffic • The installation of multiple sensors is required to monitor a specific location
Pneumatic tube	• Cheap and flexible • Easy installation • The use of two tubes makes possible the measurement of speed	• They impact traffic as vehicles contact the tubes • They wear away quickly • They need amplifiers at the receptor as the electrical signal produced is very weak
Piezoelectric sensor	• Adequate for mobile measurements • With higher endurance than rubber tubes • Thinner so with less impact on the traffic	• They impact traffic as the vehicles contact the sensors • They wear away quickly (with higher endurance than rubber tubes) • They need amplifiers at the receptor as the electrical signal produced is very weak
Fiber optic sensor	• They are not positioned on the pavement so they do not interfere traffic	• They need amplifiers at the receptor • Its installation requires working on the pavement
Geomagnetic sensor	*Simple sensors* • Less sensitive to excessive traffic than inductive loops • Some models transmit data through wireless connections • Not sensitive to inclement weather	*Simple sensors* • Its installation usually requires working on the pavement and disrupting traffic • In general they are not capable of detecting stationary vehicles

(Continued)

Table 2.3 (Continued)

Technology	Strengths	Weaknesses
	Double axles sensors • They can be used at sites where it is not possible to install inductive loops (e.g., at bridges) • Some models are installed under the pavement not requiring works on the pavement • Less sensitive to excessive traffic than inductive loops • Not sensitive to inclement weather	*Double axles sensors* • Its installation usually require working on the pavement and disrupting traffic • They diminish the lifetime of the pavement • Maintenance operations require disrupting the traffic in the associated lane

2.4.1.3 Summary of Strengths and Weaknesses of Autonomous Traffic Sensors

To finish this section of traffic sensors, a summary of the main characteristics of each type of sensor is included. In Table 2.3 the strong points and weaknesses of intrusive sensors are gathered up and in Table 2.4 are those of nonintrusive sensors, distinguishing active and passive sensors.

2.4.2 Dependant Traffic Sensors

Unlike independent traffic sensors described previously, dependent traffic sensors require the installation or presence of a device in the vehicle.

The sensors positioned in the infrastructure are capable of controlling the position of each vehicle from the detection of the equipment inside the vehicle using different technologies.

On the other hand, the sensors on board of the vehicles make possible the detection of situations in areas close to the vehicles (obstacles, meteorological conditions, state of the road, etc.).

In this section, two technologies that give the position of the vehicle based on the onboard equipment and on the equipment installed in the infrastructure will be described:

• One is based on the identification of the vehicle by radiofrequency.
• The other is Global Navigation Satellite System or GNSS (*Global Navigation Satellite System*).

Hereafter sensors on board of the vehicles will be described in a generic way.

Table 2.4 Summary table of nonintrusive sensors

	Technology	Strengths	Weaknesses
Active sensors	Microwave radar	• It is not impacted by meteorological conditions • Gives directly speed measurement • Can detect multiple lanes	• Antenna and transmitted wave shape has to be adapted being a function of the specific application • CW Doppler sensors do not detect stationary vehicles
	Laser radar	• Transmits several beams to give precise measurements of speed, type of vehicle and position • It has an operation mode for multiple lanes	• Sensors functioning can be impacted by fog if visibility is less than 6 meters • Snow also impacts its adequate performance • Its installation and maintenance (including periodic lenses cleaning) of the sensors requires a temporary traffic disruption in the associated lane
	Ultrasonic sensor	• It has an operation mode for multiple lanes • Active detection	• Some environmental conditions may affect its functioning (temperature changes and strong air turbulence). Some models include systems to compensate external temperature changes. • Detection problems may occur for vehicles driving at high speed if high repetition period for the ultrasonic pulses is used.
Passive sensors	Infrared passive sensor	• Multizone sensors are capable of measuring the speed of the vehicle • It has an operation mode for multiple lanes	• They can have low sensitivity in rain and fog conditions.

(Continued)

Table 2.4 (Continued)

Technology	Strengths	Weaknesses
Acoustic sensor	• Passive detection • Its functioning is not impacted by rainfall • It has an operation mode for multiple lanes	• Low temperatures may influence data precision. • Some specific models do not work adequately in slow traffic and stop-and-go conditions.
Video cameras	• Passive detection • Visual information, understandable by a person • Very developed artificial vision algorithms	• Its functioning is very impacted by inclement weather, lighting variations, etc. • Occlusions

2.4.2.1 Vehicle Identification by RFID (RFID Radio Frequency Identification)

Vehicle identification by radiofrequency (RFID) is a storage and remote data recovery system that uses devices known as transponders or RFID tags. The fundamental purpose of this technology is to transmit the identity and facilitate the estimation of the position of an object (similar to an unique serial number) through radio waves. Passive tags gather the energy from the radio signal, while active tags have an independent power supply (a battery), thereby allowing much greater distances (such as hundreds of meters).

Onboard equipment in the vehicle is needed (*tag*) capable of communicating in a safe and reliable way with the infrastructure so that it is possible to identify and charge each vehicle. Regarding the requirements for the infrastructure, it must include a transmitter/receptor to communicate with the vehicle (TRX)

Hereafter a brief description of the onboard equipment, the equipment of the infrastructure, and the type of communications between them can be found. No details will be given of specific models as there is a great variety of them, however an example of an application for a model of a pay-toll system in free-flow will be schematically described.

2.4.2.1.1 Onboard Equipment (Tag)

The onboard equipment consists, as mentioned before, of a RFID *tag* inserted in the vehicle that allows the system to identify univocally the vehicle and to charge the toll without requiring the vehicle to stop.

The position of a *tag* impacts the communication with the transmitter/receptor device situated in the infrastructure. To achieve an optimal communication, the device is usually positioned on the front windscreen of the vehicle offering "direct vision" with the TRX.

The *tag* has, in general, a read/write memory where the application data is stored as well as the status of the *tag* (that may include information about the existence of tag handling or any other problem in it).

2.4.2.1.2 Equipment in the Infrastructure (TRX)

The equipment in the infrastructure is a transmitter/receptor in charge of the communications with the *tags* of each vehicle. In a complete system, as mentioned before, it will work in conjunction with other equipment of vehicle detection or vehicle plate recognition, for example.

The communication is accomplished by a microwave radio-link of DSRC (DSRC *Dedicated Short Range Communications*) type. Each TRX antenna produces a "communication area" (range). When a vehicle, and thus its tag, enters into the communication area of a TRX it activates the tag and a connection Infrastructure—Vehicle is established.

Once the connection is established the TRX validates and authenticates the tag based on the information that it sends, and once validated the TX sends a report with the transaction and the charge applied as a toll. Once the process is finished the communication ends.

2.4.2.2 Bluetooth Sensing

The last technology to be described here has a similar way of functioning, at least at a conceptual level, to ETC systems, using instead of RFID devices, Bluetooth devices and antennas.

It is a relatively recent technology and its application to ITS is still being developed. However, its presence is getting greater in individual mobile devices (mobile telephones) and in the vehicles (hands-free car systems), making them a cheap system for vehicle detection.

The Bluetooth wireless technology provides short range communications focused not only on mobile devices but also on fixed devices, keeping a high security level. Between its main advantages it is possible to highlight its endurance, its low power and low cost; if we add to this its high level of deployment and current use, it becomes a very interesting technology.

2.4.3 Conclusions and Recommendations

In the current section a review of the main sensing technologies in the scope of ITS has been done. Not only have the technologies currently being used been analyzed, but also the most disruptive, technologically speaking. As a result of this analysis it has been concluded that nowadays:

- Sensing, not only as a technology, but also the associated processing algorithms, is a field that is going ahead in a very quick manner.
- The traditional *data collecting station*, mainly based on loops, are getting outdated. Development is being redirected towards cheaper, more flexible, and mobile technologies that allow the sensing of most of the infrastructure.
- ITS must *benefit* from this *technological evolution* to increase the quality of the systems.
- Roads are filling up with sensors of very different natures, to facilitate the knowledge not only of environmental and meteorological conditions, but also of traffic situation.
- At the same time, vehicle manufacturers are including in the vehicles multiple sensors to improve not only the security but as well the comfort of the drivers.
- Wireless technologies provide the ideal communication media to make possible data transmission and reception between sensors (situated in the infrastructure or in vehicle), control and processing centers, and final users.

A very interesting point to exploit and develop is to use the individual vehicles as itinerant sensors and to use their wireless communication capabilities and Big Data tools to model traffic situations with certain precision.

The current trend and the one that should be followed in the future is to use all the available information not only in the infrastructure but also in the vehicles in an intelligent manner.

2.5 DATA FUSION

Perception technologies have a key role in the modern intelligent transportation systems, whether in assisting the driver to avoid hazardous situations, i.e., Advanced Driver Assistance Systems (ADAS), or in providing environment perception for autonomous vehicles. But a single sensor does not provide the reliable results necessary to fulfill the demanding requirements that road safety application demands. Here is where Data Fusion (DF) represents a key point in order to provide reliable and trustable perception. Recent research in ITS has focused on the development of fusion applications, able to cope with several sensing units and combining them to provide advanced and reliable perception.

The section aim is to prove that DF techniques can deliver more robust and reliable road safety applications, by combining the capacities of different sensors. The sensors available in the literature are numerous, e.g., radars, ultrasonic, visible spectrum cameras, infrared cameras, inertial senses, and laser scanner. Many applications also take advantage of contextual information, which is a very recent issue in the Data Fusion researching field, that takes advantage of the information available to solve the problem (Snidaro et al., 2015), whether with a priori information such as digital maps, or online information, such as traffic information or vehicle state. Contextual information can be helpful for both increasing the accuracy of each sensor independently and providing new information sources to improve the performance of the Fusion process.

2.5.1 Data Fusion Levels

In the current section, a brief introduction to Data Fusion is given, including the key points necessary in each data fusion application.

The concept of Data Fusion dates from the period between World Wars I and II. It is a concept adopted by the United States Department of Defense (DoD) with the aim of improving Command and Control (C2) decision-making. The intention was to create a technology and scientific base that could help in C2 tasks, by adding information from several sources. From that time, Data Fusion has been one of the key elements in defense and intelligence research. Fusion has become ubiquitous in more recent decades. Now it has a key role in more than defense issues, thanks to advances in information technology. Data Fusion is no longer a term associated only with military and intelligence applications. In recent times, Data Fusion is a key element in many everyday applications. Robotics, vehicles, industry, and communications are some examples of fields where data fusion has a key role nowadays.

One of the main problems related with Data Fusion is the ambiguity of the terminology. Usually, the basic concept of Data Fusion itself is hard to define and has a wide field of action. As a consequence of this, several definitions of Data Fusion have been given as well as several models for Data Fusion procedures. The United States DoD created the Joint Directors of Laboratories (JDL) Data Fusion Group in the mid 1980s with the aim of improving communications among military researchers and system developers. The JDL tried to create a common model for data Fusion processing as well as a new lexicon. Over the years these definitions and models, despite much criticism, have become the basis of the Data Fusion. In this section, different models and definitions are going to be described, focusing on the JDL model, which currently is the basis of most Data Fusion procedures.

2.5.1.1 Data Fusion Definition

Data Fusion (DF) is sometimes referred to as sensor fusion. JDL defined DF in the 1980s as:

> *A process dealing with the association, correlation, and combination of data and information from single and multiple sources to achieve refined position and identity estimates, and complete and timely assessments of situations and threats, and their significance. The process is characterized by continuous refinements of its estimates and assessments, and the evaluation of the need for additional sources, or modification of the process itself, to achieve improved results.*
>
> **Hall, D.L., Llinas, J., 2001. Handbook of multisensor data fusion, America, The Electrical Engineering and Applied Signal Processing Series. CRC Press. Hall and Llinas (2001).**

Authors have often pointed out that the definition is very restrictive. Consequently, several other definitions have been proposed. Hall and Llinas (2001) give a wider definition, describing DF as the set of techniques and procedures that "seeks to combine information from multiple sources to achieve inferences that cannot be obtained from a single sensor or source, or wise quantity exceeds that of an inference drawn from any single source."

A similar definition is given Steinberg and Bowman (Steinberg et al., 1999): "Data fusion is the process of combining data or information to estimate or predict entity states." As it can be discerned, new definitions, which are much less restrictive, tend to give Data Fusion an open dimension, allowing it to be used in any discipline.

2.5.2 Architectures

Different models and architectures are defined. The variety of definitions and models are usually set due to the open definition of the Data Fusion and the wide varieties of situations possible to use Data Fusion processes. In this chapter we focus on the two main ones: according to the abstraction level and the localization.

1. Division according to the abstraction level

Data Fusion architectures are typically divided according to the abstraction level in which the fusion is performed. The simplicity and utility of this division makes it one of the broadest and easy to adapt to the wide variety of fields where data fusion is used:

a. Low level. Also referred as direct fusion (Fig. 2.27), it combines unprocessed information, from different sources to create a more complex set of data to be processed. Low level fusion procedures can be directly applied when both sensors are measuring the same physical phenomena (e.g., two images of the same target with thermal camera and color camera). Classic estimators such as

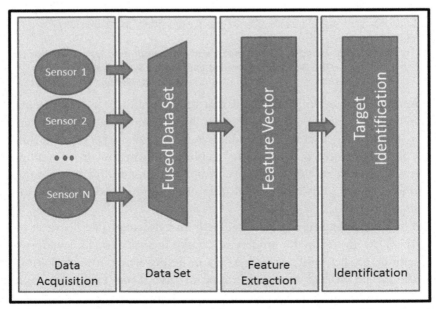

Figure 2.27 Low-level Fusion processing diagram. Fusion is performed from the raw data creating a new set of more complex information to be processed.

Kalman Filters (KF) are very common for raw data fusion for example, to provide enhanced GPS estimation (Martí et al., 2012).

In automotive applications, a stereo system is also a well-known fusion system that uses low-level fusion since it receives information from two different cameras and fuses them to create a new set of information that includes depth information—this new set of information is called disparity map. In Bertozzi et al. (2009) and Hilario et al. (2005) stereovision is used to perform pedestrian detection over this new set of information, while computer vision-based pattern matching methods are employed to give a final decision, such as active contours or probabilistic models.

b. Medium level or feature level fusion (Fig. 2.28). A procedure is defined as such when there is a first preprocessing stage per sensing device, later, based in a set of featured extracted for every sensor independently, an inference is provided accordingly. This estimation can be performed for a wide variety of approaches: machine learning algorithms, Neural Networks, State Vector Machines, etc. In Premebida et al. (2010) and Premebida et al. (2009a), features are extracted for each sensor independently and a new data set is created, and the authors present different approaches whether combining or not the different features of the different sensors comparing results.

c. High level or decision level fusion (Fig. 2.29) are procedures which combine at high levels, inferences provided by each sensor independently (sometimes, a given set of sensors can provide combined detection). Therefore, the final decision is provided according to each subsystem's decisions and the reliability of both the sensor and the inference itself. Example of these technologies can be voting schemes, decision trees, Bayesian inferences, etc.

Premebida et al. (2007) uses Adaboost vision-based pedestrian detection and Gaussian Mixture Model classifier (GMM) for laser scanner-based pedestrian detection. Finally, a Bayesian decisor is used to combine the detections of both subsystems. In Spinello and Siegwart (2008), pedestrians are detected by a laser scanner using multidimensional features, which describes the geometric properties of the detections and Histograms of Oriented Gradients (HOG) features and Support Vector Machine (SVM) for pedestrian detection using computer vision. Final fusion is performed by a Bayesian modeling approach; a similar approach is presented

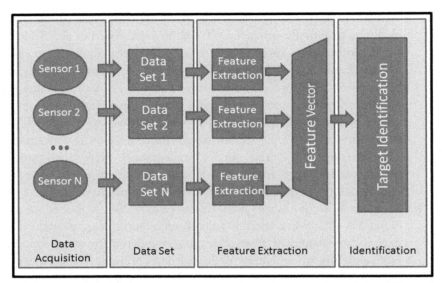

Figure 2.28 Medium-level approach diagram, preprocessing is performed for each sensor and a combined feature set of data is used combining features from all the sensors.

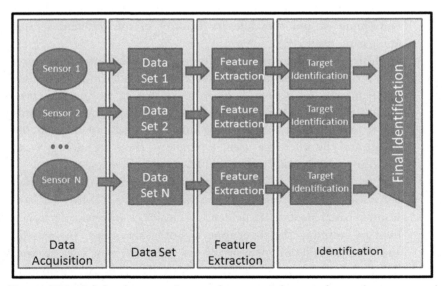

Figure 2.29 High-level approaches perform an inference for each sensor and combine them according to the certainty degree of each one.

in Premebida et al. (2009b), among some other medium level approaches. In Floudas et al. (2007), track to track fusion of obstacles, with no obstacle classification, is performed, using laser scanner, vision, and long- and short-range radars. The same sensors were used in García et al. (2014) for context enhanced pedestrian detection and tracking, in Garcia et al. (2017) for vehicles and Anaya et al. (2017) for motorcycles.

The use of any of these approaches is connected with the nature of the data fusion procedure. Each of them have some advantages and disadvantages that must be considered when making a choice:

d. Low-level procedures are complex and abstract. The new set of data created is intended to give more complex information; it may lead to more accurate estimations but with high effort costs, mainly related with data alignment. Thus, it is assumed that low-level fusion adds more information as well as more complexity to the system. Since these systems are completely dependent on the sensors used, adding new sensors to the system requires a complete revision of the procedure creating the worst possible choice when a scalable system is needed.

e. Feature level fusion has the advantage of being at an intermediate level, allowing the systems to have extra information as a result of the different sources and maintaining a medium level of complexity. The use of different features from various sensors allows the system to take advantage of the possibilities that each sensor can provide independently. On the other hand, the training processes, usually found in these approaches, make the addition of new sensors to the system a difficult task since it must to be trained using the features of the new sensors.

f. High-level fusion requires less complexity since it is based in different pre-established subsystems. The mission of the fusion process here is to add reliability and certainty to the detections and estimations, given by the different subsystems, through combining the final information provided by each subsystem. This way, high-level fusion is easy to implement even though it lacks most of the advantages of the information fusion since the information is only fused al the end of the process. On the other hand, scalability is easy. A new sensor would add more confidence and certainty to the detections in an immediate way, generally without adding complexity.

2. Centralized versus decentralized Data Fusion

Mobility is a key factor in modern applications thanks to the advances in communications and information technologies. These new challenges require different topologies, according to the way in which communications are performed. These applications are divided in nodes, each one of them is formed by one or more sensors, connected to other nodes and with a processing unit. These architectures are basically divided according to where the DF is performed:

a. Centralized Systems are those systems where each node sends all information to a central node where fusion is performed. This way the central node has all of the information from different sensors, which means that Fusion can be accomplished with more certainty.

b. Decentralized Fusion schemes are those schemes where each node performs fusion locally, with the information from the local node and, sometimes, from the adjacent nodes. There is no central node that performs global fusion. Nodes do not have information of the global topology. These schemes easily enable the scalability of the systems since a new node can be easily added or removed. On the other hand the lack of global information suggests that the real fusion procedure is usually not as effective as centralized schemes.

Some authors tend to associate these two topology schemes for Data Fusion with the previously presented Data Fusion differentiation, according to the abstraction level where fusion is performed. This is due to the fact that decentralized Data Fusion usually involves high-level Fusion since each subsystem performs a decision independently. Low-level as well as medium-level fusion schemes require unique estimators, thus it is usually associated with centralized schemes.

2.5.3 Data Fusion in Intelligent Transport Systems

Data Fusion in road safety technologies is a recent issue that tries to enhance the detection capacity of the available sensors by combining several sensor technologies. As will be detailed in the next section of this chapter, these researches are state-of-the-art and focus mainly on the detection and classification of the various objects that may be found in a road environment. Thus, the processes that are generally detailed in ITS, which use Data Fusion for detection purposes, are related with these levels:

- *Data alignment.* Sensors do not usually share a common coordinate system; it means different coordinate axes, as well as, different units and transitions. These units have to be reordered to a common frame that both sensors must share in order to combine the information provided by them. This is a very difficult task, usually too dependent on the applications and the sensor itself, however many authors have provided automatic extrinsic techniques, such as Rodriguez-Garavito et al. (2014) who provided an novel algorithm for stereo-camera and multiple layer laser scanner calibration. On the other hand, Debattisti et al. (2013) proposed using triangle patterns to provide extrinsic calibrations using also a camera and multilayer laser scanner.
- *Data/object correlation.* Previous detections should be associated with new recent detections in order to keep track of them along time. Also, detections from different sensors and subsystems have to be associated. There are multiple possibilities, such as the use of Nearest Neighbors, to associate the closes points according to a given distance definition, the use of Multiple Hypotheses Tracking along time, which allows to delay the detections of the best option for further sequences or the use of probabilistic approach such as Joint Probabilistic Approach. The work presented by Garcia et al. describes all these configurations to be used in the detection and tracking of pedestrians (Snidaro et al., 2015) and vehicles (Garcia et al., 2017).
- Position/movement estimation. Estimation techniques are used to predict the movement of targets in future detections. This prediction can be used for future associations or to help upper levels to estimate the behavior of the targets. Estimation methods are very usual in literature: Kalman Filters (Fan et al., 2013), Unscentered Kalman Filters (Wan and Van Der Merwe, 2000), particle filters (Shao et al., 2008), and in more recent years Probabilistic Hyphothesis Particle Filter (Vo and Ma, 2006) used by Garcia et al. (2014) for vehicle tracking and Meissner et al. (2013) for tracking at intersections.
- *Object/identity estimation.* Final decisions about the detections are provided, and, usually, a confidence level about the estimation is also provided.

2.5.3.1 Other Approaches

Other Data Fusion approaches, among Intelligent Vehicles research, use data from a laser scanner to detect regions of interest (ROI) in images, and computer vision to classify among different obstacles that are included in these ROIs. In Hwang et al. (2007), raw images with the SVM machine learning method is utilized. Szarvas and Sakai (2006) use Convolutional Neural Networks. Ludwig et al. (2011) use HOG features and SVM classification approach. Finally, Pérez Grassi et al. (2010) use Invariant features and SVM to perform the vision-based pedestrian detections. These approaches take advantage of the trustability of the laser scanner for obstacle detection, however, fusion is limited to speed up the process by detecting robust ROIs. Consequently, the information added by the fusion process is limited and can hardly be considered real Data Fusion.

Some fusion approaches take advantage of the properties from various sensors in a different way, which does not fit with any of the previously presented configurations:

Labayrade et al. (2005) combine information from a stereovision camera and a laser scanner. First, the application uses stereovision information to locate the road. Then it uses this information to remove those obstacles that are irrelevant for the application (i.e., outside the road). Finally, it constructs a set of obstacles using the information from both sensors. Tracking is performed using a Kalman Filter approach.

Broggi et al. (2008) use information from a laser scanner to search particular zones of the environment where pedestrians could be located and visibility is reduced, such as the space between two vehicles, and performs detections using vision approach.

Cheng et al. (2007) use a laser scanner and radar approach for obstacle detection and tracking as well as a camera to show the results. Obstacle classification only differentiates among moving and nonmoving obstacles through computing Mahalanobis among the clusters given by the laser scanner.

Some authors also presented grid-based fusion. Fusion is performed dividing the detection space into a grid space, fusing the information based on the sensor's accuracy and detection's certainty (Floudas et al., 2007; Aycard et al., 2006).

Modern technologies, based on the recent Deep Learning techniques provide data fusion with optical and LIDAR data (Kim and Ghosh, 2016).

REFERENCES

Akatsuka, H., Imai, S., 1987. Road signpost recognition system. SAE Veh. Highway Infrastructure Safety Compatibility, 189−196. Available from: http://dx.doi.org/10.4271/870239.

Aliane, N., Fernandez, J., Bemposta, S., Mata, M., 2014. A system for traffic violation detection. Sensors 14 (11), 22113−22127.

Alvarez, J.M., Lopez, A.M., 2011. Road detection based on illuminant invariance. IEEE Trans. Intell. Transp. Syst. 12 (1), 184−193.

Aly, M., 2008. Real time detection of lane markers in urban streets. In: Proceedings of the IEEE Intelligent Vehicles Symposium, Eindhoven, June 4−6, 2008, pp. 7−12.

Amemiya, M., Ishikawa, K., Kobayashi, K., Watanabe, K., 2004, Lane detection for intelligent vehicle employing omni-directional camera. In: Proceedings SICE 2004 Annual Conference, Sapporo, August 4−6, 2004, pp. 2166−2170.

Anaya, J.J., Ponz, A., García, F., Talavera, E., 2017. Motorcycle Detection for ADAS through Camera and V2V Communication, a Comparative Analysis of two Modern Technologies. Expert Syst. Appl.

Arshad, N., Moon, K.S., Park, S.S., Kim, J.N., 2011. Lane detection with moving vehicles using color information. In: Proceedings of the World Congress on Engineering and Computer Science, WCECS 2011, San Francisco, USA, October 19−21, 2011.

Assidiq, A.A., Khalifa, O.O., Islam, M.R., Khan, S., 2008. Real time lane detection for autonomous vehicles. In: Proceedings of Conference on Computer and Communication Engineering, Kuala Lumpur, May 13−15, 2008, pp. 82−88.

Asvadi, A., Premebida, C., Peixoto, P., Nunes, U., 2016. 3D Lidar-based static and moving obstacle detection in driving environments: an approach based on voxels and multi-region ground planes. Rob. Auton. Syst. 83, 299−311.

Aycard, O., Spalanzani, A., Burlet, J., Fulgenzi, C., Vu, D., Raulo, D., et al., 2006. Grid based fusion and tracking. IEEE Intelligent Transportation Systems Conference ITSC 2−7.

Ballard, D.H., 1981. Generalizing the Hough transform to detect arbitrary shapes. Pattern. Recognit. 13 (2), 111−122.

Badrinarayanan, V., Kendall, A., Cipolla, R., 2015. SegNet: A Deep Convolutional Encoder-Decoder Architecture for Image Segmentation. arXiv:1511.00561.

Bar Hillel, A., Lerner, R., Levi, D., Raz, G., 2012. Recent progress in road and lane detection: a survey. Mach. Vis. Appl. 25 (3), 727−745.

Barnes, N., Zelinsky, A., 2004. Real-time radial symmetry for speed sign detection. In: Proceedings of IEEE Intelligent Vehicles Symposium, June 14−17, 2004, pp. 566−571.

BENDIX, <http://www.bendix.com/en/products/autovue/AutoVue.jsp> (retrieved 10.01.17).

Benenson, R., Mathias, M., Timofte, R., Van Gool, L., 2012. Pedestrian detection at 100 frames per second. IEEE Conference on Computer Vision and Pattern Recognition.

Bertozzi, M., Broggi, A., Felisa, M., Ghidoni, S., Grisleri, P., Vezzoni, G., et al., 2009. Multi stereo-based pedestrian detection by means of daylight and far infrared cameras. In: Hammoud, R.I. (Ed.), Object Tracking and Classification Beyond the Visible Spectrum, Lecture Notes in Computer Science. Springer-Verlag, pp. 371−401.

Bila, C., Sivrikaya, F., Khan, M.A., Albayrak, S., 2016. Vehicles of the future: a survey of research on safety issues. IEEE Trans. Intell. Transp. Syst.1−20. Available from: http://dx.doi.org/10.1109/TITS.2016.2600300.

Borkar, A., Hayes, M., Smith, M., Pankanti, S., 2009. A layered approach to robust lane detection at night. In: Proceedings of the IEEE Workshop on Computational Intelligence in Vehicles and Vehicular Systems, Nashville, TN, April 1−2, 2009, pp. 51−57.

BOSH-1, <http://products.bosch-mobility-solutions.com/en/de/_technik/component/SF_PC_DA_Lane-Departure-Warning_SF_PC_Driver-Assistance-Systems_17856.html?compId = 2880> (retrieved 10.0117).

BOSH-2   <https://appcenter.bosch.com/details/-/app/myDriveAssist>   (retrieved 10.01.17).

Broggi, A., Cerri, P., Ghidoni, S., Grisleri, P., Jung, H.G., 2008. Localization and analysis of critical areas in urban scenarios. In: IEEE Intelligent Vehicles Symposium, pp. 1074−1079.

Brostow, G.J., Fauqueur, J., Cipolla, R., 2009. Semantic object classes in video: a high-definition ground truth database. Pattern Recognit. Lett. 30 (20), 88−89.

Burns, J.B., Hanson, A.R., Riseman, E.M., 1986. Extracting straight lines. IEEE Trans. Pattern Anal. Mach. Intell. 8 (4), 425−455.

Canny, J., 1986. A computational approach to edge detection. IEEE Trans. Pattern Anal. Mach. Intell. PAMI 8 (6), 679−698.

Chen, C., Seff, A., Kornhauser, A., Xiao, J., 2015. Deep driving: learning affordance for direct perception in autonomous driving. International Conference on Computer Vision.

Cheng, H., Zheng, N., Zhang, X., Qin, J., Van De Wetering, H., 2007. Interactive road situation analysis for driver assistance and safety warning systems: framework and algorithms. IEEE Trans. Intell. Transp. Syst. 8, 157−167.

CONTINENTAL,   <http://www.continental-automotive.com/www/automotive_de_en/themes/passenger_cars/chassis_safety/adas/ProductInfo_CMArticleslm_en.html> (retrieved 10.01.17).

Dalal, N., Triggs, B., 2005. Histograms of oriented gradients for human detection. In: Proceedings of IEEE international conference on Computer Vision and Pattern Recognition, San Diego, CA, June 20−25, 2005, pp. 886−893.

DELPHI, <http://www.autonomoustuff.com/product/delphi-ifv250/> (retrieved 10.01.17).

Debattisti, S., Mazzei, L., Panciroli, M., 2013. Automated extrinsic laser and camera inter-calibration using triangular targets. In: Intelligent Vehicles Symposium (IV), 2013 IEEE. pp. 696−701.

Dollar, P., Tu, Z., Perona, P., Belongie, S., 2009. Integral channel features. In: Proceedings of the British Machine Conference, September 2009, 91.(1-11), BMVA Press. doi:10.5244/C.23.91.

Dollar, P., Wojek, C., Schiele, B., Perona, P., 2012. Pedestrian detection: an evaluation of the state of the art. IEEE Trans. Pattern Anal. Mach. Intell. 34 (4), 743−761.

Dosovitskiy, A., Koltun, V., 2017. Learning to act by predicting the future. International Conference on Learning Representations.

Enzweiler, M., Gavrila, D.M., 2009. Monocular pedestrian detection: survey and experiments. Trans. Pattern Recognit. Mach. Anal. 31 (12), 2179−2195.

Everingham, M., Van Gool, L., Williams, C.K.I., Winn, J., Zisserman, A., 2010. The PASCAL visual object classes (VOC) challenge. Int. J. Comput. Vis. 88 (2), 303−338.

Fan, R., Prokhorov, V., Dahnoun, N., 2016. Faster-than-real-time linear lane detection implementation using SoC DSP TMS320C6678. In: Proceeding of the IEEE

International Conference on Imaging Systems and Techniques, China, October 4–6, 2016, pp. 306–311.

Fan, X., Mittal, S., Prasad, T., Saurabh, S., Shin, H., 2013. Pedestrain detection and tracking using deformable part models and kalman filtering. J. Commun. Comput. 10, 960–966.

Felzenszwalb, P., Girshick, R., McAllester, D., Ramanan, D., 2010. Object detection with discriminatively trained part based models. IEEE Trans. Pattern Anal. Mach. Intell. 32 (9), 1627–1645.

Fernandes, L.A.F., Oliveira, M.M., 2008. Real-time line detection through an improved Hough transform voting scheme. Pattern. Recognit. 41 (1), 299–314.

Finlayson, G.D., Hordley, S.D., Lu, C., Drew, M.S., 2006. On the removal of shadows from images. IEEE. Trans. Pattern. Anal. Mach. Intell. 28 (1), 59–68.

Fischler, M.A., Bolles, R.C., 1981. Random sample consensus: a paradigm for model fitting with applications to image analysis and automated cartography. Commun. ACM 24 (6), 381–395.

Floudas, N., Polychronopoulos, A., Aycard, O., Burlet, J., Ahrholdt, M., 2007. High level sensor data fusion approaches for object recognition in road. Environ. 2007 IEEE Intell. Veh. Symp136–141.

Franke, U., 2017. Autonomous Driving. In Computer Vision in Vehicle Technology: Land, Sea, & Air, Chapter 2, Edited by Lopez, A.M., Pajdla, T., Imiya, A., Alvarez, J.M.

Gavrila, D.M., 1998. Multi-feature hierarchical template matching using distance transforms. In: Proceedings of the 14th International Conference on Pattern Recognition, Brisbane, Qld, August 20-20, 1998, 439-444.

Gavrila, D.M., 1999. Traffic sign recognition revisited. 21st DAGM Symposium fuer Mustererkennung, pp. 86–93. Springer Verlag, Bonn, Germany.

García, F., García, J., Ponz, A., de la Escalera, A., Armingol, J.M., 2014. Context aided pedestrian detection for danger estimation based on laser scanner and computer vision. Expert Syst. Appl. 41, 6646–6661.

Garcia, F., Martin, D., de la Escalera, A., Armingol, J.M., 2017. Sensor fusion methodology for vehicle detection. IEEE Intell. Transp. Syst. Mag. 9, 123–133.

Garcia, F., Prioletti, A., Cerri, P., Broggi, A., Escalera, A. de la, Armingol, J.M., 2014. Visual feature tracking based on PHD filter for vehicle detection. In: Proceedings of IEEE International Conference on Information Fusion, pp. 1–6.

Geiger, A., Lenz, P., Stiller, C., Urtasun, R., 2016. Vision meets robotics: The KITTI dataset. Int. J. Robot. Res. 32 (11), 1231–1237.

Geronimo, D., Sappa, A.D., Ponsa, D., López, A.M., 2010. 2D–3D-based on-board pedestrian detection system. Comput. Vis. Image Understanding 114 (5), 583–595.

Geronimo, D., Lopez, A.M., 2014a. Vision-based pedestrian protection systems for intelligent vehicles. Springer Briefs in Computer Science. Springer, Chapter 1.

Geronimo, D., Lopez, A.M., 2014b. Vision-based pedestrian protection systems for intelligent vehicles. Springer Briefs in Computer Science. Springer, Chapter 3.

Geronimo, D., Vazquez, D., Escalera, A., 2017. Vision-based advanced driver assistance systems. computer vision in vehicle. In: Lopez, A.M., Pajdla, T., Imiya, A., Alvarez, J. M. (Eds.), Technology: Land, Sea, & Air. Chapter 5.

Gidel, S., Checchin, P., Blanc, C., Chateau, T., Trassoudaine, L., 2010. Pedestrian detection and tracking in an urban environment using a multilayer laser scanner. IEEE Trans. Intell. Transp. Syst. 11, 579–588.

Girshick, R., 2015. Fast R-CNN. International Conference on Computer Vision.

Girshick, R., Donahue, J., Darrell, T., Malik, J., 2016. Region-based convolutional networks for accurate object detection and segmentation. IEEE. Trans. Pattern. Anal. Mach. Intell. 38 (1), 142–158.

González, A., Fang, Z., Socarras, Y., Serrat, J., Vázquez, D., Xu, J., et al., 2016. Pedestrian detection at day/night time with visible and FIR cameras: a comparison. Sensors 16 (6), 820.

Goodfellow, I., Bengio, Y., Courville, A., 2016. Deep Learning. MIT Press.

Hall, D.L., Llinas, J., 2001. Handbook of multisensor data fusion, America. The Electrical Engineering and Applied Signal Processing Series. CRC Press.

Han, J., Dong, Y., Kim, H., Park, S.K., 2015. A new lane detection method based on vanishing point estimation with probabilistic voting. In: Proceedings of the IEEE Int. Conference on Consumer Electronics (ICCE), Las Vegas, NV, January 9–12, 2015, pp. 204–205.

Han, J., Kim, D., Lee, M., Sunwoo, M., 2012. Enhanced road boundary and obstacle detection using a downward-looking LIDAR sensor. IEEE Trans. Veh. Technol. 61, 971–985.

Hariharan, B., Arbeláez, P., Girshick, R., Malik, J., 2015. Hypercolumns for object segmentation and fine-grained localization. In: Conference on Computer Vision and Pattern Recognition.

Hilario, C., Collado, J., Armingol, J., La Escalera, A., 2005. Pedestrian detection for intelligent vehicles based on active contour models and stereo vision. Computer Aided System Theory –EUROCAST 2005.

Hoang, T., Hong, H., Vokhidov, H., Park, K., 2016. Road lane detection by discriminating dashed and solid road lanes using a visible light camera sensor. Sensors 16 (8), 1313–1336.

Hsiao, P.Y., Yeh, C.W., Huang, S.S., Fu, L.C., 2009. A portable vision-based real-time lane departure warning system: day and night. IEEE Trans. Veh. Technol. 58 (4), 2089–2094.

Huang, Z., Yu, Y., Gu, J., Liu, H., 2016. An efficient method for traffic sign recognition based on extreme learning machine. IEEE Trans. Cybern.1–14. Available from: http://dx.doi.org/10.1109/TCYB.2016.2533424.

Hwang, S. Lee, Y., 2016. FPGA-based real-time lane detection for advanced driver assistance systems. In: Proceeding of the IEEE Asia Pacific Conference on Circuits and Systems (APCCAS), Jeju, South Korea, October, 25–28, 2016, pp. 218–219.

Hwang, J.P., Cho, S.E., Ryu, K.J., Park, S., Kim, E., 2007. Multi-classifier based LIDAR and camera fusion. IEEE Intell. Transp. Syst. Conf. ITSC467–472.

Jiang, Y., Gao, F., Xu, G., 2010. Computer vision-based multiple-lane detection on straight road and in a curve. In: Proceeding of the International Conference on Image Analysis and Signal Processing, Zhejiang, April, 9–11, 2010, 114–117.

Jung, C.R., Kelber, C.R., 2005. Lane following and lane departure using a linear parabolic model. Image Vis. Comput. 23, 1192–1202.

Jung, S., Youn, J., Sull, S., 2016. Efficient lane detection based on Spatio-temporal images. IEEE Trans. Intell. Transp. Syst. 17 (1), 289–295.

Kang, D.J., Jung, M.H., 2003. Road lane segmentation using dynamic programming for active safety vehicles. Pattern Recogn. Lett. 24 (16), 3177–3185.

Kaur, G., Kumar, D., 2014. Performance evaluation of modified Hough transformation for lane detection. international. J. Eng. Innovat. Technol. (IJEIT) 4 (2), 74–79.

Kim, T., Ghosh, J., 2016. Robust detection of non-motorized road users using deep learning on optical and LIDAR data. In: 2016 IEEE 19th International Conference on Intelligent Transportation Systems (ITSC), pp. 271–276.

Kim, J., Lee, M., 2014. Robust lane detection based on convolutional neural network and random sample consensus. Neural Information Processing, ICONIP 2014. Lecture Notes in Computer Science. Springer, pp. 454–461.

Kim, Z., 2008. Robust lane detection and tracking in challenging scenarios. IEEE Trans. Intell. Transp. Syst. 9 (1), 16–26.

Klein, L.A., Mills, M.K., Gibson, D.R.P., 2006. Traffic detector handbook. McLean, VA, U.S. Dept. of Transportation, Federal Highway Administration, Research, Development and Technology, Turner-Fairbank Highway Research Center. <http://purl.access.gpo.gov/GPO/LPS91983>.

Kressel, U., Lindner, F., Woehler, C., Linz, A., 1999. Hypothesis verification based on classification at unequal error rates. In: Proceedings of the 9th International Conference on Artificial Neural Networks, Edinburgh, September 7–10, 1999, pp. 874-879.

Krizhevsky, A., Sutskever, I., Hinton, G., 2012. ImageNet classification with deep convolutional neural networks. In: Conference on Neural Information Processing Systems.

Kuk, J.G., An, J.H., Ki, H., Cho, N.I., 2010. Fast lane detection & tracking based on Hough transform with reduced memory requirement. In: Proceedings of the 13th International IEEE Conference on Intelligent Transportation Systems, Funchal, Portugal, September 19–22, 2010, 1344–1349.

Labayrade, R., Royere, C., Gruyer, D., Aubert, D., 2005. Cooperative fusion for multi-obstacles detection with use of stereovision and laser scanner. Auton. Robots. 19, 117–140.

Larson, J., Trivedi, M., 2011. Lidar based off-road negative obstacle detection and analysis. In: IEEE Conference on Intelligent Transportation System Proceedings, ITSC 192–197.

Laurent, J., Hérbert, J.F., Lefebvre, D., Savard, Y., 2009. Using 3D laser profiling sensors for the automated measurement of road surface conditions (ruts, macro-texture, raveling, cracks), pp. 1–8.

Lee, J.W., 2002. A machine vision system for lane-departure detection. Comput. Vis. Image Understanding 86 (1), 52–78.

<http://yann.lecun.com/exdb/lenet/>.

Li, C., Dai, B., Wang, R., Fang, Y., Yuan, X., Wu, T., 2016. Multi-lane detection based on omnidirectional camera using anisotropic steerable filters. IET Intell. Transp. Syst. 10 (5), 298–307.

Li, Q., Chen, L., Li, M., Shaw, S.-L., Nuchter, A., 2014. A sensor-fusion drivable-region and lane-detection system for autonomous vehicle navigation in challenging road scenarios. IEEE Trans. Veh. Technol. 63 (2), 540–555.

Li, Q., Yao, M., Yao, X., Xu, B., 2009. A real-time 3D scanning system for pavement distortion inspection. Meas. Sci. Technol. 21, 15702.

Lin, C.W., Wang, H.Y., Tseng, D.C., 2009. A robust lane detection and verification method for intelligent vehicles. In: Proceedings of Third International Symposium on Intelligent Information Technology Application, IITA-2009, November 21–22, 2009, pp. 521–524.

Lin, T-Y., Maire, M., Belongie, S., Hays, J., Perona, P., Ramanan, D., et al., 2014. Microsoft COCO: Common Objects in Context. In: European Conference on Computer Vision.

Lindner, P., Richter, E., Wanielik, G., Takagi, K., Isogai, A., 2009. Multi-channel lidar processing for lane detection and estimation. In: Proceedings of 12th International IEEE Conference on Intelligent Transportation Systems ITSC-09, St Louis, MO, October 4–7, 2009, 1–6.

Liu, W., Anguelov, D., Erhan, D., Szegedy, C., Reed, S., Fu, C.-Y., et al., 2016. SSD: Single Shot MultiBox Detector. European Conference on Computer Vision.

Long, J., Shelhamer, E., Darrell, T., 2015. Fully convolutional networks for semantic segmentation. In: Conference on Computer Vision and Pattern Recognition.

Lu, W., Zheng, Y., Ma, Y.Q., Liu, T., 2008. An integrated approach to recognition of lane marking and road boundary. In: Proceedings of the Workshop on Knowledge Discovery and Data Mining, Adelaide, SA, January 23–24, 2008, pp. 649–653.

Lu, X., Song, L., Shen, S., He, K., Yu, S., Ling, N., 2013. Parallel Hough transform-based straight line detection and its FPGA implementation in embedded vision. Sensors 13 (7), 9223–9247.

Ludwig, O., Premebida, C., Nunes, U., Ara, R., 2011. Evaluation of boosting-SVM and SRM-SVM cascade classifiers in laser and vision-based pedestrian detection. In: IEEE Intelligent Transportation Systems Conference ITSC, pp. 1574–1579.

Maier, G., Pangerl, S., Schindler, A., 2011. Real-time detection and classification of arrow markings using curve-based prototype fitting. Proceedings of the IEEE Intelligent Vehicles Symposium, Baden-Baden, June 5–9, 2011, pp. 442–447.

Maldonado, B.S., La-fuente, A.S., Gil, J.P., Gomez, M.H., Lopez, F.F., 2007. Road-sign detection and recognition based on support vector machines. IEEE Trans. Intell. Transp. Syst. 8 (2), 264–278.

Mallot, H.A., Bülthoff, H.H., Little, J.J., Bohrer, S., 1991. Inverse perspective mapping simplifies optical flow computation and obstacle detection. Biol. Cybern. 64 (3), 177–185.

Martí, E.D., Martín, D., García, J., de la Escalera, A., Molina, J.M., Armingol, J.M., 2012. Context-aided sensor fusion for enhanced urban navigation. Sensors (Basel) 12, 16802–16837.

Mathias, M., Timofte, R., Benenson, R., Gool, L.V., 2013. Traffic sign recognition-How far are we from the solution? In: Proceedings of the IEEE International Joint Conference on Neural Networks, Dallas, TX, August 4–9, 2013, pp. 1–8.

McCall, J.C., Trivedi, M.M., 2006. Video-based lane estimation and tracking for driver assistance: survey, system, and evaluation. IEEE Trans. Intell. Transp. Syst. 7 (1), 20–37.

Meissner, D., Reuter, S., Dietmayer, K., 2013. Road user tracking at intersections using a multiple-model PHD filter. In: Intelligent Vehicles Symposium (IV), 2013 IEEE.

Miller, T.R., 1993. Benefit-cost analysis of lane marking. Public Roads 56 (4), 153–163.

Montemerlo, M., Becker, J., Bhat, S., Dahlkamp, H., Dolgov, D., Ettinger, S., et al., 2009. Junior: the stanford entry in the urban challenge. Springer Tracts Adv. Robot. 56, 91–123.

Mori, R., Kobayashi, K., Watanabe, K., 2004. Hough-based robust lane boundary detection for the omni-directional camera. In: Proceedings SICE 2004 Annual Conference, Sapporo, August 4–6, 2004, pp. 2113–2117.

Noh, H., Hong, S., Han, B., 2015. Learning deconvolution network for semantic segmentation. In: International Conference on Computer Vision.

Oniga, F., Nedevschi, S., 2010. Processing dense stereo data using elevation maps: road surface, traffic isle, and obstacle detection. IEEE Trans. Veh. Technol. 59, 1172–1182.

Ozgunalp, U., Fan, R., Ai, X., Dahnoun, N., 2016. Multiple lane detection algorithm based on novel dense vanishing point estimation. IEEE Trans. Intell. Transp. Syst.1–12. Available from: http://dx.doi.org/10.1109/TITS.2016.2586187.

Park, D., Ramanan, D., Fowlkes, C., 2010. Multiresolution models for object detection. European Conference on Computer Vision.

Paula, M.B., Jung, C.R., 2013. Real-time detection and classification of road lane markings. In: Proceeding of the XXVI Conference on Graphics, Patterns and Images, Arequipa, August, 5–8, 2013, pp. 83–90.

Pérez Grassi, A., Frolov, V., Puente León, F., 2010. Information fusion to detect and classify pedestrians using invariant features. Inf. Fusion 12, 284–292.

Pratt, W.K., 2001. Digital Image Processing, 3ed ed. John Wiley & Sons, New York.

Petrovskaya, A., Thrun, S., 2009. Model based vehicle detection and tracking for autonomous urban driving. Auton. Robots. 26, 123–139.

Ponsa, D., Lopez, A., Lumbreras, F., Serrat, J., Graf, T., 2005. 3D vehicle sensor based on monocular vision. In: IEEE Intelligent Transportation Systems Conference.

Premebida, C., Ludwig, O., Nunes, U., 2009a. LIDAR and vision-based pedestrian detection system. J. F. Robot 26, 696−711.

Premebida, C., Ludwig, O., Nunes, U., 2009b. Exploiting LIDAR-based features on pedestrian detection in urban scenarios. In: IEEE Intelligent Transportation Systems Conference ITSC, IEEE, pp. 1−6.

Premebida, C., Ludwig, O., Silva, M., Nunes, U., 2010. A cascade classifier applied in pedestrian detection using laser and image-based features. In: IEEE Intelligent Transportation Systems Conference ITSC, pp. 1153−1159.

Premebida, C., Monteiro, G., Nunes, U., Peixoto, P., 2007. A lidar and vision-based approach for pedestrian and vehicle detection and tracking. In: IEEE International Conference on Intelligent Transportation Systems ITSC, pp. 1044−1049.

Razavian, A.S., Azizpour, H., Sullivan, J., Carlsson, S., 2014. CNN features off-the-shelf: an astounding baseline for recognition. In: Conference on Computer Vision and Pattern Recognition.

Redmon, J., Farhadi, A., 2016. YOLO9000: Better, Faster, Stronger. arXiv:1612.0824.

Ren, S., He, K., Girshick, R., Sun, J., 2015. Faster R-CNN: towards real-time object detection with region proposal networks. In: Conference on Neural Information Processing Systems.

Revilloud, M., Gruyer, D., Rahal, M.C., 2016. A lane marker estimation method for improving lane detection. In: Proceeding of the 19th IEEE International Conference on Intelligent Transportation Systems, Rio de Janeiro, Brazil, November 1−4, 2016, pp. 289−295.

Rodriguez-Garavito, C.H., Ponz, A., Garcia, F., Martin, D., de la Escalera, A., Armingol, J.M., 2014. Automatic laser and camera extrinsic calibration for data fusion using road plane. In: Information Fusion (FUSION), 2014 17th International Conference on, pp. 1−6.

Ros, G., Ramos, S., Granados, M., Bakhtiary, A., Vazquez, D., Lopez, A.M., 2015. Vision-based offline-online perception paradigm for autonomous driving. In: IEEE Winter Conference on Applications of Computer Vision.

Sermanet, P., Eigen, D., Zhang, X., Mathieu, M., Fergus, R., LeCun. Y., 2014. OverFeat: integrated recognition, localization and detection using convolutional networks. In: International Conference on Learning Representations.

Schreiber, D., Alefs, B., Clabian, M., 2005. Single camera lane detection and tracking. In: Proceedings IEEE Intelligent Transportation Systems, September 16, 2005, pp. 302−307.

Srivastava, S., Singal, R., Lumba, M., 2014. Efficient lane detection algorithm using different filtering techniques. Int. J. Comput. Appl. 88 (3), 6−11.

Schneider, L., Cordts, M., Rehfeld, T., Pfeiffer, D., Enzweiler, M., Franke, U., et al., 2016. Semantic stixels: depth is not enough. IEEE Intell. Veh. Symp.

Shang, E., An, X., Wu, T., Hu, T., Yuan, Q., He, H., 2016. LiDAR based negative obstacle detection for field autonomous land vehicles. J. F. Robot 33, 591−617.

Shao, X.S.X., Katabira, K., Shibasaki, R., Zhao, H.Z.H., Nakagawa, Y., 2008. Tracking a variable number of pedestrians in crowded scenes by using laser range scanners. In: IEEE International Conference on System Man Cybernatics.

Stallkamp, J., Schlipsing, M., Salmen, J., Igel, C., 2012. Man versus computer: benchmarking machine learning algorithms for traffic sign recognition. Neural Networks 32, 323−332.

Snidaro, L., García, J., Llinas, J., 2015. Context-based information fusion: a survey and discussion. Inf. Fusion 25, 16−31.

Spinello, L., Siegwart, R., 2008. Human detection using multimodal and multidimensional features. 2008 IEEE Int. Conf. Robot. Autom3264−3269.

Steinberg, A.N., Bowman, C.L., White, F.E., 1999. Revisions to the JDL Model. In: Proceedings of the SPIE Conference on Architectures Algorithms and Applications.

Szarvas, M., Sakai, U., 2006. Real-time Pedestrian Detection Using LIDAR and Convolutional Neural Networks. In: EEE Intelligent Vehicles Symposium, pp. 213–218.

Takahashi, A., Ninomiya, Y., Ohta, M., Nishida, M., Takayama, M., 2002. Rear view lane detection by wide angle camera. In: Proceedings of the 2002 IEEE Intelligent Vehicle Symposium, Versailles, France, June 17–21, pp. 148–153.

Teichmann, M., Weber, M., Zoellner, M., 2016. MultiNet: real-time joint semantic reasoning for autonomous driving. Roberto Cipolla, Raquel Urtasun.

Tuohy, S., O'Cualain, D., Jones, E., Glavin, M., 2010. Distance determination for an automobile environment using inverse perspective mapping in OpenCV. In: Proceedings of IET Irish Irish Signals and Systems Conference, Cork, June 23–24, 2010, pp. 100–105.

Uhrig, J., Cordts, M., Franke, U., Brox, T., 2016. Pixel-level encoding and depth layering for instance-level semantic labeling. In: German Conference on Pattern Recognition.

Uijlings, J., Van De Sande, K., Gevers, T., Smeulders, A., 2013. Selective search for object recognition. Int. J. Comput. Vis. 104 (2), 154–171.

Vaisala web. <http://www.vaisala.com > (accessed 03.2017).

Viola, P., Jones, M.J., 2001. Robust real-time object detection: Technical Report CRL 2001/01, Cambridge Research Laboratory. On-line <http://www.hpl.hp.com/tech-reports/Compaq-DEC/CRL-2001-1.pdf> (Retrieved 10.01.17).

Vo, B.-N., Ma, W.-K., 2006. The Gaussian mixture probability hypothesis density filter. IEEE Trans. Signal Process.54.

Wan, E.A., Van Der Merwe, R., 2000. The unscented Kalman filter for nonlinear estimation. Technology v153–158.

Wang, Y., Teoh, E., Shen, D., 2004. Lane detection and tracking using B-snake. Image Vis. Comput. 22 (4), 269–280.

Wang, W., Yan, X., Huang, H., Chu, X., Abdel-Aty, M., 2011. Design and verification of a laser based device for pavement macrotexture measurement. Transp. Res. Part C Emerg. Technol 19, 682–694.

Wu, Z., Shen, C., van den Hengel, A., 2016. Wider or deeper: revisiting the resnet model for visual recognition. arXiv:1611.10080.

Xu, J., Ramos, S., Vazquez, D., Lopez, A.M., 2016. Hierarchical adaptive structural SVM for domain adaptation. Int. J. Comput. Vis. 119 (2), 159–178.

Yu, F., Koltun, V., 2016. Multi-scale context aggregation by dilated convolutions. International Conference on Learning Representations.

Zhang, L., Lin, L., Liang, X., He, K., 2016. Is faster R-CNN doing well for pedestrian detection? In: European Conference on Computer Vision.

Zhang, F., Stahle, H., Chen, C., Buckl, C., Knoll, A., 2013. A lane marking extraction approach based on random finite set statistics. In: Proceedings of the IEEE Intelligent Vehicles Symposium, Gold Coast, June 23–26, 2013, pp. 1143–1148.

Zhou, H., Alvarez, J.M., Porikli, F., 2016. Less is more: towards compact CNNs. In: European Conference on Computer Vision.

Zhu, Z., Liang, D., Zhang, S., Huang, X., Li, B., Hu, S., 2016. Traffic-sign detection and classification in the wild. In: Proceedings of the 2016 IEEE Conference on Computer Vision and Pattern Recognition, Las Vegas, NV, June 27–30, 2016, pp. 2110–2118.

FURTHER READING

Bishop, R., 2005a. Arizona I-19 Wi-Fi Corridor: Assessment of Opportunities for Probe Data Operations. Report TRQS-02, prepared for Arizona Department of

Transportation, in cooperation with U.S. Department of Transportation, Federal Highway Administration.

Bishop, R., 2005b. Intelligent Vehicle Technology and Trends. Artech House.

CEN EN 12253: 2004. Road transport and traffic telematics; Dedicated Short Range Communication; Physycal layer using microwave at 5.8 GHz.

Gidel, S., Blanc, C., Chateau, T., Checchin, P., Trassoudaine, L., 2009. Non-parametric laser and video data fusion: Application to pedestrian detection in urban environment. 12th International Conference on Information Fusion, pp. 623−632.

Global Water Instrumentation, inc. <http://www.globalw.com> (accessed 03.2017).

Gossen web. <http://www.gossen-photo.de/english/> (accessed 03.2017).

Zhao, H., Cui, J., Zha, H., Katabira, K., Shao, X., Shibasaki, R., 2009. Sensing an intersection using a network of laser scanners and video cameras. IEEE Intell. Transp. Syst. Mag. 1 (2), 31−37.

Kämpchen, N., 2007. Feature-level fusion of laser scanner and video data for advanced driver assistance systems. PhD dissertation, Universität Ulm, Hannover, Germany. <https://oparu.uni-ulm.de/xmlui/bitstream/handle/123456789/409/vts_5958_7991. pdf?sequence = 1>.

Klein, L.A., 2001. Sensor Technologies and Data Requirements for ITS. Artech House, Boston.

Maerivoet, S., Logghe, S., 2007. Validation of travel times based on cellular floating vehicle data. In: Proceedings of the 6th European Congress and Exhibition on Intelligent Transport Systems and Services. Aalborg, Denmark.

Satzoda, R.K., Trivedi, M.M., 2013. Vision-based lane analysis: exploration of issues and approaches for embedded realization. In: Proceedings of International IEEE Conference on Computer Vision and Pattern Recognition Workshops, Portland, OR, June 23−28, 2013, pp. 604−609.

Sensys Networks web. <http://www.sensysnetworks.com/home> (accessed 3.2017).

Sifuentes, E., Casas, O., Pallas-Areny, R., 2011. Wireless magnetic sensor node for vehicle detection with optical wake-up. IEEE. Sens. J. 11 (8), 1669−1676.

TRB, 2000. Highway Capacity Manual. Transportation Research Board, Washington, DC. ISBN 0-309-06681-6.

Turner, S., 1998. Travel time data collection handbook. Washington, DC, Office of Highway Information Management, Federal Highway Administration, U.S. Dept. of Transportation.

CHAPTER 3

Vehicular Communications

Jorge Alfonso[1], José E. Naranjo[1], José M. Menéndez[1] and Arrate Alonso[2]
[1]Universidad Politécnica de Madrid, Madrid, Spain
[2]Mondragon Unibertsitatea, Mondragón, Spain

Contents

Traditionally, vehicles have been considered as individual elements within the road infrastructure, which, irrespective of their technological level, rely solely on the information perceived by the driver in his visual field and with that supplied by the sensors they equip. This implies that the information the driver has or the corresponding energy efficiency or safety system the vehicle equips is limited by the visual horizon of the road area in which it is located. This visual horizon will be greater or less depending on the type of road through which it is circulated, weather conditions, the road environment itself, the traffic condition, and other large number of factors. However, in any case, it is not guaranteed that the perceived environment is sufficient to avoid an accident or to optimize the efficiency of the transport.

Intelligent Vehicles
DOI: http://dx.doi.org/10.1016/B978-0-12-812800-8.00003-5

As a solution to this limitation, the wireless communications have appeared in transport, which allow the exchange of data in real time between all the elements that are part of or make use of the route. In this way, two significant advantages are obtained over unconnected vehicles: on the one hand, the limitation of the visual horizon is exceeded when planning the vehicle maneuvres, increasing traffic safety; on the other hand, the possibility of cooperating with other vehicles or following the indications of the infrastructure itself is improved, improving the efficiency of road transport. In any case, the introduction of wireless communication systems technology in road transport involves the enabling of a new set of systems and services for the improvement of safety and efficiency, and these have become the basis of Cooperative Systems, culminating in Cooperative Autonomous Driving (OCDE, 2015).

Thus, vehicular communications can be considered to be one of the keys to the Intelligent Transport Systems; the integrated view of technologies and ICTs into the transport environment. This view encompass aspects related to the data collection, processing, communications and user services, amongst others, and results in a number of service categories, such as:

- Safety critical applications.
- Traffic efficiency applications.
- Infotainment applications.
- Road-charging applications.

Each of these categories will impose a number of operational and functional requirements on the data to be collected, and the means to do so, the processing and storage of such data, the underlying communications supporting data exchanges amongst all the parts of the applications/services, and finally the way in which processed information will be given to the user.

In terms of vehicular communications, the goal is to provide connectivity in scenarios requiring high-speed mobility and very short latency times. The basic scenarios to be covered are the links between vehicles (V2V) or between vehicles and infrastructure (V2I).

Throughout recent years, efforts have been made by different initiatives to first develop, and then integrate, technologies, and mechanisms that would allow the communications between mobile entities and infrastructure entities at different levels. These levels would include, for example, a radio link between a nomadic onboard device antenna and its corresponding base antenna, or the fiber between an optical enabled

CCTV camera and the nearby optical switch. This is what is referred to as the access layer of the communications stack. At a different level, we might be interested in describing the mechanisms that would allow sending end-to-end packets of data: we would be talking here about the network layer of the communications stack. At yet another level, we start considering information exchanges between applications, or between applications and users. A part of these information exchanges and applications are what the ITS services developments are focused upon. Another critical topic has been the integration of these technologies in a common framework or reference architecture, the final component integration specifications of which are still undergoing tasks in standardization groups. Privacy and security are also two open topics in the area of these communications, which, up until now, do not have a clear solution that has been widely accepted by the entire sector. Finally, there is still another additional uncertainty: the communications service providers. The trend up to the present is that the transport communications service is part of the vehicles and the road infrastructure, and its operation as such is free, without the need to pay fees to any communications service provider and enabling access to all types of cooperative and information service providers. In other words, each vehicle (and access point in the infrastructure) is a node of a decentralized communications network. However with the enabling of the 5th generation of mobile telephony (5G), this paradigm is undergoing a radical change, appearing the figure of the network operators, communications service providers that would perform the task of providing access to the vehicular network with the same performance characteristics as decentralized systems, similarly than mobile telephony. These two operating philosophies will be discussed in depth in this chapter.

As these different aspects of the vehicular communications were developed, and deployment stages were getting closer, additional aspects such as interoperability and scalability gained importance. Along with these, also the mechanisms for the standardization, certification, and homologation at different local and international level started their work on the integration and common specification of a global framework for all these aspects and components of the vehicular communications. Therefore, standardization in the field of vehicular communications will be introduced first, giving a clear idea of the main groups involved, and the major advancements achieved in standardizing technologies in these groups, together with the overall proposed reference architecture. The second part of the chapter will focus on more specific aspects of selected technologies.

3.1 STANDARDIZATION IN VEHICULAR COMMUNICATIONS

3.1.1 Introduction

There are mainly two types of standards: industry standards or open standards produced by standardization organization such as ISO, ETSI, Internet Engineering Task Force (IETF), Institute of Electrical and Electronics Engineers (IEEE), or European Committee for Standardization (CEN).

- *ISO*: Is an independent, nongovernmental organization, the members of which are the standards organizations of the 164 member countries. It is composed of a number of Technical Committees, with the one relevant to vehicular communications being ISO/TC204. The scope of the committee is the standardization of information, communication, and control systems in the field of urban and rural surface transportation, including intermodal and multimodal aspects thereof, traveler information, traffic management, public transport, commercial transport, emergency services, and commercial services in the intelligent transport systems (ITS) field.

- *CEN*: Is a public standards organization whose mission is to foster the economy of the European Union in global trading, the welfare of European citizens and the environment by providing an efficient infrastructure to interested parties for the development, maintenance, and distribution of coherent sets of standards and specifications. It is composed of a number of Technical Committees, with the one relevant to vehicular communications being ISO/TC278, with a scope similar to that of the ISO/TC204.

- *IEEE*: The IEEE is a professional association formed in 1963, with the objectives of the educational and technical advancement of electrical and electronic engineering, telecommunications, computer engineering, and allied disciplines. Not limited to standardization activities, it includes 39 Technical Societies, acting as a major publisher of scientific journals and organizer of conferences, workshops, and symposia. Its IEEE Standards Association is a leading standards development organization for the development of industrial standards in a broad range of disciplines.

- *IETF*: The IETF develops and promotes voluntary Internet standards, in particular the standards that comprise the Internet Protocol suite (IP). It is an open standards organization, with no formal membership. The IETF started as an activity supported by the US federal government, but since 1993 it has operated as a standards development function under the auspices of the Internet Society, an international membership-based nonprofit organization.

- *ETSI*: The European Telecommunications Standards Institute is a nonprofit organization whose mission is to produce the telecommunications standards applicable in Europe. The Technical Committee ETSI TC ITS was established in 2007, and its work includes aspects of DSRC, CALM communications, architecture, and security. It is one of the bodies usually appointed to carry out European Commission mandates or directives on standardization.

Relevant to CEN and ISO, the Vienna Agreement signed by both institutions in 1991 aims to avoid duplication of standards between CEN and ISO, so that joint work and adoption of standards have become the norm in the past years.

Other relevant players in the standardization arena are the governmental institutions and public authorities, which actively contribute to push standardization bodies to specific actions towards integration and support to deployment of ITS solutions. The European Commission and the US department of transport have both taken throughout the years a number of actions to ensure that the standardization efforts of the different entities were consistent and produced results that would benefit the citizens and the society. In the case of the European Commission, these efforts often come in the form of Directives or Mandates which address a specific area of application, providing guidelines and broad objectives, together with the specific standardization bodies which should be involved in the activities.

3.1.2 The ISO CALM Framework
3.1.2.1 The ISO CALM Communications Reference Architecture
An overall view of the integration effort in vehicular communications research initiatives and standardization activities, and the resulting framework can be seen in Fig. 3.1. This is known as the ISO CALM (Communications Access for Land Mobiles, in its latest definition). It is a multilayered diagram for the different technologies relevant in vehicular communications, as considered in the most recent ISO 21217 ITS Reference Architecture. The layered description follows roughly a simplified OSI stack specification, with physical (access) components at the bottom, and increasingly application-oriented components higher up in the structure.

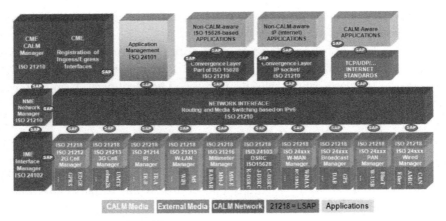

Figure 3.1 CALM system architecture.

It is important to note that the overall approach to the specification of this framework is modular. That is, components have clear specifications in terms of intended objective and scope of the technologies on which they are based. In some cases, this is more difficult, for example those components specifying data and information flow control from the applications, or those related with security of the communications. But the goal is to try to make components modular as it will facilitate not only interoperability aspects, but also the replacement of these components as technology evolves.

ISO 21217:2014 describes the communications reference architecture of nodes called "ITS station units" designed for deployment in ITS communication networks. The ITS station reference architecture is described in an abstract way. While ISO 21217:2014 describes a number of ITS station elements, whether or not a particular element is implemented in a ITS station unit depends on the specific communication requirements of the implementation. Thus, the standard defines the frame, including provisions for the most significant technologies in different levels to provide the tools for the bigger number of particular implementations possible.

ISO 21217:2014 also describes the various communication modes for peer-to-peer communications over various networks between ITS communication nodes. These nodes may be ITS station units as described in ISO 21217:2014 or any other reachable nodes.

ISO 21217:2014 specifies the minimum set of normative requirements for a physical instantiation of the ITS station based on the principles of a bounded secured managed domain.

In the following pages, some of the components of the communications framework will be highlighted, focusing on the access technologies and network technologies, as these provide the critical first link between the ITS entities, and the basic end-to-end capabilities for ITS services and applications respectively. In this section, these technologies are introduced in relation to their reference technical standards, whereas section 3.2 will address more in detail their technical specifications.

3.1.2.2 The ISO CALM Access Media
3.1.2.2.1 IEEE WAVE

Wireless Access in Vehicular Environments (WAVE) is a set of standards addressed at ensuring a homogeneous access for communications between automotive manufacturers. It defines an architecture and a set of services and interfaces that collectively enable secure V2V and V2I wireless communications. As a complete architecture, it covers the complete communications stack and additionally addresses overall issues such as security and management.

The set of standards which specify the WAVE architecture are developed by the IEEE standardization body under the title of IEEE 1609 family. In relation to the access media, IEEE 1609 WAVE relies on the IEEE 802.11p standard which defines the physical and medium access layers of the communications stack (the two lowest levels of the OSI stack) (Fig. 3.2).

3.1.2.2.2 CEN DSRC

Dedicated Short Range Communications (CEN DSRC) is the European implementation of a short-range wireless communications system, and is

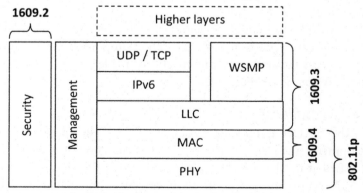

Figure 3.2 IEEE WAVE standards family and 802.11p standard scope.

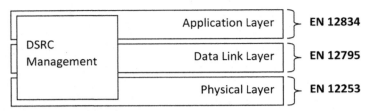

Figure 3.3 CEN DSRC protocol stack.

the one traditionally used for electronic fee collection applications. In a similar way to the American WAVE, CEN DSRC is a family of standards which specify the whole DSRC stack, including PHY, DLL, and application layers, and additionally considering management issues. Compliance with CEN DSRC allows multiple simultaneous RTTT (Road Transport and Traffic Telematics) applications in parallel, but the particular constraints of the application scenarios in which DSRC is likely to be used, and particularly regarding the low latency requirements, led CEN to assume a simplified architecture for DSRC (Fig. 3.3).

3.1.2.2.3 ETSI ITS G5

ETSI ITS-G5 includes the specification on functionalities providing communications focused on the ETSI definition of ITS services and architecture, and based in the 5 GHz band. As the ISO CALM M5 developments, it is based on the work done in IEEE 802.11p, but adopts a reduced set of the services (at different layers) described in IEEE 802.11p.

The ITS-G5 standard includes also specific requirements for different spectrum bands, in this case explicitly addressing those subbands in the document, as a direct response to the EC Directive on harmonized spectrum allocation in the 5.9 GHz.

The ETSI ITS-G5 standard specifies several aspects of the ITS Station reference architecture, as well as the corresponding SAPs (Service Access Points) and, particularly, the modifications made to the base IEEE 802.11p equivalent parts (Fig. 3.4).

3.1.2.2.4 ISO CALM M5

Specifies the CALM architecture access in the 5 GHz microwave range. Its development was done in parallel with IEEE 1609 WAVE and is in fact also based at this level on the work done in IEEE 802.11p.

A CALM M5 communication interface can be integrated with CEN DSRC as a way to ensure currently deployed payment solutions (this

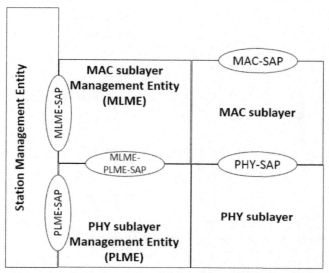

Figure 3.4 Scope and specification of the ETSI ITS-G5 standard. *Adapted from ETSI ES 202 663 v1.1.0.*

integration expects the CEN DSRC equipment to be compliant with the CEN EN 12253:2004, CEN EN 12795:2003 and CEN EN 12834:2003 standards which specify several DSRC layers).

The specification of M5 relies in the concept of Communication Interface (CI), which in a similar way of that of the ETSI ITS-G5 describes an entity which includes aspects of the PHY, MAC, and management layers of the overall ISO CALM structure and defines the corresponding access points to this entity and the compliance requirements for these SAPs (Fig. 3.5).

The compliance requirements of the M5 CI are extensive, with other CALM standards and with IEEE 802.11, amongst others. The global concept of M5 CI is a direct implementation of the overall CALM concept of achieving interoperability at the network and transport level through the use of the Communication Adaptation Layers which isolate particular access technologies from upper layers for data transmission and CM and CI management entities to make the overall CALM system aware of the capabilities of a given CI.

3.1.2.2.5 IEEE 802.11

The IEEE 802.11 family of standards in general specifies the characteristics of the lower layers of the wireless Local Area Network (LAN)

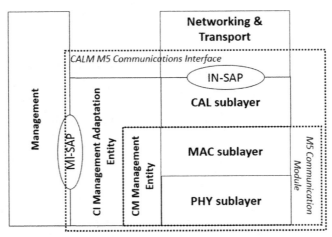

Figure 3.5 ISO CALM M5 CI architecture. *Adapted from ISO 21215:2010.*

communications in local and metropolitan area networks. The first 802.11 standard was released in 1999, and already considered two different implementations, 802.11a and 802.11b, operating at the 5 GHz and 2.4 GHz bands, respectively. The original 802.11 standard has been under development since then, with a number of amendments being published by IEEE, although in several cases there has not been commercial implementation of the standard documents, with 802.11a, 802.11b, 802.11g, 802.11n, and 802.11p being the only ones which have been widely implemented. Originally detailed in IEEE annexes to the initial 802.11 standard, these have since been integrated into the main document, so nowadays there is a single reference IEEE 802.11 specification. For the purposes of clarity, and giving support to existing documentation, the suffixes of the annexes will be maintained in this document.

3.1.2.2.6 IEEE 802.11p

IEEE 802.11p access standard specifies the technical characteristics of the radio link of WAVE, and can be seen in turn as an IEEE validation of the ASTM E2213–03 specification for telecommunications and information exchange between roadside and vehicle systems using the 5 GHz band. It is to be noted, though, that the 11p amendment applies to a band starting also at 5 GHz, but specifies particular radio requirements (maximum transmit power, spectrum masks, etc...) for the 5.85–5.925 GHz band, probably as a result of a combined development effort with European standardization to ensure interoperability at this level and the US FCC allocation of that band to ITS services,

such as traffic light control, traffic monitoring, travellers' alerts, automatic toll collection, traffic congestion detection, emergency vehicle signal preemption of traffic lights, and electronic inspection of moving trucks through data transmissions with roadside inspection facilities.

From the standard's specification and deployment, the primary use of IEEE 802.11p compliant 5.9 GHz equipment in the WAVE environment is the provision of ITS services, with the main characteristics of high speed (27 Mbps), short range (up to 1000 m), and low latency. Integration with existing CEN DSRC PHY 5.8 GHz applications is possible by implementing the corresponding spectrum masks specified in the IEEE 802.11p standard to avoid interferences with these applications. Current development of commercial IEEE 802.11p compliant devices is in fact slowed down by the difficulty in correctly implementing these power masks, and it might be a factor to be considered in relation to the equipment to be used.

As of September 2009, IEEE 1609 approach to Electronic Fee Collection (EFC) was that it considered a part of a broader Electronic Payment Service concept (EPS), and that whenever possible, standards already specified (at higher levels, such as the CEN/ISO 14906 DSRC-based EFC application interface definition) should be used and in agreement with the established IEEE 1609 WAVE standards structure (maybe in the shape of application profiles, for example).

3.1.2.2.7 IEEE 802.16 WiMAX

Worldwide Interoperability for Microwave Access (WiMAX) is a communication technology trying to fill the gap between 3G and WLAN standards, thus fitting in the concept of Metropolitan Area Network (MAN). WiMAX is actually the commercial name of the IEEE 802.16 set of wireless broadband standards.

As is the case with IEEE 802.11 WiFi standards, 802.16 implementations have also evolved with time. The first release of the standard was published in 2001, specifying a point-to-multipoint broadband wireless transmission in the 10−66 GHz band, with only direct line-of-sight capability. The IEEE 802.16a evolution, released in 2003 extended the specifications for the 2−11 GHz band. However, arguably the most relevant evolution of the standard in terms of applicability to the mobile environment is the 802.16e iteration.

3.1.2.3 The ISO CALM Network Layer
3.1.2.3.1 IETF IPv4

IPv4 is the fourth revision in the development of the Internet Protocol (IP) and the first version of the protocol to be widely deployed. Together with IPv6, it is at the core of standards-based internetworking methods of the Internet. IPv4 is described in the IETF RFC 791 from 1981.

IPv4 is a connectionless protocol for use on packet-switched Link Layer networks (e.g., Ethernet). It operates on a best effort delivery model; in that it does not guarantee delivery, nor does it assure proper sequencing or avoidance of duplicate delivery. These aspects, including data integrity, are addressed by an upper layer transport protocol, such as the Transmission Control Protocol (TCP).

As one of the key aspects of the protocol, IPv4 uses 32-bit addresses, with some address blocks reserved for special purposes such as private networks and multicast addresses. As addresses were assigned to end users, an IPv4 address shortage was developing, prompting both network addressing mechanisms and the development of IPv6 to delay and eventually overcome the address exhaustion.

IPv4 addresses may be written in any notation expressing a 32-bit integer value, but for convenience, they are most often written in dot-decimal notation, which consists of four octets of the address expressed individually in decimal and separated by periods.

```
192.0.2.235
```

3.1.2.3.2 IETF/ISO IPv6 Networking and Mobility

Internet Protocol version 6 (IPv6) is the current version of the Internet Protocol (IP), and is currently in the process of being deployed together with existing IPv4, to eventually reach an IPv6-only network scenario.

Arguably, the most important reason for the development of the IPv6 was the issue of address exhaustion caused by the relatively small addressing space of IPv4, but with the increased knowledge of the networking scenario to be faced by IPv6, its design and specification include some significant specific improvements.

- Expanded addressing capabilities
  IPv6 increases the IP address size from 32 bits to 128 bits, to support more levels of addressing hierarchy, a much greater number of addressable nodes, and simpler autoconfiguration of addresses. The scalability of multicast routing is improved by adding a "scope" field to multicast

addresses. And a new type of address called an "anycast address" is defined, used to send a packet to any one of a group of nodes.

A full IPv6 address can be, for example:

```
2001:0DB8:C003:0001:0000:0000:0000:F00D
```

- Header format simplification.
  Some IPv4 headers have been dropped or made optional, to reduce the common-case processing cost of packet handling and to limit the bandwidth cost of the IPv6 header.
- Improved support for extensions and options.
  Changes in the way IP header options are encoded allows for more efficient forwarding, less stringent limits on the length of options, and greater flexibility for introducing new options in the future.
- Flow labeling capability.
  A new capability is added to enable the labeling of packets belonging to particular "traffic" flows for which the sender requests special handling, such as nondefault quality of service or "real-time" service.
- Authentication and Privacy capabilities.
  Extensions to support authentication, data integrity, and optionally data confidentiality are specified for IPv6.

Even though the IPv6 specification was released (officially) in 1998, the truth is that its implementation is far from extended. There may be several reasons for this slow deployment, the assessment of which is out of the scope of this document. The fact is that very likely service deployment scenarios will include both IPv4 and IPv6 devices, and therefore, the introduction of some ideas about the coexistence of IPv4 and IPv6 are necessary.

The differences between IPv4 and IPv6 go beyond the extended addressing space of IPv6, and include improved and added capabilities related to self-configuration of nodes (both fixed and mobile), multicasting, and both network level security and mobility facilities. Some of the solutions developed for the coexistence of IPv4 and IPv6 reduce the impact of these changes in order to maintain compatibility with the older IPv4 protocol.

3.1.2.3.3 Mobility in IPv6 Networks

Mobility for IP networks is the implementation of a concept by means of which nodes remain reachable while moving around in an IP-based network. Each mobile node is identified by its home address, regardless of its current point of attachment to the Internet. While away from its home network, a mobile node is also associated with a Care-of Address (CoA), which provides information about the mobile node's current location. The mobility extensions implement the necessary protocols so that IP nodes cache the binding of a mobile node's home address with their CoA, sending packets destined to the mobile node to this CoA, with the sender only knowing the home address of the mobile node.

It is to be noted that Mobile IP protocols solve specifically the problems related to the mobility at the network layer protocol, trying to provide upper layers with a single destination IP address independently of the particular network location of the destination device. Mobile IP does not solve mobility issues at lower layers, such as possible access technology changes or cell-to-cell handovers.

However, there is also a need to support the movement of a complete network that changes its point of attachment to the fixed infrastructure, maintaining the sessions of every device of the network. This is basically the NEMO IP Network Mobility concept, developed by the IETF.

A Mobile Network is a network segment or subnet that can move and attach to an arbitrary point in the routing infrastructure. A Mobile Network can only be accessed via specific gateways called Mobile Routers that manage its movement. Mobile Networks have at least one Mobile Router serving them. A Mobile Router does not distribute the Mobile Network routes to the infrastructure at its point of attachment (i.e., in the visited network), instead, it maintains a bidirectional tunnel to a Home Agent that advertises an aggregation of Mobile Networks to the infrastructure. The Mobile Router is also the default gateway for the Mobile Network (Fig. 3.6).

3.1.2.3.4 IEEE 1609.3 WAVE WSMP

One of the main goals of the WAVE development was to provide an implementation framework optimized for the particular requirements of vehicular environments, thus focusing mainly in air interface efficiency and low latency. While a WAVE compliant implementation of Network Services could be based on the IPv6/TCP/UDP stack, the potential of WAVE arguably lies on its WAVE-specific WAVE Short Message Protocol.

Air interface efficiency is tightly related to the signal transmission parameters, and although it can be said that these are issues which belong to lower layers of the stack, WAVE exploits the particular operating environment of the communication links and specifies a number of physical

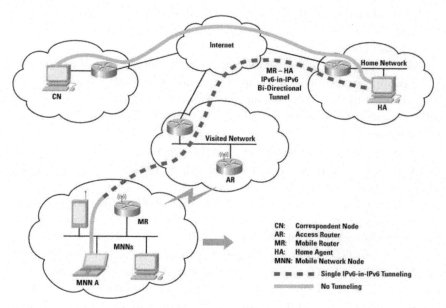

Figure 3.6 Example of NEMO basic support protocol operation.

parameters in the transmission of network-related data packets. Even in an IP-based stack, WAVE uses a transmitter profile containing these data. The WSMP is based on a number of primitives which allow higher layer entities to send and receive WSMs.

3.1.2.3.5 GeoNetworking

GeoNetworking is a practical development of a concept in which network nodes can be reached not only by means of their identified or known address, but also by means of geographic information. With GeoNetworking, it would be possible to request or send information to "the vehicles 200 m behind me in this lane and direction," "one vehicle approaching one intersection in front of me," or "all the HGVs in a given geographical area."

The GeoNetworking communication modes considered are:

- *Unicast* Communication between a single entity and an identified destination entity at a given geographical position.
- *Anycast* Communication between a single entity and a single arbitrary entity from a predefined group within a geographical area.
- *Broadcast* Communication between a single entity and all entities within a given geographical area.

In combination with IPv6 Unicast, Multicast, and Anycast modes, GeoNetworking + IPv6 can integrate any communication link mode for fixed or mobile scenarios.

The development of the GeoNetworking concept started at the C2C-CC consortium, after which the main effort was carried out by the GeoNet project, with the final results being incorporated into ETSI and ISO as a series of standards defining the integration of the GeoNetworking concept into the existing ITS framework.

In its more general specification, GeoNetworking is achieved by means of a C2CNet network sublayer sitting on top of the access layer and servicing basic transport protocols. In practice, however, GeoNet and the standardization efforts have been addressed to exploit the advanced addressing capabilities of IPv6 and the foreseeable future widespread support to IPv6-based applications to push the integration of GeoNetworking concepts with the IPv6/TCP-UDP stack communications.

Within this integrated framework, geonetworks can be seen as ad hoc subnets of a larger IP-based environment, and GeoNetworking capabilities are called upon only at certain links of the communication (C2CNet domain) when necessary depending on what the application wants to do.

3.1.3 Vehicular Communications in a Mobile Communications Scenario

Vehicular communications, and specifically the ITS-focused approaches described in this section consider mainly the so-called medium range access technologies. These are particularly suited for vehicular communications for their low-latency performance, and more importantly, their latency management capabilities. It cannot be ignored, however, that ICT technologies have also been developing rapidly in recent years, and that concepts like Internet of Things (IoT), Cloud Computing, and in general, Future Internet (FI) elements are advancing rapidly and getting increasingly accepted by users and industrial sectors.

The current ITS development and deployment scenarios presents several issues and challenges. Some of them are related to the completion of the components to make different implementations of the ITS station reference architecture fully interoperable. Others are related with the fact that it is possible that a significant amount of data relevant for traffic management purposes and ITS services is not currently used by ITS services. It is generally acknowledged that ITS should make use of all the available data sources, including those coming from mobile applications and Future

Internet related sources. But there has not been any detailed approach on how to integrate these into the ITS technological environment. Alfonso et al. (2014) goes over the potential benefits of this integration and the possible approach to its implementation.

One of the main characteristics of the ITS Station reference architecture is its design as a secure bounded environment. This has facilitated the design and development of ITS safety critical services. On the other hand, the strength of the Future Internet approach is based on the ideas of ubiquitous connection of devices and sensors and extended connectivity to the users, creating a virtual pool of data and service that can be accessed in different ways. The key aspect when combining these two approaches lies in the realization of the limitations and constraints that each imposes on the data exchanges at different levels, and the importance of the management in handling these limitations.

A potentially valid approach would be to consider the Cooperative ITS and the more strict implementations of the ITS station reference architecture as particular cases of a FI implementation. This is due to the facts that the FI architecture in principle seems to facilitate integration of different systems and networks, with C-ITS applications being the ones imposing tighter requirements on data and processes. Thus the integration proposal is based on the exploitation of these requirements and their implementation in the Interfaces To Networks and Devices (I2ND) block of the FI architecture. Specifically, provisions would have to be made in the Service Capability, Connectivity and Control (S3C) component to extend the flow management and application interfaces capabilities to include ITS requirements and resource handling (Fig. 3.7).

The first step for this integration focuses on the data collection stage of the ITS applications, implementing a quality metadata management and control subsystem, which would handle the procedures of data quality measurement and assignment, security aspects, and advanced flow control. Subsequent steps could include processing and presentation of service data to the users, particularly making use of distributed data and processing resources.

Integrating FI and Cooperative ITS resources will facilitate in turn the management of ITS service quality requirements, including Quality of Experience together with Quality of Service concepts, supported by a harmonized conceptual and technological framework related with data quality. This can be considered to be an important idea in the development of the Mobility as a Service (MaaS) concept, in its integrated view of interests, technologies, and goals.

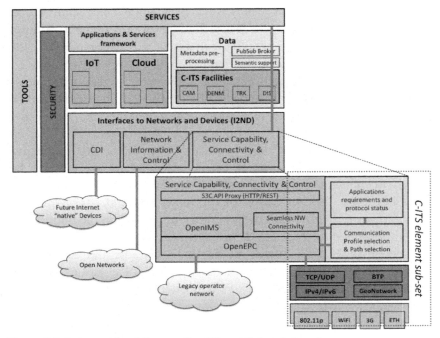

Figure 3.7 An example of Cooperative ITS and Future Internet convergence.

3.1.4 Conclusions

We have seen in this section some of the aspects related with the vehicular communications capabilities as described in the main international standards. We shall repeat here however one of the basic ideas behind vehicular communications standards: the modularity of the design of the overall support framework and architecture. Technology evolves, and it is critical that the impact on the structure of mobility and vehicular applications and services already deployed is minimized whenever a new WiFi specification appears on the market, or when a new annex for network protocols is defined. This is a way to guarantee that existing deployments on the roads are future-proof, while also allowing for the development of existing technologies.

Another aspect to take into account is the increasing importance—and performance—of long-range mobile technologies, such as 4G, and the already underdrafting 5G. Safety-Critical vehicular applications will rely on medium-range technologies such as 802.11(p) for the next few years, but it is necessary to consider in the supporting framework of vehicular communications initiatives more related to mobile communications. The promises of QoS management and latency control will make future mobile communications more important in transport, but again, interoperability with existing and developing ITS systems will make it necessary to address standardization in these areas.

3.2 TECHNOLOGY

3.2.1 Introduction

In the previous section in this chapter, some relevant work lines of standardization in the field of vehicular communications were presented, together with a brief introduction to the main actors involved in these activities. While Section 3.1 focused in the standardization work, and the scope of the technical specifications related with different vehicular communication technologies, this Section 3.2 will briefly describe more detailed technical aspects of these technologies.

3.2.2 Reference Architecture

The wireless communications in vehicular environments are generically called Vehicular Ad hoc Network (VANETs). A VANET consists of an ad hoc wireless network where its nodes correspond to the different elements of the road, vehicles (cars, trucks, buses ... etc), access points in the infrastructure, or the road users themselves. These nodes form a high mobility communications network (e.g., while driving on a motorway) and therefore, the network is formed by a set of nodes that move arbitrarily, communicating with each other in a decentralized way. Then, the data circulates directly through the nodes of the network (hence the concept of ad hoc), without any centralizing element that is responsible for data management (network operator). This implies that the vehicular network behaves as a mesh network, where each node is an access point to the network that, in addition, acts as a router for the retransmission of the messages that it receives, enabling the so-called multihop capability. This implies that a message can be sent from a sender to a receiver or a set of receivers that are out of reach; that message will be transmitted from the origin to the destination through several hops in a set of intermediate nodes, which route the message to its destination or to the next node. This network structure and constitutive nodes are automatically reconfigured according to the position of the various vehicles, allowing a high mobility and guaranteeing the access and fluid availability of the information.

In terms of nomenclature, it can be said that a VANET is a type of Mobile Ad hoc Network (MANET), for the reasons given above; although it is possible to differentiate that MANET describes mainly a field of academic research, whereas the term VANET is more focused to a specific application of these.

Currently, everything that involves the topic of VANETs is in development and research. In fact, there are several working groups, both on the part of the universities and the governments, as well as of industry, that are investigating this field due to the multitude of possible applications that could support its use.

For VANETS networks to be operational, it is necessary that all the nodes that compose them are compatible with each other and speak the same language in order to share information and access cooperative services. This implies that these networks must follow two types of normalization, at low and high level. On the one hand, at a low level, they must have standardized their stacks of layers of communications protocols, so that their operation for data exchange is compatible and interoperable; this will be seen in the section on operative technologies (Section 3.2.3). On the other hand, at a high level, communications must be organized on the basis of a global architecture that guarantees the provision of cooperative services and access to data, encompassing all available operational technologies. This reference architecture is the aforementioned Communications Access for Land Mobiles (CALM) architecture.

CALM is composed of a series of international standards in this field, and it supports the continuity of applications through different physical interfaces and media such as IEEE 802.11, IEEE 802.11p, IEEE 802.15, IEEE 802.16e, IEEE 802.20, 2G/3G/4G mobile telephony, or the proprietary systems of national ITS programs. More specifically, the main physical means used to support the CALM architecture are:

- *ISO 21212*: 2G Cellular (GSM).
- *ISO 21213*: 3G Cellular (UMTS).
- *ISO 21214*: InfraRed.
- *ISO 21215*: M5 (802.11p).
- *ISO 25112*: WiMAX (802.16e).
- *ISO 25113*: HC-SDMA (802.20).
- Bluetooth (IEEE 802.15).
- Ethernet (IEEE 802.3).
- DSRC.

The fundamental applications of CALM are: to support Internet services in mobile environments; to support national ITS applications; to support the next generation of ITS applications, i.e., communications systems for vehicle safety and new commercial applications based on their high bandwidth and wide range capability.

This architecture also defines a series of cooperative services that are associated to the different interfaces and communications requirements in order to support the entire design cycle.

The great contribution of the CALM architecture is that for the first time all the actors involved in the field of transport (automobile companies, infrastructure construction companies, social agents, public administration, etc.) have collaborated to build a framework that unifies all the aspects of the communications in the transport, in respect to physical means as well as in respect to applications that give use to the mentioned interfaces. However, CALM is a framework architecture, which is based on standards of both physical media and services, some of which have not yet been fully implemented and, in some cases, there is no consensus in the community to adopt a specific technology. Because of this, one of the major challenges facing vehicular communications is to ensure absolute interoperability in communications between vehicles of different manufacturers, as well as between vehicles and different types of infrastructure. Furthermore, there are a number of nontechnical obstacles, such as user acceptance, data security and privacy, system access and interoperability, risk and liability, public order needs, cost/benefit and business models, and deployment plans for implementation.

In addition, the installation in Europe of such a series of communications networks and services faces a serious problem of a higher order, such as the difficulty in the acceptance and harmonization of regulations by the different EU member states, which maintain the competences in terms of transport and infrastructures and that in some cases have conflicting regulations that make it impossible to put them into operation.

3.2.3 Operative Technologies

Once the framework architecture that makes use of wireless communications for the transmission of information in the field of transport has been defined, it is necessary to define which technologies are currently available as well as which technology fits best with each one of the proposed applications. It is clear that all these sets of services associated with transport require a specific type of communication systems, emphasizing concepts such as real time, privacy, connectivity between vehicles and infrastructures, etc. None of these technologies are purely new, but have been adapted to a greater or lesser extent from existing wireless communications technologies, developed to operate with high mobility or not. That is why, although some of these wireless technologies work perfectly when installed in a vehicle, they have not been designed specifically for this and in some cases present problems and incompatibilities that have had to be improved. Even in most cases, it is necessary to combine several systems to cover all needs, even in order to have redundancy and guarantee access to data in any situation.

Depending on the scope and range of communications systems, wireless communication networks can now be divided into four types depending on their scope. The Personal Area Network (PAN), which serves local communication needs, with a maximum range of 10 meters, which mainly supports V2P services. The most characteristic application that uses this type of communications and with which we are all familiar are the hands-free devices, i.e., talking with the mobile phone. However, the range of applications available in the automotive industry is not yet very wide, for example, wireless GPS devices, music reproduction devices, electronic payment, identification of traffic signals, and of course, installed communication devices as part of the vehicle's electronics. However, it is expected that the number of applications for the intravehicular Personal Area Network will increase in the coming years, mainly due to the replacement of current wired communications buses by this type of wireless technologies that can lead to a decrease in cost and the weight of the vehicle, while maintaining the same characteristics of safety and reliability. The technologies available for this type of application are Bluetooth (IEEE 802.15), Zigbee (IEEE 802.15.4) networks, and RFID identification systems (ISO/IEC 10536 family of standards).

The Local Area Network (LAN) is the interconnection of several devices with limited range to an environment up to 1000 m. Logically in the automotive and wireless communications environment, the types of networks most used are the Wireless Local Area Network (WLAN), which allow the exchange of information between different devices located inside the vehicle, between several vehicles, and with road infrastructure at broadband and while several of the nodes of the network are in motion. Although these types of networks were not designed originally for mobility but simply to avoid the installation of network wires in offices, they have now evolved to support short-range embedded communications systems. The protocol family on which the physical and MAC level of Mobile WLANs is based is IEEE 802.11p (ETSI-ITS G5 in Europe), being the basis of Dedicated Short Range Communications (DSRC) devices. The services associated with this type of network are those related to the area of intelligent vehicles, such as V2V and V2I communications for the transmission of safety information, incidents on the road, emergency or hazard maneuvers, etc. This type of connection, just like the PAMs are ad hoc, so there is no connection provider as such and, therefore, data exchange has no cost to the user and no third-party dependencies exist except, logically, the provider of cooperative services where appropriate.

Increasing the range of the network, Metropolitan Area Networks (MAN) are defined as high-speed networks with capacity for services requiring broadband (70 Mbps), which, covering a large geographic area (around 48 km), provides capacity of integration of multiple services through the transmission of data, voice, and video. This type of network is used in the automotive field mainly for the interconnection of systems located in the infrastructure, in areas where a high-speed cabling is not available, for example on many roads, where transmission of information from sensors, such as surveillance cameras or radars, is necessary and there is not a wired means to transmit it. The technology used to support this type of service is mainly based on IEEE 802.16 WiMAX, under an extension of the original protocol known as IEEE 802.16e for applications with high mobility, up to 200 kmh. However, WiMAX technology has suffered a slowdown in its development in recent times as it requires a specific deployment of communications infrastructure and its capabilities have been matched and surpassed by the mobile telephony networks from UMTS (3G) technology.

Finally, the type of network with the greatest coverage is the so-called Wide Area Network (WAN), capable of covering unlimited distances and providing broadband Internet access in any physical location, with coverage reaching 99% of the territory of the developed countries. In the case of automotive, the available WANs are those based on mobile telephony, based on 3G protocols, and 4G, considering also the future development of 5G. Given the global coverage of these types of networks, they are ideal for the transmission of information between vehicle and infrastructure and even between vehicles. Its fundamental limitation lies in the delays in the transmission and the establishment of the connection, the saturation of the cells by a large number of users, and in the cost of the connections, all provided by an access provider. They are used basically for emergency applications such e-Call or OnStar.

Given the importance of the technologies described in this section, the following sections describe them in detail in order to relate their communication capabilities with the cooperative services they enable, depending on the requirements of each application.

3.2.3.1 Dedicated Short Range Communications

Dedicated Short Range Communications are ad hoc (decentralized) short- and medium-range data transmission systems that support public and private security operations in vehicle-to-infrastructure and vehicle-to-vehicle communications environments or vice versa. The DSRC are standardized to guarantee their interoperability independently of the manufacturer of the media access devices, following the protocol layer stack of ISO Model of Architecture for Open Systems Interconnection, comprising five layers (Physical, MAC and Link, GeoNetwork, Transport, and Application), where we can highlight three differentiating characteristics: the IEEE 802.11p (ITS G5 in Europe) specification is followed at the physical level and MAC, which allows the transmission of data in the dedicated 5.9 GHz channel through spread spectrum technique, and the sending of MAC-level broadcast packets. The network level includes the geographical location of the information handled by the communications device, enabling the so-called GeoNetworking. Finally, the transport level enables the multihop capability for the retransmission and routing of the packets of the vehicular network.

Such networks provide very high data transfer rates in circumstances where it is important to minimize latency times in channel establishment and communications zones are relatively small and isolated from other broadband access. Data communication between stationary or moving vehicles and fixed equipment on the road is used in applications involving payments and information transfer for security or monitoring. Among others, at the application level the most important services are Cooperative Awareness Message (CAM), Decentralized Environmental Notification Message (DENM), Road Topology (MAP), Signal Phase And Timing (SPAT), and Service Announcement Message (SAM), standardized by ISO for ITS communications, which include roadwork warning, postcrash warning, icy road, emergency vehicle warning, electronic brake light (CAM/DENM), and other messages related to traffic sign states (SPAT/MAP), vehicle signage, and vehicle probe data (Santa et al., 2014).

They are able to broadcast information in broadcast mode (GeoBroadcast) or to a specific user in a range of between 300 and 1 km, depending on the needs and the nature of the information by multiple hops, integrating all the vehicles that are nearby in a reduced zone of road in the same network, which allows the exchange of information quickly and reliably, with a minimum time to connect to the network and without delays in accessing the information. Specifically, DSRC-based networks offer maximum packet transmission time between source and destination of <5 ms and end-to-end latency (<100 ms), including security (Festag et al., 2008). The use and specification of this technology involves several car manufacturers, including Volkswagen or Volvo.

DSRC technical description:

The communication modes of DSRC allow V2V and V2I communication.

1. **V2V Communications:** includes multihop geographic routing, using other vehicles as relays for the message delivery.
   a. *GeoUnicast*: provides packet delivery from an emitting vehicle to a receiving vehicle that is located in a fixed geographic position, via multiple hops.
   b. *GeoAnycast*: provides packet delivery to a vehicle (node) that is in a specific geographic area as a function of set conditions (i.e., nearer).
   c. *GeoBroadcast*: provides packet delivery in broadcast mode to all the vehicles that are in fixed geographic area.
   d. *Topollogically-scoped broadcast (TSB)*: provides packet delivery to every vehicle that is in a range of n-hops from the emitting vehicle.
2. Vehicle to Infrastructure (uplink) V2I and infrastructure to Vehicle I2V (downlink) Communications: they have an equivalent behavior to V2V but involve DSRC modules installed in the roadside:
   a. One vehicle to beacon (Geounicast).
   b. Beacon to one vehicle (Geounicast).
   c. Beacon to many vehicles (GeoBroadcast, TSB).
   d. Beacon to selected vehicles (GeoAnycast).

3.2.3.2 3/4G Mobile Telephony.

In 1985, the first generation of mobile telephony (1G) emerged in Europe after adapting the American Mobile Phone System (AMPS) system to European requirements, and was named (Total Access Communications System (TACS). TACS encompasses all analog mobile communications technologies. It can transmit voice but not data.

Due to the simplicity and the limitations of the first generation of mobile telephony, the Global System for Mobile Communications system (GSM) was enabled, which marked the beginning of the second generation (2G). Its main feature is the ability to transmit data in addition to voice, at a speed of 9.6 kbit/s.

In 2001, the so-called second and second dot five generation (2.5G) emerged in the United States and Europe. In this generation were included those technologies that allowed a greater capacity of data transmission and that emerged as a previous step to 3G technologies. The most notorious technology of this generation is the General Packet Radio System (GPRS), capable of coexisting with GSM, but offering a more efficient service for accessing IP networks such as the Internet. The maximum GPRS speed is 171.2 kbit/s. In Japan this generation did not exist since the direct jump from 2G to 3G was made.

In 2005 the 3G technologies of mobile telephony were deployed. Third-generation technologies are categorized within the International Telecommunication Union (ITU) standard International Mobile Telecommunications-2000 (IMT-2000), which guarantees interoperability among all 3G networks independently of the telephony operator and technology provider. The most widely used 3G technology is the Universal Mobile Telecommunication System (UMTS). The services offered by 3G technologies are basically: Internet access, broadband services, international roaming, and interoperability. But fundamentally, these systems allow the development of multimedia environments for the transmission of video and images in real time, encouraging the emergence of new applications and services such as videoconference or electronic commerce with a maximum speed of 2 Mbit/s in optimal conditions, such as, for example, in the interior of buildings.

In 2007, High Speed Downlink Packet Access (HSDPA) technology, corresponding to the 3.5G mobile phone, was available to users, allowing wireless broadband access over high speed UMTS to a maximum bandwidth of 14.4 Mbps.

HSDPA technology was surpassed in 2010 by the Long Term Evolution (LTE), enabling the 4th generation of mobile telephony (4G). LTE is the standard for high-speed wireless data communications for mobile phones and data terminals, with transmission speeds of up to 75 Mbit/s for high mobility (200 km/h) and 300 Mbit/s for low mobility, with latencies between 50 and 150 ms. In 2014, the Long Term Evolution Advanced (LTE-A) technology, a 4G evolution, was developed, enabling transmission speeds up to 500 Mbit/s for high mobility (200 km/h) and 1 Gbit/s for low mobility, with latencies between 10 and 20 ms.

Mobile telephony technology applied to vehicular environments is currently in addition to the DSRC networks, the only one that is fully developed, operational, and available for all types of applications. While DSRC networks focus primarily on short/medium-range V2V communications, data exchange via mobile telephony allows operations with the infrastructure and even with other vehicles when DSRC networks are unavailable. Additionally, its implementation in road and automotive environments is much more deployed than any other technology and, in some cases, mobile telephony is used as the only system for all types of communications. However, there are two clear limitations regarding the use of mobile telephony-based communications in vehicular environments. On the one hand, given the characteristics of cell-based communications, a massive deployment in vehicles could lead to saturation of communications in areas with few nodes in the infrastructure. On the other hand, since the service is provided by telephony provider companies, it is essential to pay a fee for use, which represents an increase in costs for the user that, in some cases, they may not be willing to assume, even in case of providing services for safety applications. In addition to that, mobile telephony may have certain delays in establishing the connection that would imply an inability to use it in some critical systems. The other important factor is the lack of quality coverage in some stretches of the road, mainly in rural or mountain roads that are, at the same time, areas with a high accident probability. The most commonly used mobile-based systems are the e-Call and its commercial version General Motors' OnStar, already implemented in several European countries and USA respectively. However, the general trend today is not to choose between DSRC and 3/4G, but an integrated vision is sought, using the advantages of both technologies in what are called hybrid communication systems. In this way, DSRC would be fundamentally dedicated to V2V communications security and efficiency and 3/4G would be dedicated to V2I communications and applications with high demand for bandwidth as infotainment.

3.2.3.3 5G Mobile Telephony

The new generation of mobile telephony is expected for 2021 and known as 5th generation—5G (Kljaic et al., 2016). It was based on the early developments of the European Project FP7 METIS (Mobile and wireless communications Enablers for the Twenty-twenty Information Society, 2012—15), which highlights the participation of BMW as s manufacturer of cars to contribute to the definition of requirements in the field of road transport. Based on these initial results, the European Commission defined the 5G Public Private Partnership (PPP) within the H2020 program for the purpose of developing 5G technology and the Internet of the future. 5G technology is expected to be a hybrid of 3G, 4G, and WiFi-WLAN technology, which, when applied to the transport sector, unifies the advantages of mobile telephony and DSRC, including direct communication between multihop devices and device-to-device (ERTICO, 2015).

The preliminary 5G technical capabilities are:
- *Capacity*: 50 to 100 times 4G.
- *Quality of Service*: Ultra reliable communication for many critical applications.
- *Transmission time*: 50—100 times faster than 4G LTE.
- *Latency*: 1 ms.
- *Bidirectional*: Direct communications between devices (device to device: D2D). In case of road transport, V2V.
- *Broadcast*: Enabled.

3.2.3.4 RFID

Radio Frequency IDentification (RFID) is a data storage and retrieval system that uses devices called tags, transponders, or RFID tags. The fundamental purpose of RFID technology is to transmit the identity of an object (similar to a unique serial number) using wireless data transmission.

An RFID tag is a small device, similar to a sticker that can be attached or incorporated into a product, animal, or person. They contain antennas to enable them to receive and respond to radiofrequency requests from an

RFID transceiver. One of the advantages of using radiofrequency is that direct vision is not required between transmitter and receiver.

The tags can be active, semipassive (or semi-active, battery-assisted), or passive. Passive tags do not require any internal power supply and are in fact purely passive devices (only activated when a reader is nearby to supply them with the necessary power). The other two types need power, typically a small battery.

Depending on the frequencies used in RFID systems, cost, range, and applications are different. Systems employing low frequencies also have low costs, but also low usage distance. Those employing higher frequencies provide longer reading distances and faster read speeds. Thus, low frequency is commonly used for animal identification, goods tracking, car key for vehicles, pallet tracking and packaging, and tracking of trucks and trailers on shipments.

Another important application of RFID in transport applications is the electronic toll collection. This technology has been used in many deployments in Spain, Mexico, USA, France, and Germany. In this case, a RFID tag installed in each vehicle connects and exchanges information with the infrastructure when the car enters onto the ramp of a highway, charging the costs of this access automatically.

Another common application is the use of RFID in smart keys, available in models from most car manufacturers. In this case, the key is replaced by a card with an active RFID circuit that allows the car to recognize the presence of the key within 1 m of the sensor.

Another proposed application is the use of RFID for road traffic signals (Road Beacon System). It is based on the use of floor-embedded RFID transponders (radio beacons) that are read by a vehicle-carrying unit (OBU) that filters the various traffic signals, warning the driver if necessary.

3.2.3.5 Bluetooth

Bluetooth (Bluetooth, 2016) is the specification for so-called Personal Area Wireless Networks (WPANs) that enables data transmission between different devices through a radio frequency link in the 2.4 GHz band. The Bluetooth specification has been designed to enable the development of low-cost, low-power, and short-range communications devices (up to 100 m).

The reason for the creation of this specification is to obtain a single digital wireless protocol that is capable of interconnecting multiple devices very simply and solving classic problems such as the synchronization between them. Similar to WiFi networks, Bluetooth uses Frequency Hopping Spread Spectrum (FHSS) technology for data transmission, using the 2.4 GHz band.

Bluetooth networks support up to 1 Mbps band rate in basic transfer mode and 3 Mbps in the enhanced data transfer mode.

The normal operation of Bluetooth networks follows the master–slave scheme. One of the devices in the network, called master, provides the reference values for the connection, such as synchronization and frequency hopping sequence. The other devices in the network are called slaves and exchange data with the master. This network consisting of short-range devices is called a piconet (μNet). One of the fundamental characteristics of this type of network is that the information can circulate between the master and any other device; however, different devices can change their roles among themselves and, in this way, a master can be transformed into a slave and vice versa, depending on the needs of applications that support communications.

The Bluetooth specification also allows the interconnection of two or more piconets, thus forming a scatternet, in which some of the slave devices act as gateways between two networks, being master in one and slave in another.

At present, Bluetooth communications are widely used in the automotive field in many applications and systems that use connections between different devices of the vehicle and mobile phones, MP3 players, or GPS. The common feature of these applications is that the Bluetooth connections are aimed at entertainment, not safety systems. However, the master–slave scheme of Bluetooth networks is very similar to the scheme of multiplexed buses that equip vehicles and may well appear in the not too distant future applications to replace these wired networks.

3.2.4 Hybrid Communication Approach

Cooperative Intelligent Transport System (C-ITS) messages will be transmitted for a wide range of services, in different transport situations. End-users do not care about the specific communication technology used to transmit C-ITS messages, but will expect to receive all information on traffic and safety conditions seamlessly. This can only be achieved through a so-called hybrid communication approach, i.e., by combining complementary communication technologies.

As mentioned in *A European strategy on Cooperative Intelligent Transport Systems, a milestone towards cooperative, connected and automated mobility,* to support all C-ITS services both On-Board Units (OBUs) and Road-Side Units (RSU) have to be compliant. On the vehicle side the full hybrid communication mix needs to be onboard. On the infrastructure side the choice of communication technology will depend on the location, the type of service, and cost-efficiency. C-ITS messages should be unaware of, and thus flexible about the communication technology used, easing the inclusion of future technologies (e.g., 5G and satellite communication).

Currently, the best option for the hybrid communication mix is a combination of IEEE802.11p/ETSI ITS-G5 and next-generation cellular networks (5G). This ensures the best possible support for deployment of all Day 1 C-ITS services. It combines low latency of ETSI ITS-G5 for time-critical safety-related C-ITS messages with wide geographical coverage and access to large user groups of existing cellular networks.

3.2.5 Services

Innovation in transport comes hand in hand with the development of enabling technologies and suitable business models. As a result of that transformation, socio-economic opportunities will be created, and the societies ready to seize them will benefit from them.

There have been several road test demonstrators, which have shown that services enabled by CALM operative technologies are not only possible but that they perform efficiently in real-life vehicular scenarios. Following CALM architecture, Table 3.1 reviews some of these applications considering: 5G as WAN-enabling technology; DSRC as VANET-enabling technology; and RFID and Bluetooth as PAN-enabling technologies.

Digital technologies are said to be the strongest driver and enabler of this process. Exchanging real-time data amongst actors:

- Leads to a more efficient use of the resources (e.g., a shared car or a container or a rail network);
- Facilitates anticipation so that human error is reduced, thereby diminishing the greatest source of accidents in transport;

Table 3.1 Typical communications services in real-life vehicular scenarios

Service example	Enabling technology	Project reference
Route time collection systems	Bluetooth	Chitturi, M.V., et al., 2014. Validation of Origin-Destination Data from Bluetooth Re-Identification and Aerial Observation. TRB 2014 Annual Meeting
Electronic toll systems	RFID	Salunke, P., et al., 2013. Automated toll collection system using RFID. IOSR J. Comput. Eng. (IOSR-JCE)
Location data offloading service	Hybrid communication approach (IEEE802.11p-5G)	Katsaros, K., et al., 2013. Effective implementation of location services for VANETs in hybrid network infrastructures. IEEE ICC 2013 Workshop on Emerging Vehicular Networks, Budapest; 9–13 June 2013.

- Facilitates the creation of a real multimodal transport system, by which people can reach their destination using various means of transport during an itinerary; and
- Induces social innovation and the emergence of collaborative economic solutions.

The C-ITS platform from the European Commission has identified that from the end-user point of view, the way towards the deployment of C-ITS *services* is to provide *continuous service* (source: *Final Report of the C-ITS Platform*) on both sides, infrastructure and vehicle. Therefore a prioritization of the ITS-services to be deployed is enforced: *Day 1 C-ITS Services* and *1.5 C-ITS Services*.

According to the aforementioned report, *Day 1 C-ITS services in Europe*, when deployed in an interoperable way, will produce a benefit cost ratio of up to 3 to 1 based on cumulative costs and benefits from 2018 to 2030. Rapidly deploying as many services as possible will also mean that they will more quickly break even and will lead to higher overall benefits, mainly due to the network effects (which means that slow initial uptake would result in relatively long periods with few benefits).

On the other hand, the full specifications or standards for large-scale deployment of Day 1.5 C-ITS services are expected to be completed by 2019, even though they are considered to be generally mature. A detailed description of those C-ITS services is included in Chapter 7, Cooperative Systems.

Infotainment is the other pillar that supports the structure of vehicular communication based services. Infotainment supports all the services not directly related with safety or efficiency, mainly focused on the provision of resources and applications based on wideband Internet access in mobility. This set of services is relevant twofold; first, from the point of view of users (driver and passengers) through the provision of comfort, navigation, audio/video streaming and Internet applications. Second, from the point of view of network operators and digital contents providers, enabling new business possibilities related to road transport. Mobile apps, vehicle embedded apps and in-car Internet access through mobile telephony, are the actual basis of infotainment.

3.2.6 Security and Privacy

The way towards the deployment of C-ITS *security* relies on digitized transport systems, which are more vulnerable to hacking and cyberattacks. Cybersecurity is critical and requires actions at global level (e.g., European or US national). Clear rules are needed in order to accelerate the C-ITS deployment. Fragmented security solutions will put interoperability and the safety of end-users at risk.

A common security and certificate policy for C-ITS deployment, such as that proposed by the European Commission, needs to be developed. *Political support* is needed in order to develop a uniform and widely-accepted security solution for cooperative and connected vehicles.

Technology-wise, the security framework is based on Public Key Infrastructure technology, a combination of software, asymmetric cryptographic technologies, processes, and services that enable an organization to secure C-ITS communications. A challenge will be to set up the necessary governance at institutional and industry levels involving main stakeholders, including public authorities, road operators, vehicle manufacturers, C-ITS service suppliers and operators.

And finally *users* have to be aware of the importance of *privacy and data protection*. They have to understand that personal data are not a commodity, and know they can effectively control how and for what purposes their data are being used.

Safety-related CAM and DENM data traffic is broadcasted by C-ITS infrastructure and vehicles. The data flow coming from the vehicles will, in principle, qualify as personal data as will relate to an identified or identifiable natural person. Thus, the implementation of C-ITS requires compliance with the applicable data protection legal framework. Processing of such data is only legal if it is based on one of the laws listed therein, for example the consent of users.

Data protection principles and data protection impact assessments are basic C-ITS system layout and engineering, especially in the context of the applied communication security scheme. When these conditions are met the willingness of end-users to give consent to broadcast data is not a barrier, in particular if the data is to be used to enhance road safety or improve traffic management.

3.2.7 Interoperability

In order to have an integrated transport system, its components have to be interoperable. In other words, the systems need to be able to interact with each other, across borders and considering different transport modes. And this interoperability is necessary at all levels: infrastructure, data, services, applications, and networks. While standardization activities are important, they are not sufficient to ensure interoperability. EU and worldwide deployment specifications therefore have to be defined and agreed upon.

To this end, C-ITS deployment initiatives should define and publish the technical C-ITS communication profiles needed to ensure the interoperability of Day 1 C-ITS services. They should also develop test procedures to check the interoperability of these profiles. Granting mutual access to communication profiles will ensure that best practices and lessons learned from real life operation are shared. It should also lead to a gradual convergence of profiles, creating the conditions for interoperability. The aim is to enable a single market for C-ITS services based on common communication profiles, which, however, leave space for future innovative services.

In 2016, Member States and the Commission launched the C-Roads Platform to link C-ITS deployment activities, jointly develop and share technical specifications, and to verify interoperability through cross-site testing. Initially created for C-ITS deployment initiatives cofunded by the EU, C-Roads is open to all deployment activities for interoperability testing.

REFERENCES

Alfonso, J., Sánchez, N., Menéndez, J.M., Cacheiro, E., 2014. Cooperative ITS architecture — the FOTsis Project and beyond. IET Intell. Transport Syst. http://dx.doi.org/10.1049/iet-its.2014.0205, Online ISSN 1751-9578.
Bluetooth Special Interest Group, 2016. Specification of the Bluetooth System.
ERTICO, 2015. Guide about technologies for future C-ITS service scenarios.
Festag, A., Baldessari, R., Zhang, W., Le, L., Sarma, A., Fukukawa, M., 2008. CAR-2-X communication for safety and infotainment in Europe. NEC Tech. J. 3 (1), 21—26.
Kljaić, Z., Škorput, P., Amin, N., 2016. The challenge of cellular cooperative ITS services based on 5G communications technology, 39th International Convention on Information and Communication Technology, Electronics and Microelectronics (MIPRO), Opatija, 2016, pp. 587—594.
OCDE, 2015. Automated and Autonomous Driving Regulation under uncertainty, International Transport Forum, Corporate Partnership Board Report.
Santa, J., Pereñíguez, F., Moragón, A., Skarmeta, A.F., 2014. Experimental evaluation of CAM and DENM messaging services in vehicular communications. Transp. Res. Part C Emerg. Technol. 46, 98—120.

RELATED STANDARDS

A European strategy on cooperative, intelligent transport systems, a milestone towards cooperative, connected and automated mobility (http://ec.europa.eu/energy/sites/ener/files/documents/1_en_act_part1_v5.pdf).
CEN EN 12253:2004: "Road transport and traffic telematics; Dedicated Short Range Communication; Physical layer using microwave at 5.8 GHz."
CEN EN 12795:2003: "Road transport and traffic telematics; Dedicated Short Range Communication (DSRC); DSRC data link layer; medium access and logical link control."
CEN EN 12834:2003: "Road transport and traffic telematics; Dedicated Short Range Communication (DSRC); DSRC application layer."
ETSI TS 102 636-3 (V1.1.1): "Intelligent Transport Systems (ITS); Vehicular Communications; GeoNetworking; Part 3: Network architecture," 2010-03.
ETSI TS 102 637-1 (V1.1.1): "Intelligent Transport Systems (ITS); Vehicular Communications; Basic Set of Applications; Part 1: Functional Requirements," 2010-09.
ETSI ES 202 663 (V1.1.0): "Intelligent Transport Systems (ITS); European profile standard for the physical and medium access control layer of Intelligent Transport Systems operating in the 5 GHz frequency band," 2009-11.
ETSI EN 302 571 (V1.1.1): "Intelligent Transport Systems (ITS); Radiocommunications equipment operating in the 5 855 MHz to 5 925 MHz frequency band; Harmonized EN covering the essential requirements of article 3.2 of the R&TTE Directive," 2008-09.
ETSI EN 302 665 (V1.1.1): "Intelligent Transport Systems (ITS); Communications Architecture," 2010-09.
Final Report of the C-ITS Platform (January 2016): http://ec.europa.eu/transport/themes/its/c-its_en.
IEEE Std 802.11-2007 (Revision of IEEE Std 802.11-1999): "IEEE Standard for Information technology; Telecommunications and information exchange between systems; Local and metropolitan area networks; Specific requirements; Part 11: Wireless LAN Medium Access Control (MAC) and Physical Layer (PHY) Specifications," 2007-06.

IEEE Std 802.11p-2010: "IEEE Standard for Information technology; Telecommunications and information exchange between systems; Local and metropolitan area networks; Specific requirements; Part 11: Wireless LAN Medium Access Control (MAC) and Physical Layer (PHY) Specifications; Amendment 6: Wireless Access in Vehicular Environments," 2010-07.

IEEE Std 1609.2-2006: "IEEE Trial-Use Standard for Wireless Access in Vehicular Environments; Security Services for Applications and Management Messages," 2006-07.

IEEE Std 1609.3-2010 (Revision of IEEE Std 1609.3-2007): "IEEE Standard for Wireless Access in Vehicular Environments (WAVE); Networking Services," 2010-12.

IEEE Std 1609.4-2010 "Revision of IEEE Std 1609.4-2006): "IEEE Standard for Wireless Access in Vehicular Environments (WAVE); Multichannel Operation," 2011-02.

IETF RFC 791: "Internet Protocol DARPA Internet Program Protocol Specification," Information Sciences Institute, University of Southern California, 1981-09.

IETF RFC 2460: "Internet Protocol, Version 6 (IPv6) Specification," IETF Network Working Group, 1998-12.

IETF RFC 2766: "Network Address Translation – Protocol Translation (NAT-PT)," IETF Network Working Group, 2000-02.

IETF RFC 2784: "Generic Routing Encapsulation (GRE)," IETF Network Working Group, 2000-03.

IETF RFC 3024: "Reverse Tunneling for Mobile IP, revised," IETF Network Working Group, 2001-01.

IETF RFC 3315: "Dynamic Host Configuration Protocol for IPv6 (DHCPv6)," IETF Network Working Group, 2003-07.

IETF RFC 3596: "DNS Extensions to Support IP Version 6," IETF Network Working Group, 2003-10.

IETF RFC 3775bis: "Mobility Support in IPv6," IETF Mobile IP Working Group, 2011-09 (Expiring date for approval).

IETF RFC 3963: "Network Mobility (NEMO) Basic Support Protocol," IETF Network Working Group, 2005-01.

IETF RFC 4213: "Basic Transition Mechanisms for IPv6 Hosts and Routers," IETF Network Working Group, 2005-10.

IETF RFC 5177: "Network Mobility (NEMO) Extensions for Mobile IPv4," IETF Network Working Group, 2008-04.

IETF RFC 5454: "Dual-Stack Mobile IPv4," IETF Network Working Group, 2009-03.

IETF RFC 5555: "Mobile IPv6 Support for Dual Stack Hosts and Routers," IETF Networking Group, 2009-06.

IETF RFC 5944: "IP Mobility Support for IPv4, Revised," IETF Standards Track, 2010-11.

In Europe: Respecting the Radio Equipment Directive 2014/53/EU (http://eur-lex.europa.eu/legal-content/EN/TXT/?uri = celex:32014L0053).

In Europe: COM(2016)588: 5G for Europe: An Action Plan and accompanying Staff Working Document SWD(2016)306 (http://eur-lex.europa.eu/legal-content/EN/TXT/?qid = 1479301654220&uri =CELEX:52016DC0588).

ISO/DIS 21210.2: "Intelligent Transport Systems; Communications access for land mobiles (CALM); IPv6 Networking," 2011-03 (Final Draft to be formally voted and approved).

ISO 21215:2010: "Intelligent Transport Systems; Communications access for land mobiles (CALM); M5," 2010-11.

ISO/DIS 21217: "Intelligent Transport Systems; Communications access for land mobiles (CALM); Architecture" (2010, currently under revision).

ISO 21218:2008: "Intelligent Transport Systems; Communications access for land mobiles (CALM); Medium service access points," 2008-08.

ISO 24102:2010: "Intelligent Transport Systems; Communications access for land mobiles (CALM); Management," 2010-11.

ISO 29281:2011: "Intelligent Transport Systems; Communications access for land mobiles (CALM); Non-IP networking," 2011-03.

The common security and certificate policy documents will for instance define the European C-ITS Trust model based on Public Key Infrastructure. They will, amongst others define legal, organizational and technical requirements for the management of public key certificates for C-ITS services based on the structures identified in (IETF) RFC 3647.

The C-Roads Platform is cofunded under the Connecting Europe Facility (CEF)— https://www.c-roads.eu/platform.html.

CHAPTER 4

Positioning and Digital Maps

Rafael Toledo-Moreo[1], José M. Armingol[2], Miguel Clavijo[3], Arturo de la Escalera[2], Javier del Ser[4], Felipe Jiménez[3], Basam Musleh[5], José E. Naranjo[3], Ignacio (Iñaki) Olabarrieta[6] and Javier Sánchez-Cubillo[6]

[1]Universidad Politécnica de Cartagena, Cartagena, Spain
[2]UC3M, Leganés, Spain
[3]Universidad Politécnica de Madrid, Madrid, Spain
[4]TECNALIA, University of the Basque Country (UPV/EHU and Basque Center for Applied Mathematics (BCAM), Bizkaia, Spain
[5]ABALIA, Madrid, Spain
[6]TECNALIA, Bizkaia, Spain

Contents

Intelligent Vehicles
DOI: http://dx.doi.org/10.1016/B978-0-12-812800-8.00004-7

4.1 POSITIONING BASED SYSTEMS FOR INTELLIGENT VEHICLES

It seems obvious that positioning is a fundamental element in road transport, as transportation could not be understood without knowing the position of the road vehicle, either because the driver knows the area, there are relevant road signs, or simply because the navigation system says so. When positioning technology is applied to make smarter vehicles, the benefit may be three-fold. They apply for both people and good transportation:

1. For safety, allowing the deployment of Advanced Driver Assistance Systems (ADAS) that can avoid collisions or support their mitigation.
2. For efficiency, improving the navigation times/fuel consumption, with a positive impact on the road traffic conditions.
3. For comfort, being essential in a good number of Location-Based Services (LBS) that make the driving experience more pleasant.

Today's most outstanding technology for vehicle positioning is the Global Navigation Satellite System (GNSS), led by the US system Global Positioning System (GPS). However, satellite-based positioning suffers from relevant flaws that linger on in the implementation of applications and services. This is particularly crucial in safety-critical, where human life is in play, and liability-critical applications, in which a guaranteed performance is a must for accepting the system deployment (as in the GNSS-based toll collect). GNSS performance problems, such as lack of coverage, multipath errors, or jamming/spoofing, cannot be overcome without the support of additional sources of information. For this reason, key technologies for GPS-aid positioning (and how to use them) will also be presented in this Chapter.

Moreover, GNSS systems provide absolute positioning referred to a global frame (for instance, the WGS84, standing for World Geodetic System 1984). Despite the fact that this may be useful per se in some applications and services, in most of the cases, a local reference, namely a map, is needed. Indeed, for the large majority of ADAS and LBS, the positioning system will feature a map. The fundamentals of digital maps and the process of map-matching the positioning output deserve a dedicated section.

4.1.1 Definitions

This section is intended to be a glossary of the key terms used along this chapter. The definitions, which are oriented to the specific field of intelligent vehicles, neither pretend to follow any standard, nor to dispute with them the meanings. It must be simply understood as a handy list of descriptions to support the later discussion (Table 4.1).

Table 4.1 Positioning and digital maps key terms definitions

Term	Description
Position	Spatial location of the vehicle
Positioning performance	Quality of the positioning outputs
Trajectory	Time series of position values of a vehicle
Position availability	Capability to provide a position solution at a given time and place NB: the availability of position does not imply that its value is correct
Position accuracy	Difference between the position output and the real value
Position integrity	Level of trust that can be laid in the position solution
Digital map	Digital description of the geometry and topology of the road
ITS	Intelligent Transportation Systems: application of IT technologies in the transportation domain
Map-matching	Algorithm to match the position into a map reference
Data fusion	Process of fusing information coming from different sources (sensor, map, etc.) in a common solution that benefits the overall position performance
Inertial sensor	Motion sensors that provide acceleration and rates of turn measurements
Odometry	Estimate of traveled distance based on the fraction of wheel rotation
PVT	Position Velocity Time
V2X communications	Communications Vehicle to Any (Vehicle, Infrastructure, Person)
IEEE 802.11p	Standard of the physical layer of vehicular communications (mobile WiFi)
Hybrid communications	Communications that involves Mobile WiFi and cellular communications
GeoNetworking	Vehicular communications that includes the GPS Positioning in the data packets routing
RSU	Road Side Unit. Each one of the roadside infrastructure components of a V2X system
OBU	On-Board Unit. Each one of the in-vehicle components of a V2X system
OSI	Open Systems Interconnection
ISO	International Standardization Organization
DSRC	Dedicated Short Range Communications
CDMA	Code Division Multiple Access, the basis of the IEEE 802.11p communications

(Continued)

Table 4.1 (Continued)

Term	Description
Trilateration	A method of determining the relative positions of objects using the geometry of triangles
Ephemerides or ephemeris	Information about the position of astronomical objects as well as artificial satellites in the sky at a given time or times
Urban canyon	A place where the street is flanked by buildings on both sides creating a canyon-like environment
LOS	Line-Of-Sight (between a satellite and the receiver's antenna)
NLOS	Non-Line-Of-Sight (between a satellite and the receiver's antenna)

4.1.2 Location Based Services and Applications Based on Position

Position-based location services onboard a vehicle can be used to improve the travel experience. Their number is really unlimited, as new apps may arise everyday that exploit the location of our smartphones to provide information relevant to the user. Just to name a few of them:

- Navigation
- Traffic management
- Incident alarm
- Fleet management
- Emergency call
- Intelligent speed control
- Collision avoidance and collision avoidance support
- Lane departure and lane change warnings
- Find my friends
- Etc.

4.2 GNSS-BASED POSITIONING

4.2.1 Motivation, Requirements and Working Principles

A GNSS is a system offering the users a service to determine their three-dimensional position worldwide. It involves a constellation of satellites orbiting at an altitude of about 20,000 km over the earth's surface, continuously transmitting positioning, navigation, and timing signals.

Two GNSS constellations are fully operational in 2017: the American NAVSTAR Global Positioning System (GPS) and the Russian GLONASS (European Space Agency, 2017).

4.2.1.1 Motivation

GPS is the most widely used GNSS in the world and became fully operational in 1993. It was first promoted by the U.S. military for national defense and homeland security which made it necessary to geolocate their position around the globe. Although service availability was also provided for civilian use, at the beginning of operation an artificial degradation in the civilian signal was added. It was called *selective availability*. This artificial degradation was switched off in May 2000 enabling civilian users, commercial, and scientific needs to take advantage of full positioning performance of the system. GPS currently provides two levels of service, each using different codes to transmit the information. The civilian service is called Standard Positioning Service (SPS) and is available to all users on a continuous, worldwide basis, free of any direct user charges. The restricted level is called Precise Positioning Service (PPS) and has restricted access through cryptography to US Armed Forces, US Federal agencies, and selected allied armed forces and governments.

GLONASS became fully operational in 2011, after the Government of the Russian Federation acknowledged in 2001 that the Russian navigation satellite system was a top priority.

Galileo is the GNSS of the European Union, providing a highly accurate and guaranteed global positioning service under civilian control. It was designed to be interoperable with GPS and GLONASS. Galileo started offering its first services, called Early Operational Capability, on 15th December 2016, and is expected to reach Full Operational Capability in 2019. The complete 30-satellite Galileo system, comprising 24 operational satellites and six active spares, is expected to be deployed and running by 2020 (European Space Agency, 2016; European Commission, 2015). Fig. 4.1 illustrates the satellites' constellations of GPS, Galileo, and GLONASS.

Additionally, China is creating its own GNSS system called COMPASS/Beidou, offering complete worldwide coverage and enhanced regional coverage in China. Similarly to GPS, Beidou is expected to provide a free localization service for civilian users and will also have a licensed service with higher accuracy, only for authorized and military users.

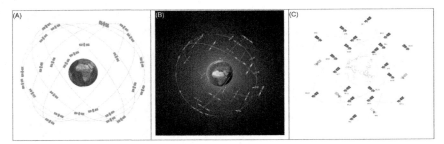

Figure 4.1 Illustration of space segments of (A) NAVSTAR GPS, (B) Galileo, and (C) GLONASS. *From (A) NASA.gov. (B) ESA. (C) ESA Navipedia.*

Table 4.2 Space segment differences of active and future GNSS

Space segment	GPS	GLONASS	GALILEO	COMPASS
Number of satellites	32	24	27 + 3	35
Orbital planes	6	3	3	7
Inclination (degrees)	55	64.8	56	55
Altitude (km from earth surface)	20,180	19,140	23,222	35,786/21,500
Period	12 h	11.25 h	14.08 h	23.93 h/12.85 h

4.2.1.2 Requirements

All four GNSS systems are based on the same architecture requirements and working principles. A complete GNSS system is composed by (European Space Agency, 2003):

1. Satellites that define the *space segment* or satellite constellation in outer space.
2. A *ground segment* (also referred to as *control segment* or *operational control system*), responsible for the proper operation of the GNSS system.
3. A *user segment* that consists of a suitable radio receiver with the processors and antennas needed, which receive the signals, and determine the navigation data (i.e., position as latitude and longitude, altitude, orientation, velocity) through the received information and internal calculations.

Table 4.2 shows the main differences in the space segments of GNSS systems GPS (US), GLONASS (Russia), Galileo (Europe), and COMPASS (China). For example, the GPS constellation consists of 24 satellites in six orbital planes with 55 degrees inclination to the equatorial plane. Each plane holds four satellites, orbiting at around 20,000 km from the earth.

4.2.1.3 Working Principle

The positioning principle is based on solving a geometric problem. From the measured ranges to four or more satellites, the receivers on earth are able to compute worldwide their own position. This principle is called *trilateration* and assumes that the receiver knows with high accuracy its distance to the visible satellites. Since the satellites are continuously orbiting the earth, not all of them are visible to the receiver at each moment in a certain place. With the distance information to one satellite, a receiver could only determine its own position on a sphere, with the radius of that sphere being the distance to the satellite. With the information of a second satellite, the intersection of the two spheres would determine the position on a circle, and with a third satellite the positioning reduces to just two possible points in space. One of the intersected points normally falls in space, far away from earth, so it can automatically be discarded from the calculations. The other intersection point is the position of the receiver on earth. This is only true if the coordinates of the three satellites are known with a high accuracy. As this is not the case for physical reasons, an additional timing variable has to be taken into account to precisely solve the problem. Thus, a fourth GNSS satellite is needed in the field of view of the receiver for providing the right positioning calculations. A unified terrestrial reference system for position and vector referencing is required to correctly position the calculated coordinates. For instance, GPS uses the World Geodetic System WGS-84 developed by the US Defense Department.

The distance to the satellites is computed by a GNSS receiver by extracting the propagation time of the incoming signal sent by each satellite. The signal is detected and tracked and once this is done, the receiver application decodes the *navigation message* and estimates the user position. The navigation message sent by the satellites includes:

- Ephemeris parameters, needed to compute the satellite's coordinates.
- Time parameters and precise clock corrections, even taking relativistic effects into account, to compute satellite clock offsets and time conversions.
- Service parameters with satellite health information.
- Ionospheric parameters model needed for single frequency receivers. GNSS signals get degraded when crossing the earth ionosphere, so these models help reconstructing the original signal.
- Almanacs that allow computing the position of all satellites but with a lower accuracy than the ephemeris.

With the received information and additional physical information, called *observables*, the receiver solves the navigation equations and obtains the satellites' coordinates. A highly accurate time is also provided. The basic observable in a GNSS system is the time required for a signal to travel from the satellite (transmitter) to the receiver. This traveling time, multiplied by the speed of light, provides a measure of the relative or apparent distance between them. This relative distance is called *pseudor-ange* or *pseudodistance*, as it is an "apparent range" between the satellite and the receiver. The values do not exactly match with its geometric distance due to, among other factors, synchronism errors between receiver and satellite clocks. In order to provide accurate timing information that can be correlated at the receivers, GNSS satellites are equipped with atomic oscillators to ensure that such clocks are stable. They are one of the most critical components of a GNSS system, providing very high timing precision and high daily stabilities. The small satellite clock offsets accumulated through time are continuously estimated by the Ground Segment and transmitted in the navigation message to the users to correct the measurements. The receivers, on the other hand, are equipped with quartz-based clocks, much more economical but with a poorer stability, but this is overcome by estimating its clock offset together with the receiver coordinates.

Additional information about the working principle of GNSS systems can be found in (European Space Agency, 2017).

4.2.2 Performance Parameters

The performance of a navigation system is usually measured in terms of the positioning *accuracy* and the *availability* of the system. The accuracy of an estimated or measured position of a vehicle, aircraft, or vessel at a given time is the degree of conformance of that position with the true position, velocity, and/or time of the mobile. Since accuracy is a statistical measure of performance, a statement of navigation accuracy is meaningless unless it includes a statement of the uncertainty in position that applies. For example, GPS provides accuracies of 5−10 m for SPS and better accuracies in the range of 2−9 m for PPS. GLONASS is able to provide around 5 m accuracies, and Galileo is expected to provide positioning accuracies in ranges of 1−4 m depending on the receivers' location. Beidou is designed to achieve positioning errors lower than 10 m worldwide, and an even better performance in China.

The availability of a navigation system is the percentage of time that the services of the system are usable by the navigator and can be used for the intended application. Availability is an indication of the ability of the system to provide usable service within the specified coverage area. Signal availability is the percentage of time that navigation signals transmitted from external sources are available for use. It is a function of both the physical characteristics of the environment and the technical capabilities of the transmitter facilities.

The current performance in accuracy and availability is sufficient for most noncritical applications in Intelligent Transport Systems today. However, for some critical applications additional performance parameters are important. An unwarned large solution error can seriously increase the risk of an accident, for instance, when a navigation system is used for air or maritime navigation, or in a near future with the introduction of automated guided vehicles in roads. Such errors can occur even without violating the accuracy specification, and this is why the concept of integrity has been defined. *Integrity* is the measure of the trust that can be placed in the correctness of the information supplied by a navigation system. It includes the ability of the system to provide timely warnings to users when the system should not be used for navigation.

A fourth performance parameter called *continuity* is important when the navigation system must remain available, without interruptions, during a certain period of special navigation criticality, like for instance in the approach and landing flight phases. The continuity of a system is the ability of the total system (comprising all elements necessary to maintain craft position within the defined area) to perform its function without interruption during the intended operation. More specifically, continuity is the probability that the specified system performance will be maintained for the duration of a phase of operation, assuming that the system was available at the beginning of that phase of operation.

Apart from these general performance metrics, the receiver performance can be influenced by local factors such as:

- *Location of the user*: different locations yield different performances since different satellites will be visible at different positions in the sky.
- *Time of day*: despite the fact that different views of constellation will be visible at different times of day, some errors (such as the errors caused by ionospheric delay) have different impacts during different times of the day.

- *Surrounding environment*: surrounding buildings and vegetation can lead to masking of the sky leading to less visible satellites, signal attenuation effects or multipath effects coming from the reflection of the GNSS signal on buildings or other landmarks.

Additional information on performance parameters for GNSS can be found in Federal Aviation Administration (2011) and GPS.gov (2008, 2007).

4.2.3 Satellite Positioning in ITS Domain and Applications

The Intelligent Transport Systems (ITS) are becoming more and more popular in current society for their capacity for improving transportation issues, such as traffic accidents, congestion, and increasing emissions. They make use of existing information and communication technologies, and also take advantage of sensors, actuators, and other control technologies for proper operation. This is the direction embraced by modern vehicles built by car manufacturers, who are continuously improving the onboard ITS technology.

ITS include applications that use GNSS systems for its functionality. The GNSS receivers can collect position, velocity, and time information to be used by the ITS application, not only for its use for localization services in transportation systems, but also for an increasing number of ITS applications, further buttressed by the fact that GNSS is today embedded in a wide variety of nomadic devices such as smart phones, smart watches, or hiking positioning systems. This has unchained useful positioning services also for pedestrians or goods, and is playing a crucial role in some rescue operations.

Current GNSS positioning accuracy is very well suited for several ITS applications, like fleet management and vehicle tracking, vehicle scheduling and control, and improved "just-in-time" delivery and tracing processes for goods. However, it might not be enough for the future ITS applications coming in the next years. Automated and cooperative vehicles will need to very accurately know their position, velocity, and direction not only to properly stay in the correct driving lane, but also as an implicit requirement for the adoption of new radio access technologies for V2V communications with strong spatial directivity (e.g., mmWave). GNSS information shall be used to actively contribute to the control loop of the automated vehicles by providing inputs to the control algorithm and by giving feedback to the control algorithm after an action is taken. If the location information is shared among cars, GNSS can be a part of a short-range situation awareness system of other vehicles in the road for collision avoidance.

4.2.4 Future of GNSS in ITS

Improved accuracy in positioning, attitude determination, and velocities will be required for ITS in the forthcoming years. Experimental use of automated vehicles today makes use of differential GPS (DGPS), which is an enhancement to primary GNSS constellation(s) information by the use of a network of ground-based reference stations. It enables the broadcasting of differential (correcting) information to the receivers to improve the accuracy of his position. In order to provide with better accuracies, even in the centimeter range, there are several techniques, such as the classical DGPS, real-time kinematics (RTK), wide-Area RTK (WARTK), or augmentation services like satellite based augmentation systems (SBAS), and the Precise Point Positioning (PPP). In case of using GPS in a differential mode, the performances that can be achieved are 0.7−3 m for the civilian codes and 0.5−2 m for the restricted ones. With RTK, performance can be improved up to the 2-cm level.

In a similar manner to how it is today for critical flight phases or complex maneuvers in the aviation or maritime sectors (i.e., approach or landing phases), the availability, integrity, and continuity performance parameters from GNSS will become increasingly important in the future for certain critical applications, such as automated driving. The problems of GNSS not properly working in urban canyons, tunnels, or other difficult environments through the lack of GNSS signal will be resolved with sensor fusion techniques, gathering and fusing information from map-matching technologies, inertial systems, and other available sensors and information technologies. Merging information from different sources, including GNSS as a positioning sensor, will provide ITS with continuous, accurate positioning information regardless of the location of the vehicle.

Some issues may appear in the future regarding the radionavigation satellite service portion of the radio frequency used for GNSS, since some of the mostly used bands are starting to be overcrowded. Even the ones that have not intensively been used yet will certainly be used and shared by ITS systems in the future so it is most probable that in the next years other free frequency resources will have to be searched for use with

GNSS. Also a worldwide interoperability of different GNSS systems will have global benefits for ITS and will thus be promoted. It is the case of Galileo with full compatibility with GPS since its inception. Additionally, it is planned that new GLONASS satellites will transmit GPS and Galileo compatible signals in addition to the ones being used today. This means that in a positive scenario of worldwide GNSS compatibility, and once all the currently planned GNSS are fully deployed, the users might benefit from multiconstellation receivers with a total of over 100 active satellites. This will significantly improve the performance of the GNSS systems and ITS services even in urban canyons in cities.

4.3 GNSS AIDING AND HYBRIDIZED POSITIONING SYSTEMS

As seen in the previous section, GNSS-only systems cannot provide accurate and integer positioning at all times and in all circumstances. However, there are means to compensate for the GNSS flaws. In this section, we will present two key technologies for GNSS support positioning, such as inertial systems and odometry. The way the information from difference sources may be fused to provide a single system solution is also introduced.

4.3.1 Technologies for GNSS-Aided Positioning and Navigation

GNSS positioning can be supported by means of different technologies. Sensors, odometers, magnetometers, etc. can provide information about the motion of the vehicle that may be combined with the GNSS for a better solution.

An odometer counts the number of wheel rotations (or a fraction of it) and, using a conversion rate related to the wheel diameter, transforms the count into information on the traveled distance. A similar concept is exploited in steering encoders to provide the angle of the steering wheel. This information may be used to calculate the angle of the front wheels, which can be transformed into a heading rate. Also, by applying the proper model, the steering wheel information may be combined with the front wheels' speed coming from odometers to predict the next position of the vehicle (Toledo-Moreo et al., 2007). Wheels may each have their dedicated encoders, and the vehicle heading may be calculated based on the difference between the velocities of each frontal wheel, too (Carlson et al., 2004).

Accelerometers can also provide useful information on the longitudinal motion that, contrary to the one provided by encoders and odometers, is not subject to wheel slides and slippages, nor to changes in the wheel diameter. Another form of inertial sensor, the gyroscope, measures turn rates. To do so, a gyroscope can use different technologies, such as MEM, fiber optic, or laser.

A traditional way to estimate the vehicle heading is the use of a compass. Because electronic compasses suffer when exposed to changes in the magnetic field, they are not considered in the community of intelligent vehicles as a sound technology (Abbott and Powell, 1999).

Beacon detection, such as in visual odometry, or the geometry and topology of the road stored in digital maps are also some of the most relevant ways to do relative positioning.

Among all of them, the ones that are available in all cars that are manufactured today and most commonly used for GNSS-aid are the odometry and the inertial sensors. For this reason, we will focus now on the hybridization of these two with GNSS, while the possibilities of visual odometry, LIDAR, wireless networks, and RFID technologies will be presented in Section 4.5, dedicated to alternative technologies for positioning. Also, Section 4.4, Digital Maps, will have a specific subsection for map-assisted GNSS positioning.

4.3.2 GNSS/DR Positioning

4.3.2.1 Principle

Dead-reckoning (DR) is the process of calculating the current pose (position and heading) based on previous ones by integrating velocity and acceleration. Usual dead-reckoning sensors are encoders, accelerometers, and gyroscopes, providing traveled distance, acceleration, and turn rate, respectively. The fact that these sensors provide measurements of the first or second order derivative of the pose implies that the position and heading estimates are subject to cumulative errors. The magnitude of these errors depends on the sensors' accuracy (noise and bias), but also on the techniques employed for the PVT calculation, including the modeling of the sensor errors and the vehicle models. In the case of hybridized dead-reckoning-assisted GNSS-based positioning systems, the fusion architecture and the fusion algorithm have a relevant impact, as well.

The use of a GNSS-based multisensor positioning brings different benefits. For a usual configuration with encoders and inertial sensors, the most relevant features may be enumerated as follows:

1. Immunity to usual GNSS problems, such as signal blockage, jamming, weather conditions, etc. thanks to the nature of inertial navigation.
2. Provision of PVT at a higher rate, as inertial sensors usually work 1 or 2 orders of magnitude faster than GNSS receivers.
3. Complementary information, such as acceleration and attitude (roll, pitch, yaw).
4. Redundant positioning estimates coming from different sources.

Feature 1 supports the availability and continuity of the position, particularly when GNSS satellites visibility is reduced. The combination of 2 and 3 allows that navigation systems can track the vehicle dynamics. Item 4 gives redundancy, and the possibility to improve both accuracy and integrity.

4.3.2.2 Vehicle Models

Vehicle models provide the means to integrate the DR sensors. A common assumption when slippage can be neglected is that the vehicle motion is nonholonomic, so that the heading equals the track angle, as in Fig. 4.2. The literature shows many examples of how a good choice of

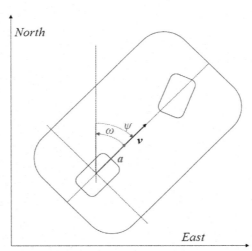

Figure 4.2 A simplified bicycle model in which velocity (v) and acceleration (a) are assumed to be aligned. ψ is the angle with respect to the North and ω the rate of turn.

the vehicle model for a dead-reckoning system leads to important reduction of the positioning error (Julier and Durrant-Whyte, 2003, Toledo-Moreo et al., 2007, Sukkarieh et al., 1999). Vehicle models become of special interest in the case of a system with low-cost sensors, as the worse the sensor performance is, the more convenient it is to constrain the positioning problem to avoid the rapid accumulation of the positioning error (Goodall et al., 2006).

4.3.2.3 Architecture

The purpose of data fusion is the combination of all the navigation information into a single hybridized output that benefits from the complementary nature of the different sources, such as GNSS and dead-reckoning (GNSS/DR). To do so, it is necessary to choose a fusion architecture, on top of which the aimed fusion technique can be applied. Although there is an uncountable number of possible data fusion architectures, it is in general agreed that they can be split in two categories: *Loosely coupled* and *tightly coupled fusion architectures*, depending on the measurements used in the fusion algorithm.

1. In loosely coupled fusion architectures, the GNSS fix is merged with the position estimate of the system. The simplest state vector will consist of the vehicle position and heading, and the observation vector can be the GNSS fix, being both variables defined in the same navigational frame. In this case, the state covariance matrix represents the confidence of the algorithm on the pose estimate.

2. In tightly coupled architectures, the fusion is not carried out with positioning variables, but with GNSS pseudoranges and Doppler measurements. At the expense of a higher complexity of the architecture, this approach allows three remarkable benefits:

    a. The GNSS pseudorange measurements errors are, a priori, less correlated than the errors of the calculated positioning variables, which matches better the assumptions of the most usual filtering techniques, such as the Kalman Filter (Goodall et al., 2006).

    b. It provides a mean to benefit from the GNSS measurements even when the GNSS fix cannot be calculated (when there are less than four satellites in view).

    c. It makes it possible to exploit dead-reckoning measurements for detecting faulty satellite measurements, enhancing the integrity of the position (Le Marchand et al., 2008).

4.3.2.4 Fusion Techniques

The literature of the field presents a good number of different fusion techniques for GNSS/DR fusion, with the most common algorithm being the Extended Kalman Filter (EFK). With an EKF, the filtering problem is linearized around the current estimates. EKF gives the optimal (with Minimum Mean Square Error) solution to the linearized problem, when the error sources are uncorrelated and zero-mean. The EKF cycle can be summarized like:

1. Prediction of the state vector by applying the state transition model.
2. Prediction of the covariance of the error of the state vector taking into consideration the system modeling error and the sensor noise characteristics.
3. Calculation of the innovation of the observation and its covariance, taking into consideration the error characteristics of the measurements.
4. Calculation of the Kalman gain. The gain will be in favor of, either the prediction, or the observation, depending on how much confidence can be laid on each of them.
5. Update of the state estimate and its covariance by using the gain calculated in 4.
6. Return to step 1.

However, when applied to the GNSS-based positioning problem, the EKF strategy presented before has several drawbacks:

1. The optimal solution stands for a linearized problem, and not for the original problem. The more nonlinear the problem, the worse the EKF.
2. The assumption of centered and uncorrelated noises is very rough for standalone GNSS receivers, especially in urban areas subject to multipath and diffraction errors.

Due to these two problems, the literature of the field presents many alternatives. Mostly, these can be categorized in two groups:

1. Fusion algorithms that exploit more refined a priori statistical information about the noises and errors.
2. Intelligent Fusion techniques that do not use an explicit model of the system and often demand a learning phase (e.g., Neural networks, fuzzy logic).

Among those of category 1, shaping filtering of correlated noise can be applied thanks to the use of the aiding sensors. It is also worth mentioning the Particle Filter (PF) which is capable of dealing with non-Gaussian errors, at the expense of increasing the computational complexity (Carvalho et al., 1997).

4.4 DIGITAL MAPS

4.4.1 Importance and Utility

A Geographic Information System (GIS) is an organized integration designed to store, manipulate, analyze, and display geographically referenced information. The system allows separating the information in different thematic layers.

Although its initial use was oriented to navigation systems, digital maps have become an essential tool to support numerous road vehicle assistance systems. With them, it is intended to extend the visual horizon of the driver to what could be called an electronic horizon (Njord et al, 2006). This greater knowledge of the conditions that the vehicle will find in points far from its location results in improvements of the safety and the efficiency of the traffic. Thus, it is possible to anticipate potential risk situations well in advance of what would be possible with only the visual horizon. Similarly, knowledge of dynamic variables allows the optimization of traffic, improving traffic efficiency.

In this sense, the map becomes a relevant piece for the deployment of assistance systems (McDonald et al, 2006), and the digital map can be understood as an additional sensor and its use can be classified in two main categories:

- Primary sensor, when the information of the map is essential for the system functioning. In other word, it is not possible to develop a similar task without that information, but this fact is completely independent of required positioning accuracy. Some examples are Intelligent Speed Adaptation systems, Predictive Powertrain Control System, Hot spots warning, ecall, almost every cooperative application, etc.
- Secondary sensor, when map information is used to validate data from other sensors or to enable more efficient detection of the primary sensors of the system. Some examples are adaptive cruise control, lane keeping assistance, lane change assistance, etc.

However, in order to support previous applications, conventional digital maps used for navigation are not precise enough, and they must be improved in two aspects: detail of the information contained in the map and precision. Thus, safety applications tend to be the most demanding in these aspects and it is usually indicated that a move from a navigation map to a safety map is required (T'Siobbel et al., 2003), which has clear repercussions in the form of constructing the map, as will be discussed below.

In order to show the relevance of digital maps, some international initiatives are now cited. Within the strategic agenda given by the European Road Transport Research Advisory Council (ERTRAC), in which four major areas of activity are established (Urban Mobility; Energy, Resources, and Climate Change; Long Distance Freight Transport; Road Transport Safety), the horizontal priority of "development of universal digital maps for navigation and positioning with integrated real-time updating" is identified (European Road Transport Research Advisory Council, 2008).

Similarly, on the strategic research agenda presented at the eSafety Forum (2006), in which the main lines of research are mobility services for people and for goods, intelligent vehicle systems, cooperative systems, and field operational tests, the development of enhanced digital maps and enhanced position systems is specifically cited as an important horizontal activity. Thus, it is stated that innovative, enhanced digital maps need to be developed and maintained in a cost-efficient way to support systems and services. The main objectives are accuracy and up-to-dateness of the maps. The same organization created a Working Group on digital maps in order to promote the availability of safety-related attributes in digital maps. In turn, the Road Map Working Group indicated the need to develop this enhanced digital map and to create public—private partnerships to produce, maintain, verify, and distribute this map database.

The relevance of the maps is also evidenced by the numerous research projects that have been developed in this area, such as, for example, NextMAP that proposed the establishment of compulsory requirements for digital maps in terms of the detail and precision to be used in new applications for driver assistance, HIGHWAY that defines the architecture and specifications to offer integrated safety and added-value services, in which the role of digital maps is central, MAPS&ADAS that presents an extensive list of data to be included in digital maps for ADAS for present and future applications, and EDMAP.

Finally, it should be noted that the new applications that are included or will be included on the vehicles involve not only more detailed and precise maps but also a positioning in those maps that is much more precise and reliable and, in many cases they surpass the current state of the technology at an affordable cost in standard vehicles.

4.4.2 Specifications

The specifications of the digital maps include the aspects of contained information and the accuracy of the data. For precise navigation, the attributes used by the system include street names, restrictions of movement, particularities in certain sections depending on the time or day of the week, location of points of interest, expected speed, etc. However, safety attributes imply much more complete information. More specifically, in the framework of MAPS&ADAS subproject of the European Integrated Project PREVENT the attributes of the map were classified into four major categories (geometry, feature, attribute, and relationship). Some of the most important ones are now listed:

- Characteristics of road geometry, including data such as grades, super-elevation, radius, etc. (Three-dimensional information of the geometry with a precision of one or two orders of magnitude higher than maps currently used for navigation)
- Fixed and variable speed limits
- Road signaling in terms of signs of prohibition, danger, recommendation, etc.
- Information about the lanes in terms of number, width, elements that delimit them, maneuvers allowed in each one, etc.
- Typology, signaling, and maneuvers allowed at intersections
- Characterization of potentially hazardous areas
- Visibility distance
- Information and positioning of other elements in order to improve location and facilitate real obstacle detection and identification of hotspots.

In terms of precision, it is important to note that, for the positioning of vehicles, three levels of quality have been distinguished (EDMap Consortium, 2004):

- In which road the vehicle is, means the level reached with current technology except in very adverse operating conditions
- In which lane the vehicle is, a state that is estimated to be achievable over the next years and many solutions are now under development. In this case, a error total budget of 1.5 m is accepted (1 m is the allowable limit for positioning error because half a meter error is assumed to be included in the digital map properties)
- Laterally where in the lane the vehicle is, a state that presents greater difficulties and a longer time horizon to achieve. Now, a total error budget of 0.5 m and 0.3 m in vehicle positioning should be assumed

Finally, it should be borne in mind that these levels of detail and accuracy have implications for construction, operation, maintenance, and distribution of digital maps.

4.4.3 Digital Map Development

The construction of digital maps can be made from detailed topographic maps, aerial photographs, or digitized conventional maps. However, the new information demanded by the new systems and the level of precision required imply that new means of construction of these maps must be used, since the previous ones are insufficient. Thus, the use of instrumented vehicles is a solution that allows reaching the levels of detail and precision required. The instrumented vehicles allow obtaining the geometry of the road and acquiring additional information. For this purpose, satellite positioning receivers and/or inertial systems are usually used to measure the geometry, complementing it with vision systems for the detection of traffic signals, estimation of the width of lanes, etc. Satellite positioning systems provide the Cartesian coordinates directly after the corresponding transformations of the latitude and longitude information. However, inertial systems require a prior calculation to obtain the Cartesian coordinates from the speed and rotation angles. In addition, to refer to a global and predefined coordinate system, a rotation and a translation are made.

Besides the Cartesian coordinates of the path, for some driver assistance systems, like curve warning systems or intelligent speed adaptation systems, it is useful to know the radius. This variable is usually calculated using the chord–offset method from the 3 point positions.

Although information is not yet demanded in digital maps, road auscultation systems complement previous equipment with road surface measuring systems, to obtain roughness values or to identify irregularities. In addition, external sources of information such as databases (street maps, road maps, etc.) that contain data can be merged with those acquired by the instrumented vehicle. This task entails high interoperability between diverse sources, which may come from public or private sectors or even private vehicles.

However, the use of instrumented vehicles for the measurement of digital maps has a number of problems among which the following ones can be highlighted:

- The GNSS positioning signal is subject to loss or deterioration and even differential correction does not guarantee the precision required for many applications in all circumstances. These deviations are even greater in the vertical coordinate.
- Inertial systems do not provide an absolute reference, although this aspect is easy to solve if it is complemented by an absolute positioning system. On the other hand, they provide variables such as ramps and,

above all, superelevation directly, contrary to GNSS receivers that cannot provide or do not do that directly. Furthermore, they do not have signal losses. However, the greatest limitation of inertial systems lies in cumulative deviations that are more significant as the distance traveled increases.

- In order to reduce some of the problems of previous systems, sensor fusion is often used, for example, using tools such as Kalman filters. However, those correction systems that work in real time must be differentiated from those that carry out a postprocessing after the measurements have been carried out. This postprocess, which is acceptable in the construction of digital maps, but not for the positioning of vehicles, allows the reduction of errors mainly in critical areas such as those with long GNSS signal losses.

- In addition, inertial systems are subject to the influence of vehicular dynamics, mainly in the determination of grade and superelevation rates, but also depending on the trajectory described. Thus, there are algorithms oriented to correct this influence and to separate the contributions in the measurement of the vehicle (rotations of the sprung mass, mainly) and the geometry of the road.

- High cost of the equipment, since high quality GNSS receivers, if possible with differential correction, and high performance inertial sensors with small drift are used. However, this high cost is not an obstacle since it is a vehicle specifically designed for this purpose, distinguishing this case from the equipment of standard vehicles that must keep the cost within a market range.

On the other hand, the information contained in the digital map cannot be stored continuously, but is usually discretized in stretches and is usually simplified for a more efficient storage. In this sense, maps for navigation have important simplifications that obviate details that may be relevant to other applications. These simplifications are not acceptable in many applications oriented to improving safety, but it is possible to accept other minor simplifications to reduce the volume of data that must be stored. Thus, in the literature works that approximate sections of road using splines can be found, reducing the characterization of geometry by points to the analytical representation of complete sections. However, this solution does not fit exactly to the reality and, therefore, in some works a segmentation is realized in straight sections and curves considering the variation of the yaw angle. In the same way, segmentations of the geometry of the road are proposed considering the classic geometric elements of the road design: straight lines, circular curves, transition curves, parabolic alignments, etc.

Finally, once the map is constructed and, depending on the application to which it is oriented, the information update is required. In this sense, the static and dynamic information is distinguished according to the intervals of time in which variations can be found. Within the ACT-Map project the requirements that must be met in the update process are defined:

- To avoid duplication of data.
- It should be possible to update part of the digital map.
- It should be possible to filter data in response to various criteria.
- In the updating process, permanent, temporary, and dynamic updates should be accepted.

As a means of capturing variable information, sensors can be used in the infrastructure (traffic, meteorology, etc.) or floating vehicles, with or without specific sensors.

4.4.4 Map Quality Assessment

Although the precision of a digital map oriented to navigation applications or even efficiency is not critical and some deviations are acceptable, it is not the case for maps for safety applications where some attributes and variables must be collected with high precision. It is possible to observe how the uncertainty in the measurement of geometric variables of the road can have an influence. As an example, the radius of the curve or the superelevation grade influence the estimation of the safe speed of circulation. In particular, it is observed that a 10% error in the radius estimation leads to errors of 5% in the calculation of the safe speed, which justifies the need for quantifying the uncertainty in the measurement of the map and, above all, of its geometric characteristics.

Regarding the evaluation of the accuracy of the geometry stored in the digital map, different methods have been proposed, which depend, fundamentally, on the method used for the construction of the map. Hence, the most common and general quality indicator of a map's accuracy is the one that can be obtained by following the procedure included in the Transportation Research Board (2002) where the positioning of random points on the map is compared to the actual locations obtained by means that are more accurate than what is required.

Unlike the previous method based on a comparison of Cartesian coordinates, other authors propose the use of the mean and standard deviation of the lane width measurement when measuring the same route using different lanes, as an indicator of accuracy, but the main disadvantage of this procedure being that results are highly conditioned by the path followed by the instrumented vehicle.

Furthermore, in order to choose the proper method, it should be taken into account that the nature of the error that occurs when working with inertial measurement error is different from that which appears with satellite positioning systems. Therefore, given the cumulative nature of the error, there is a need to establish a methodology that considers the uncertainty prior to the completion of the measurements. The methodology proposed by Jimenez et al. (2009) is based on the law of propagation of uncertainty and applies to all variables which are obtained indirectly. The method also allows estimating the maximum distance that can be traveled without correcting the measurements of an inertial measurement system and without the accumulated error exceeding a certain preset tolerance given by a driving assistance application.

Finally, within the EDMap project, the relative effort for the construction of digital maps in terms of the final application was analyzed, since not all systems require the same information. The conclusions of the study point, for example, to a stop sign assistant requiring much less effort than a curve speed assistant with a control function or an adaptive cruise control. Additionally, not all sections of the road require the same detail and precision. Therefore, the realization of hybrid maps has been proposed, where the quality of the map is variable and the level of detail and precision increases in critical road sections, such as bifurcations and curves. This approach is valid for most systems and significantly reduces the development of map construction.

On the other hand, in the digital map, different errors can be found in the definition of attributes:

- *Omission Error*: an attribute exists in reality, but is not on the map.
- *Commission Error*: an attribute does not exist in reality, but is included in the map, which is usually due to failure in updating the map after a change.
- *Classification Error*: an attribute is misidentified or classified (e.g., confusion in the information contained in a traffic sign).
- *Position Error*: an attribute of the map is not located in the correct position on the map.

4.4.5 Map-Matching

The map-matching algorithms seek to correlate the position generated by a sensor in the vehicle with the geographic information of a map. This problem is not trivial when handling inaccurate information on the digital map and the positioning itself. Hence, providing a specific location of a vehicle on a roadway presents difficulties in complex environments such as urban environments, highway junctions, and roads that are near and parallel to each other, where it is necessary to use other signals such as the ones provided by inertial sensors and to consider how the position has evolved over the preceding instants, which involves implementing complex and reliable algorithms.

Several map-matching algorithms have been developed up to the present and they can be classified into four categories:

- The geometric algorithms are the simpler ones but they do not perform well at junctions, roundabouts, and parallel roads. There are three different approaches to geometric algorithms, i.e., point-to-point matching, point-to-curve matching, and curve-to-curve matching. However, their simplicity means they do not usually perform well in complex situations.

- The topological algorithms use the relationship between entities of the digital map such as adjacency, connectivity, and containment to select the correct link. Hence, they take into account the previous information of the route to locate the vehicle into a link and allow the algorithm to keep the continuity of the trajectory.

- Probabilistic algorithms generate a confidence region around a position fix obtained from a GNSS receiver, and are used mostly in the decisions concerning junctions.

- Advanced algorithms use Kalman Filters (KF), Extended Kalman Filters (EKF), Fuzzy Logic models (FL), Dempster-Shafer's mathematical theory of evidence, particle filters, or Bayesian inference and outperform the other algorithms in terms of correct link identification and accuracy. Nevertheless, they require more input data and could be relatively slow and difficult to implement in real-time applications.

Another classification is provided by Gustafsson et al. (2002) and distinguishes simple method algorithms, weighted algorithms, and advanced algorithms. The most sophisticated algorithms reach percentages of correct link identification of between 85% and 99%. Nonetheless, their performance is not enough to support some ITS applications. A complete review of the state-of-the-art of map-matching algorithms can be found in Hashemi and Karimi (2014) and Quddus et al. (2007).

On the other hand, most of the algorithms consider the possibility of recalculating previous positioning decisions to check if errors have occurred. This solution is valid for navigation applications, but not for safety-related applications. Thus, these errors in the positioning assignment in the map increase the difficulty and limitations of using positioning in many ADAS applications.

Finally, the difference between real-time algorithms and postprocessing algorithms should be pointed out. The former, such as those focused on in-vehicle navigation systems should be simple and not computationally intense. The latter use all the position fixes generated by a GNSS along a route as an input, and compare them with the digital map data to produce the whole route traveled as an output. In this last scenario, rapid computer processing is not essential, so the algorithms can be more complex and, therefore, the results could outperform those executed in real-time.

4.4.6 Map-Assisted GNSS Positioning

As we have seen, digital map databases are key components in the majority of transport and mobility systems, since the map is normally the reference system in which the final positioning based service is delivered. To do so, the absolute position has to be map-matched, with the latter being the only thing of interest for the end-user of most services.

However, map data can also benefit directly the positioning process, at the level of the positioning system. The two main principles of map-aided GNSS positioning are:

1. Use of 2D map data in the data fusion loop. In this case, the map information describing the road shape and also possibly the relations between the road segments, or lanes, is used by the data fusion algorithm as a redundant source of information that can, either constrain the feasibility of the computed positioning solution (the vehicle is assumed to be on the road and to follow the traffic rules stored in the map) or support the estimates by using the information such as orientation, speed limit, etc. Examples of this are Cui and Ge (2003) and Toledo-Moreo et al. (2010).

2. Use of 3D map data for Fault Detection and Exclusion (FDE) of ill-conditioned satellite measurements, improving the overall quality of the solution. This way, the knowledge of the environment is used

to analyze whether a certain satellite is in Line-Of-Sight (LOS) with the receiver, and its information can be employed in the positioning computation, as shown in Fig. 4.3. In this image, by means of comparing the angles between the antenna and the surrounding buildings, with the one between the antenna and the satellite SV3 (highlighted in red), it can be stated that SV3 is out of the direct visibility, and therefore the pseudorange measurement detected by the receiver is biased. Detecting NLOS satellites is of particular interest for multipath detection and correction, when possible (Peyraud et al., 2013; Groves et al., 2012).

Both 2D and 3D map-assisted positioning may be also combined, as in Piñana-Díaz et al. (2012).

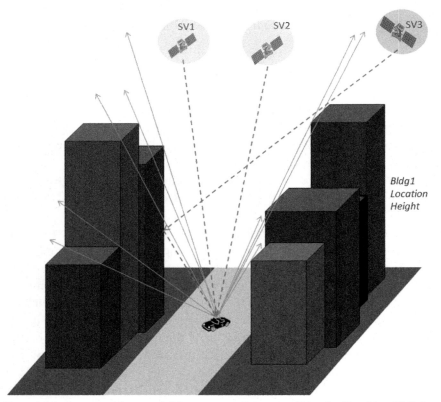

Figure 4.3 Concept of use of 3D map to detect satellites in Non-Line-Of-Sight (NLOS).

4.5 ALTERNATIVES TO GNSS POSITIONING

4.5.1 Visual Odometry as Vehicle's Movement Estimator

The demanding needs of autonomous or semi-autonomous vehicles, both in roadways and urban environments, encourage the use of systems capable of estimating the vehicle motion (velocity, acceleration, and positioning) by means of several sensors acting concurrently. To support this, the use of computer vision and LIDAR algorithms has increased considerably during recent years, with this approach being known as visual odometry (Nistér et al., 2004). Visual odometry algorithms are based on relative movement estimation by means of a vision system mounted on the vehicle or platform. However, every relative movement estimation method presents the same inconvenience: the emergence of an integral error over time, thus producing a drift from the vehicle's real location and its estimation.

As it has been discussed in Section 4.3, fusion architectures and algorithms allow for merging the information collected from absolute and relative sensors, in order to minimize the overall error. Together with the use of encoders, inertial sensors, and maps, presented in Sections 4.3 and 4.4, cameras can also work when gaps appear in the GPS information due to the vehicle passing through GPS-denied environments, especially in urban environments (see Fig. 4.4A). Likewise, the information

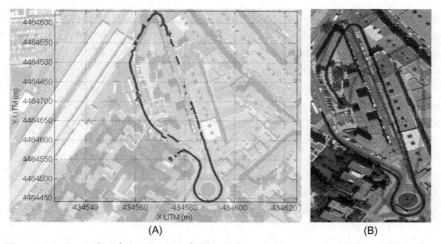

(A)   (B)

Figure 4.4 Example of a trajectory followed by a vehicle in an urban environment. (A) Data supplied by a differential GPS. (B) Data supplied by three different visual odometry algorithms.

provided by GPS may be employed to reduce the lack of accuracy due to the integral error typical of the positioning systems based on relative sensors, as is the case of visual odometry (see Fig. 4.4B).

This section focuses on the visual odometry algorithms for the estimation of a vehicle's movement. The use of visual odometry has important advantages over some other systems, for instance, its robustness to wheel drift compared to wheel odometry. Furthermore, it is based on the use of passive sensors, thus removing the possibility of interference with other vehicles' sensors. Finally, the cost of using vision sensors is relatively low when compared to GPS systems that provide similar accuracy rates, adding also the possibility to use the images information for other tasks, like obstacle detection or signal recognition. Nowadays, this advantage is not shared by LIDARs but these sensors are also used simultaneously for other purposes in the vehicle.

4.5.1.1 Visual Odometry Algorithms Using Computer Vision

The term visual odometry was introduced by Nistér et al. (2004), in contrast to the traditional wheel odometry terminology. However, there are previous works which deal with the general problem of estimating the relative position of the camera from an image set captured from different points of view, known as Structure from Motion (SFM). Visual odometry has being extensively applied, with its use on the Mars Rovers (Johnson et al., 2008) being one of the most popular examples. Visual odometry has a broad application range, starting from indoor and outdoor robotics to autonomous vehicles, both for structured environments and off-road driving, being also widely used in ADAS. It is important to highlight that the visual odometry is a subproblem of a more general problem, the Simultaneous Localization and Mapping (SLAM). The SLAM consists of estimating the vehicle ego-motion, tracking its localization and reconstructing the map of the unknown environment incrementally; furthermore, there is no information given about the initial pose. The location and map of the environment are estimated in each time step given a number of observations by the sensors. As not all the measures provided by the localization sensors are perfect, these data and all the environment-modeling elements could be treated as probabilistic variables.

A common classification for visual odometry methods is based on the sensor type which is used. The main problem derived from the usage of monocular systems for visual odometry systems is the lack of knowledge of the absolute scale (Hilsenbeck et al., 2012). Nevertheless, the main advantage of these systems is the fact of needing no complex calibration process as in stereo rigs, in addition to having a lower cost. The vast majority of monocular implementations of visual odometry are based on that presented by Nistér et al. (2004) at present. In this work, the authors make use of the five-point relative pose problem (Nistér, 2004) and RANSAC (Fischler and Bolles, 1981) in order to avoid possible outliers.

The use of stereo systems in place of monocular ones provides more accurate results despite its higher complexity during the calibration phase and its greater computational cost. The general methodology of the visual odometry based on stereo system was established by Moravec (1980), where a sliding camera was used to capture nine images from different points of view. Following research with stereo systems achieved better results (Matthies and Shafer, 1987). One of the disadvantages of the visual odometry algorithms based on stereo vision is the lack of accuracy in the depth estimation, bringing a nonisotropic error about in the triangulation (Trucco and Verri, 1998). As might be expected, the greater the depth the larger the error. This is the reason why it is not advisable to make use of far points from the vehicle (Otsu and Kubota, 2014). Another source of error is when the baseline of the stereo system is very small compared with the depth of where the elements are located in the environment (Scaramuzza and Fraundorfer, 2011).

Visual odometry implementations comply with the following scheme (see Fig. 4.5): The first stage or task encompasses the detection of feature points (P_n) in the image (I_{k-1}) that are liable to be matched in the next image (I_k) (and with the other image that make up the pair, in the stereo case). This matching process constitutes the second stage of the implementation. Once the feature points' positions along two consecutive images are computed $(fp_{n,k})$, it is possible to estimate the relative movement $(T_{k-1,k})$ the vehicle has performed between the two different locations (C_k, C_{k-1}) where the respective images (image pairs for the stereo case) were captured. Finally, an optimization stage may be used in order to reduce the estimation drift. For a more detailed description of the implementation stages of a visual odometry system, it is advised to refer to (Scaramuzza and Fraundorfer, 2011; Fraundorfer and Scaramuzza, 2012).

4.5.1.2 Visual Odometry Algorithms Using LIDAR

LIDAR can also be used for solving the SLAM problem. For example, in Zhang and Singh (2014) the Lidar Odometry and Mapping (LOAM)

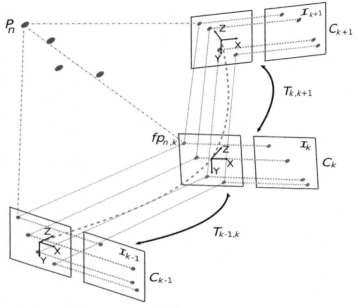

Figure 4.5 Schema of the stereo visual odometry.

algorithm is described. It is a Visual SLAM application where using only LiDAR data it is possible to estimate, with low drift and real time execution, the 6-dof trajectory of the vehicle and it achieves accuracy in real time at the level of state-of-the-art offline batch methods using the KITTI benchmark. Mainly, the algorithm performs two stages, the first one at a high frequency with low accuracy and then, a second step, slower for a fine matching and registration of the point cloud.

In this case, the SLAM stages are quite similar to the ones performed with computer vision and can be summarized as follows:

• Characteristic features (landmarks) extraction:

This is the first step to be performed by the environmental sensor, in this case the LiDAR. Characteristic points like lines, edges of planes, planes, or changes of orientation in neighboring points, are extracted. Usually, different strategies to increase the density of points are applied such as the ICP algorithm. Then, the algorithms used for this extraction could be the same described for features extraction for obstacles detection or estimating the path boundaries (e.g., RANSAC, geometric fitting, etc.).

- Data Association:
  Once, the features extraction has been done, it is necessary to perform a tracking with the features previously identified. A number of observations are stored with the location and mapping of the vehicle and also, the landmarks detected in that time step. The landmarks are seen from different perspectives, that is, in different observations, and the matching relates each set of landmarks estimating the ego-motion of the vehicle. To perform an accurate matching, good quality landmarks must be extracted. It is considered a good quality landmark when the feature has been observed in several time-steps. Some issues that may appear in data association are: not observing the same landmarks as in previous time steps due to the shadows in the LiDAR data; or a bad association because of the bad quality identification when the landmark is further from the LiDAR. A wrong data association results in a divergence issue because the vehicle begins to integrate wrong positions.
- Extended Kalman Filter or Particle Filter:
  An extended Kalman Filter or a particle filter is used to upload and integrate the results from the vehicle odometry and the data association. This is the usual solution when using several sensors. When GPS/INS or IMUs are included in the sensor fusion, the SLAM algorithm performance improves and reduces the motion estimation drift (Montemerlo et al., n.d).

4.5.2 Wireless Networks

Wireless networks may be exploited, not only for the exchange of positioning information, but also for estimation. This approach is becoming more and more popular, in particular for indoor environments. Depending on the measurements used in the positioning calculation, we may find wireless network positioning based on:

- Power, such as the Received Signal Strength (RSS).
- Angle of arrival (AoA).
- Propagation time, such as Time of Arrival (ToA), Time Difference of Arrival (TDoA), or Differential Time Difference of Arrival (DTDoA).

For the moment, their development is not remarkably linked with vehicle positioning (at least, when compared to other approaches presented in this Section). Some typical complications of wireless networks systems are:

- RSS systems demand a good knowledge of the signal propagation;
- AoA degrades when direct line-of-sight is not guaranteed; and
- ToA techniques imply outstanding synchronization among transmitters and with receivers.

4.5.3 RFID

RFID, introduced in Section 4.3, also may be used for positioning estimation of tagged vehicles. As for today, their application to intelligent vehicles is very limited. Some examples are toll-collect gantries, traffic monitoring, or as aiding positioning systems in GNSS-neglected areas, such as tunnels or parking lots.

REFERENCES

Abbott, E., Powell, D., 1999. Land-vehicle navigation using GPS. Proceedings of the IEEE 87, 145−162.

Carlson, R., Gerdes, J., Powell, J., 2004. Error sources when land vehicle dead reckoning with differential wheelspeeds. Navigation 51, 13−27.

Carvalho, H., Del Moral, P., Monin, A., Salut, G., 1997. Optimal nonlinear filtering in GPS/INS integration. IEEE. Trans. Aerosp. Electron. Syst. 33.3, 835−850.

Cui, Y., Ge, S.S., 2003. Autonomous vehicle positioning with GPS in urban canyon environments. IEEE Trans. Robot. Autom. 19, 15−25.

EDMap Consortium, 2004. Enhanced digital mapping project. Final report. EDMap Consortium, Washington.

eSafety Forum, 2006. Strategic Research Agenda. ICT for Mobility. European Commission, Brussels.

European Commission, 2015. Galileo's contribution to the MEOSAR system. <http://ec.europa.eu/growth/sectors/space/galileo/sar/meosar-contribution> (accessed 17.03.03).

European Road Transport Research Advisory Council, 2008. ERTRAC Research Framework. 'Steps to Implementation'.

European Space Agency, 2003. Galileo System Requirement Document (GSRD), Issue 3.0, ESA-APPNS-REQ-00011 [Ref.3] Space System Engineering Standard (ECSS-E-10-A).

European Space Agency, 2016. Galileo begins serving the globe <http://www.esa.int/Our_Activities/Navigation/Galileo_begins_serving_the_globe> (accessed 17.03.03).

European Space Agency, 2017. Navipedia. <http://navipedia.net> (accessed 17.03.03).

Federal Aviation Administration, 2011. FAA Global Positioning System (GPS) Standard Positioning Service (SPS) Performance Analysis Report. <http://www.nstb.tc.faa.gov/reports/PAN72_0111.pdf> (accessed 17.03.03).

Fischler, M.A., Bolles, R.C., 1981. Random sample consensus: a paradigm for model fitting with applications to image analysis and automated cartography. Commun. ACM. 24 (6), 381−395.

Fraundorfer, F., Scaramuzza, D., 2012. Visual odometry: Part II: Matching, robustness, optimization, and applications. IEEE Robot. Autom. Mag. 19 (2), 78−90.

Goodall, C., Syed, Z., El-Sheimy, N., 2006. Improving INS/GPS navigation accuracy through compensation of kalman filter errors. IEEE Veh. Technol. Conf. 1—5.

GPS.gov, 2007. Performance Standards & Specifications, GPS Precision Positioning Service (PPS) Performance Standard, first ed. <http://www.gps.gov/technical/ps/2007-PPS-performance-standard.pdf> (accessed 17.03.03).

GPS.gov, 2008. Performance Standards & Specifications, GPS Standard Positioning Service (SPS) Performance Standard, fourth ed. <http://www.gps.gov/technical/ps/2008-SPS-performance-standard.pdf> (accessed 17.03.03).

Groves P., Wang L., Ziebart M., 2012. Shadow matching: Improved GNSS accuracy in urban canyons. GPS World. Available online: <http://gpsworld.com/wirelesspersonal-navigationshadow-matching-12550/>.

Gustafsson, F., Gunnarsson, F., Bergman, N., Forssell, U., Jansson, J., Karlsson, R., et al., 2002. Particle filters for positioning, navigation, and tracking. IEEE Trans. Signal Process. 50, 425—435.

Hashemi, M., Karimi, H.A., 2014. A critical review of real-time map-matching algorithms: current issues and future directions. Comput. Environ. Urban. Syst. 48, 153—165.

Hilsenbeck, S., Möller, A., Huitl, R., Schroth, G., Kranz, M., Steinbach, E., 2012. Scale-preserving long-term visual odometry for indoor navigation. Int. Conf. Indoor Positioning Indoor Navigat. 1—10.

Jiménez, F., Aparicio, F., Estrada, G., 2009. Measurement uncertainty determination and curve fitting algorithms for development of accurate digital maps for Advanced Driver Assistance Systems. Transp. Res. Part C Emerg. Technol. 17 (3), 225—239.

Johnson, A.E., Goldberg, S.B., Cheng, Y., Matthies, L.H., 2008. Robust and efficient stereo feature tracking for visual odometry. IEEE Int. Conf. Robot. Autom. 39—46.

Julier, S., Durrant-Whyte, H., 2003. On the role of process models in autonomous land vehichle navigation systems. IEEE Trans. Robot. Autom. 19, 1—14.

Le Marchand O., Bonnifait P., Bañez-Guzman J., Peyret F., Betaille D., 2008. Performance evaluation of fault detection algorithms as applied to automotive localisation. In: European Navigation Conference - GNSS 2008, Apr 2008, Toulouse, France <hal-00445170>.

Matthies, L., Shafer, S., 1987. Error modeling in stereo navigation. IEEE J. Robot. Autom. 3 (3), 239—248.

McDonald, M., Keller, H., Klijnhout, J., Mauro, V., Hall, R., Spence, A., et al., 2006. Intelligent Transport Systems in Europe. Opportunities for Future Research. World Scientific.

Montemerlo, M., Thrun, S., Koller, D., Wegbreit, B., n.d. FastSLAM: a factored solution to the simultaneous localization and mapping problem.

Moravec, H.P., 1980. Obstacle avoidance and navigation in the real world by a seeing robot rover (No. STAN-CS-80-813). STANFORD UNIV CA Department of Computer Science.

Nistér, D., 2004. An efficient solution to the five-point relative pose problem. IEEE. Trans. Pattern. Anal. Mach. Intell. 26 (6), 756—770.

Nistér, D., Naroditsky, O., Bergen, J., 2004. Visual odometry. Proc. 2004 IEEE Comput. Soc. Conf. Comput. Vis. Pattern Recognit. (1), 652—659.

Njord, J., Peters, J., Freitas, M., Warner, B., Allred, K.C., Bertini, R., et al., 2006. Safety applications of intelligent transportation systems in Europe and Japan. Federal Highway Administration, Department of Transportation, U.S.

Otsu, K., Kubota, T., 2014. A two-point algorithm for stereo visual odometry in open outdoor environments. IEEE Int. Conf. Robot. Autom. 1042—1047.

Peyraud, S., Bétaille, D., Renault, S., Ortiz, M., Mougel, F., Meizel, D., et al., 2013. About Non-Line-Of-Sight satellite detection and exclusion in a 3D map-aided localization algorithm. Sensors (Basel) 13 (1), 829–847.

Piñana-Díaz, C., Toledo-Moreo, R., Toledo-Moreo, F.J., Skarmeta, A., 2012. A two-layers based approach of an enhanced-map for urban positioning support. Sensors (Basel) 12 (11), 14508–14524.

Quddus, M.A., Ochieng, W.Y., Noland, R.B., 2007. Current map-matching algorithms for transport applications: state-of-the art and future research directions. Transp. Res. Part C Emerg. Technol. 15, 312–328.

Scaramuzza, D., Fraundorfer, F., 2011. Visual odometry. IEEE Robot. Autom. Mag. 18 (4), 80–92.

Sukkarieh, S., Nebot, E., Durrant-Whyte, H., 1999. A high integrity IMU/GPS navigation loop for autonomous land vehicle applications. IEEE Trans. Robot. Autom. 15, 572–578.

T'Siobbel, S. 2003. The road to safety maps. In: Proceedings of 10th World Congress and Exhibition on Intelligent Transport Systems and Services. Madrid.

Toledo-Moreo, R., Bétaille, D., Peyret, F., 2010. Lane-level integrity provision for navigation and map matching with GNSS, dead reckoning, and enhanced maps. IEEE Trans. Intell. Transp. Syst. 11, 100–112.

Toledo-Moreo, R., Zamora-Izquierdo, M., Ubeda-Minarro, B., Gómez-Skarmeta, A., 2007. High-integrity IMM-EKF-based road vehicle navigation with low-cost GPS/SBAS/INS. IEEE Trans. Intell. Transp. Syst. 8, 491–511.

Transportation Research Board, 2002. Collecting, Processing and Integrating GPS Data into GIS. Transportation Research Board, Washington D.C.

Trucco, E., Verri, A., 1998. Introductory Techniques for 3-D Computer Vision, 201. Prentice Hall, Englewood Cliffs.

Zhang, J., Singh, S., 2014. LOAM: Lidar odometry and mapping in real-time. Robot. Sci. Syst.

CHAPTER 5

Big Data in Road Transport and Mobility Research

Sergio Campos-Cordobés[1], Javier del Ser[2], Ibai Laña[1],
Ignacio (Iñaki) Olabarrieta[1], Javier Sánchez-Cubillo[1],
Javier J. Sánchez-Medina[3] and Ana I. Torre-Bastida[1]
[1]TECNALIA, Bizkaia, Spain
[2]TECNALIA, University of the Basque Country (UPV/EHU) and Basque Center for Applied Mathematics (BCAM), Bizkaia, Spain
[3]Universidad Las Palmas de Gran Canaria, Las Palmas de Gran Canaria, Spain

Contents

Ubiquitous computing devices have changed the acquisition of mobility data, with two aspects contributing: the high penetration rate (90% population has a Smartphone, 15 billion users in 2016) and their ability to capture and share information on a continuous basis. This applies to

Intelligent Vehicles
DOI: http://dx.doi.org/10.1016/B978-0-12-812800-8.00005-9

geolocation information, GPS position; the operational mobile phone data, exchanged frequently with the network, and also social network crowdsourced information.

Additionally, under the umbrella of the Internet of Things (IoT) trend, the deployment of the Connected Vehicle (or Car-as-a-sensor) concept, supported by advanced V2X communications, provides massive data volume. For all these cases, data is geolocated, opening never before seen opportunities to analyze and predict individual and aggregated mobility patterns. This all allows managers to enrich existing traffic and mobility sensor infrastructure and deploy more reliable decision support tools, improving the quality of the service or optimizing the operation. In addition, it supports the provision of advanced personalized information and services for citizens, such as "Google Now" does.

This scenario defines the needs of new technologies able to capture, store, and elaborate a relevant and heterogeneous quantity of mobility data. Big Data refers to this explosion in the amount (and sometimes the quality) of available and potentially relevant data as a result of recent advances in such capture, data recording, and storage technologies.

This chapter will review the most relevant sensors and data, introduce the underlying techniques supporting the BigData paradigm and, finally, provide a list of some relevant problems in the transport and mobility domain.

5.1 DATA AND INFORMATION SOURCES

One of the key aspects of BigData versus previous approaches (e.g., data analytics, massive processing, and parallel computing) is the diversity and heterogeneity, including both structured data and also unstructured data.

At present, a great variety of sensing technologies exist that can be used to obtain information on the current state of the road. In fact, new vehicle and traveler detection and monitoring technologies are constantly being improved and developed, which can facilitate traveler and displacement characterization.

Several conclusions can be observed:
- Each detection device, however, has its pros and cons, making it appropriate for some purposes, but not for others, according to the acquired data, reliability, and accuracy.
- No detector is appropriate to meet all needs.

- Generally, sensor technologies are usually classified into two categories: "On-site" and "Onboard."
- "On-site" technologies correspond to conventional sensors installed directly in the infrastructure.
  - *Invasive sensors*: sensors that need installation tasks prior to commissioning. These devices are installed directly on the surface of the pavement, making grooves with a saw or on the surface of the road:
    - *Magnetic loops*: this is the classic technology for collecting traffic data. The information is relayed to a counting device.
    - *Passive magnetic*: they count the number of vehicles, type, and speed. However, in operating mode, they often present difficulties in the classification of very close vehicles.
    - *Pneumatic road tubes*: tubes located along the road in order to detect vehicles based on changes in pressure. The major drawback of this technology is its limited road coverage and its reliability is heavily dependent on weather, temperature, and traffic conditions. This system is inefficient in the measurement of low speed flows and retentions.
    - *Piezoelectric sensors and WIM (Piezoelectric sensors) systems*: the underlying principle is the conversion of mechanical energy into electrical energy. This system can be used in the measurement of weight and speed, variables customary when classifying a vehicle.
- *Noninvasive sensors*: sensors do not invade the traffic area of the vehicles, but are installed on the rails to be controlled or to the sides of the road, assuming a condition to the minimum traffic during the installation and maintenance.
  - *Active and passive infrared (Passive and active infrared) sensors*: the presence, speed, and type of vehicles are detected based on the infrared energy emitted in the detection area.
  - *Microwave radar*: this technology can detect moving vehicles and speeds (Doppler radar). It records the number of vehicle, speed, and classification data (approximate) and is unaffected by weather conditions.
  - *Ultrasonic and passive acoustic sensors*: record vehicle counts, speed, and classification of data. They can also be affected by poor weather conditions (e.g., low temperatures, snow, ice).
  - *Video image detection (VIP)/Video vision detection*: video cameras record the number of vehicles, type, and speeds by means of video techniques: trip line and tracking. Systems are sensitive to weather conditions.

- In the specific context of the speed of a vehicle, we find the cinemometers, devices designed to measure the speed of a mobile. There are different models of cinemometers but they are all based on the measurement of time: radar, laser, LIDAR (Light Detection And Ranging), laser barrier, and based on image recognition.

- "Onboard" (off-roadway) technologies, which collect certain information of interest from the vehicles and/or their occupants and send it to a device on the road or a receiving center for their process.
  - *Mobile Phone information.* Two kinds of events occur when a mobile phone is moving: network-triggered, automatic operational communications, and event-triggered, associated user activities.
  - *Floating Car Data (FCD).* GPS information allows the collection of a detailed spatiotemporal trace describing the mobility of an individual. All smartphones are usually equipped with GPS sensors able to calculate position by using trilateration methods.
  - Automatic Fare Collection systems collect transactional data for each public-transport travel by using different technologies: RFID, magnetic cards, plate recognition, etc. In all cases, the transaction or transit is assigned univocally to a user. Different fare structures allow different exploitation opportunities (boarding and alighting information, when mandatory).

It is important to be aware, that for crowdsourced data (GPS, mobile phone data, transactional, social networks) the main challenge is the robustness of such sparse and noisy measures to extract individual mobility sequences with semantic meaning (Widhalm et al., 2015). For these sources, the key concept is the "spatial trajectory," a trace composed by a series of chronologically ordered points where each point is represented by the object/user/traveler Id, its geographical coordinates, a time stamp, and any additional information (e.g., speed, semantic labels, altitude, orientation, accuracy, and heading, when available). Specific techniques have been defined to process trajectories under the denomination of trajectory mining (Feng and Zhu, 2016; Zheng, 2015), according to the following categories: preprocessing, data management, query processing, data mining tasks, and privacy protection.

Additionally, from the perspective of open data, the following data format standards must be considered for using traffic information and enable the scalability of services: DATEX II, SIRI, Geography Markup Language (GML), GTFS, among other relevant ad hoc standards (Section 5.3.2).

5.2 DATA PREPROCESSING

5.2.1 Feature Engineering

Machine learning techniques rely on extensive datasets, that in many cases are formed by hundreds or even thousands of attributes. Abundance of features can provide rich information for models to better adapt to each problem's characteristics. On the other hand, an excessive number of attributes presents two main issues: overfitting and running time (Blum and Langley, 1997). Overfitting occurs when the ratio of attributes over training instances is too high, creating a solution hyperspace that adjusts extremely well to training data, but fails to adapt for new unseen instances, providing poor performance. A fair selection of features, along with other techniques helps to mitigate this problem. A large amount of features usually produces increases of machine learning modeling time, as features represent dimensions of the solution hyperspace. Thus, reducing of the size of the feature set is a common practice in high dimensionality problems that cannot be addressed lightly. A typical case of high dimensionality for which feature selection and engineering is applied is text categorization and document classification, in which each instance is formed by all the words of a document (Yang and Pedersen, 1997). Other typical cases are those in which attributes are sensor readings that conform to a state, or physical and chemical characteristics that shape kinds of organisms. Big data problems can belong to similar situations, although each problem has particularities for which specific solutions can be found in the literature. Feature selection and feature engineering techniques are not mutually exclusive (in a manner, feature selection is a kind of feature engineering), and they are often used combined in sequential or iterative processes to refine and enhance machine learning model capabilities.

Feature selection. Feature selection techniques are used to trim dimensions of the features vectors. The general underlying philosophy for dimensionality reduction is to remove those features that have little relevance for the target function, or in other words, select more relevant features. For this, several techniques have been applied through the years, with some of the most important being the following (Kohavi and Jhon, 1997):

- *Manual preprocessing*: knowing the domain of application is a key aspect of data science, and intervenes strongly in initial preprocessing.

Datasets frequently contain features that are combinations of other subsets of features, or features that the data scientist knows are irrelevant for the problem (or even worse, could provide spurious correlations), and could be removed. Features with a high percentage of invalid values through the dataset are also removable.

- *Principal component analysis* (Uguz, 2011): PCA is a statistical method that represents the target variable as a linear combination of input variables. This allows the identification of the most relevant variables.
- *Heuristic approaches*: which means searching with different techniques (exhaustive, filters, etc.) combinations of variables that produce better performances. A well-known heuristic technique is RELIEF algorithm (Kirak and Rendell, 1992). This method consists of assigning weights to variables according to their ability to discriminate between neighboring patterns, and it is considered an effective and simple way to select features. In this context, evolutionary computation has brought in the last decade an assortment of powerful tools that allow data scientist building optimization wrappers that can be configured to obtain subsets of optimal features in few iterations.

This is a small sample of feature selection techniques, an area of profuse research that is gaining focus again in the big data era.

Feature engineering. Feature engineering implies modifying existing features or creating new ones to obtain more efficient models. Although creating new features might seem contrary to the aforementioned techniques, both techniques are often complementary, and new features are usually focused on replacing subsets of previous features, or rendering them more manageable for processing, like changing a text attribute into a numerical one. Besides, it is not unusual to find the opposite situation, where too few features are available for the machine learning model (with the opposite effect of underfitting the model). In this case, creating some variables can help an algorithm to fit to data. Some of the most common transformations are listed below:

- Normalization or other kinds of numerical preprocess: used (Schadt et al., 2001) when ranges of features are highly discordant, especially for machine learning methods that perform comparisons among different feature sets.
- Domain changing of the variables, transforming continuous variables into discrete ones, or text variables into numerical. Handling numbers is easier for machine learning methods than handling text, as they can perform operations and represent them in Euclidean spaces. For

example, a field like "Month," can be easily changed from "September" to 9.

- Replacing features by combinations of them or by equivalents. When there are a huge amount of features, combining subsets of them into individual features is a common practice. This kind of preprocessing is completely dependent on the problem: a text classification problem usually replaces phrases for independent words with the same meaning, or synonyms, but an image processing problem could combine features representing pixels with similar values.

- Generating new variables that are not present in the original dataset but might add explanatory capacities to each instance. This generation can be manual, parting from the knowledge of the problem, and it can be automated (Leather et al., 2009), which usually relies on heuristic searches of combinations and replacements of existing variables.

Feature preprocessing is a key element in the development of machine learning systems. An optimal set of features grants performance enhancements and speed improvements, and therefore should not be taken lightly when modeling data (Table 5.1).

5.2.2 Dimensionality Reduction

Datasets are constantly increasing, nowadays there are between a few dozen terabytes to many petabytes of data in a single data set. Morais (2015) reported that 80% data is geographic. The explosion of data set size, in number of records and attributes, has triggered the development of a number of big data platforms as well as parallel data analytics algorithms. At the same time though, it has pushed for the usage of data dimensionality reduction procedures.

Here we present a report about the most common dimensionality reduction techniques from a paper by Kdnuggets (2017).

- *Missing Values Ratio.* Data columns with a number of missing values greater than a given threshold can be removed.

- *Low Variance Filter.* Data columns with little changes in the data carry little information and data columns with variance lower than a given threshold are removed.

- *High Correlation Filter.* Data columns with very similar trends are also likely to carry very similar information. We calculate the correlation coefficient, for example Pearson's Product Moment Coefficient. Pairs of columns with a correlation coefficient higher than a threshold are reduced to only one.

Table 5.1 Comparison of dimension reduction techniques

Dimensionality reduction	Reduction rate (%)	Accuracy on validation set (%)	Best threshold	AuC (%)	Notes
Baseline	0	73	–	81	Baseline models are using all input features
Missing values ratio	71	76	0.4	82	–
Low variance filter	73	82	0.03	82	Only for numerical columns
High correlation filter	74	79	0.2	82	No correlation available between numerical and nominal columns
PCA	62	74	–	72	Only for numerical columns
Random forrest/ensemble trees	86	76	–	82	–
Backward feature elimination + missing values ratio	99	94	–	78	Backward Feature Elimination and Forward Feature Construction are prohibitively slow on high dimensional data sets. It becomes practical to use them, only if following other dimensionality reduction techniques, like here the one based on the number of missing values.
Forward feature construction + missing values ratio	91	83	–	63	

- Principal Component Analysis (PCA) is a statistical procedure that orthogonally transforms the original n coordinates of a data set into a new set of n coordinates called principal components. As a result of the transformation, the first principal component has the largest possible variance; each succeeding component has the highest possible variance under the constraint that it is orthogonal to (i.e., uncorrelated with) the preceding components. Keeping only the first $m < n$ components reduces the data dimensionality while retaining most of the data information, i.e., the variation in the data.
- *Random Forest.* Decision Tree Ensembles, also referred to as random forests, are useful for feature selection, in addition to being effective classifiers.
- *Backward Feature Elimination.* In this technique, at a given iteration, the selected classification algorithm is trained on n input features. Then we remove one input feature at a time and train the same model on $n-1$ input features n times. The input feature whose removal has produced the smallest increase in the error rate is removed, leaving us with $n-1$ input features. The classification is then repeated using $n-2$ features, and so on. Each iteration k produces a model trained on $n-k$ features and an error rate $e(k)$. Selecting the maximum tolerable error rate, we define the smallest number of features necessary to reach that classification performance with the selected machine learning algorithm.
- *Forward Feature Construction.* This is the inverse process to the Backward Feature Elimination. We start with one feature only, progressively adding one feature at a time, i.e., the feature that produces the highest increase in performance. Both algorithms, Backward Feature Elimination and Forward Feature Construction, are quite time and computationally expensive. They are practically only applicable to a data set with an already relatively low number of input columns.

Those techniques are compared into the article using the smaller data set of the 2009 KDD challenge in terms of reduction ratio, degrading accuracy, and speed. Results of this comparison are reported in Table 5.1.

In the context of Big Data, there are numerous works related to reduction dimensionality of datasets for the task of preprocessing and thus enable the subsequent analysis. The most relevant are the following ones: Xu et al. (2014), where in addition to other techniques of Big Data mining is treated the reduction of dimensionality in large volumes of data; Snášel et al. (2017), where geometrical and topological methods are used to analyze highly complex data and to reduce its dimensionality; Maier et al. (2017) present a Big Data specific reduction techniques in the smart factory environment; Huang et al. (2016) present a new approach for Big Data dimensionality reduction technique in traffic data.

5.3 DATA NORMALIZATION

The process of data normalization has different definitions depending on the field in which we locate it:

- In Databases, Beeri et al. (1978) refer to the process in charge of refining tables, keys, columns, and relationships to create an effective database. Normalization is not only applicable to relational files, it is also a common design activity for indexed files.
- In Statistics, Lévy-Leduc and Roueff (2009) report that it is understood to normalize by the process of transforming a random variable that has some distribution in a new random variable with normal or approximately normal distribution. Also it is called typing or standardization. In practice it consists of applying a simple calculation to the variable object of study, with it, we get that the data of the normalized distribution have a value of the arithmetic mean of 0 and a standard deviation of 1.
- In machine learning, Lakhina et al. (2005) report that it is usually identified with the concept known as "features scaling," which is a method used to standardize the range of independent variables of the data. In data processing it is called normalization, and it is a very important phase of data preprocessing before analysis.

In this section we focus on the Big Data normalization and therefore associate it with the preprocessing lifecycle stage. At this phase, the concepts of data cleaning and healing, as well as integration and modeling, play a key role and are therefore given special attention in the following subsections.

5.3.1 Data Cleaning

It is the process of detecting and correcting (or removing) corrupt or inaccurate records from a record set, table, or database and refers to identifying incomplete, incorrect, inaccurate, or irrelevant parts of the data and then replacing, modifying, or deleting the dirty or coarse data (Wu, 2013). This field comprises the following subareas: Data Parsing, Data transformation, Duplicate elimination, Data integration, Data editing,

Data mining, Record linkage, and Data curation (Lord et al., 2004). In the case of data obtained from GPS sensors, carried by the users, algorithms such as Map–Matching are responsible for adjusting to valid positioning data in the road network.

5.3.2 Formats and Standards

Format and standards are a way of sharing a model conceptualization between the different stakeholders. This section is focused on the main formats and standards around traffic information and ITS systems. The most important ones are cited below, along with European project and initiatives detected by IBM (2017):

- *ASTM E2665−08 Standard Specification for Archiving ITS-Generated Traffic Monitoring Data* (http://www.astm.org/Standards/E2665.htm): Standard of the ASTM (United States standardization body) for the specification of the data format of the traffic monitoring.
- *DATEX* (http://www.datex2.eu/): Set of specifications that allow the exchange of traffic information in a standard format between different systems.
- *IFOPT—Identification of Fixed Objects In Public Transport* (http://www.dft.gov.uk/naptan/ifopt/): It allows the definition of a model and the characteristics that identify the main objects related to public transport (stops, stations, service areas, interconnection areas, entrances ...)
- *GDF—Geographic Data Files* (http://www.iso.org/iso/catalogue_detail.htm?csnumber=30763): It is a European standard used for the description and transfer of data related to the road network.
- *Advanced Travel Information system (ATIS)* (http://www.standards.its.dot.gov/): Advanced traveler information systems are geared to provide users of the transportation system with more information to make decisions about route options, estimated travel times, and avoid congestion.
- *Advanced Traffic Management (ATM)* (http://www.standards.its.dot.gov/): The systems that integrate the technology in urban environments with the aim of reducing traffic congestion, increasing road safety, and improving traffic flow of vehicles. These systems use solutions to congestion problems through the deployment of simple and reliable sensors, communications, and data processing technologies.
- *Institute of Transportation Engineers (ITE)* (http://www.ite.org/): An international educational and scientific association of transport

professionals, including engineers, transportation planners, consultants, educators, and researchers. Founded in 1930, the ITE facilitates the application of the principles of technology and scientists to research, planning, functional design, implementation, operation, policy development, and management for any mode of ground transportation. More information on ITE can be found on the website.

- *Location Referencing Message Specification (LRMS)* (http://www. standards.its.dot.gov/): Describes a set of standard interfaces for the transmission of location references between the different Intelligent Transportation Systems (ITS) components. The LRMS facilitates the movement of ITS data in a transport network, providing a common language for the expression of its location among the different components. LRMS interfaces define standard meanings for the content of location reference messages and standard public domain formats for the presentation of location references in application software.

- *National Transportation Communications for Intelligent Transportation System (ITS) Protocol (NTCIP)* (http://www.ntcip.org/): A family of communications standards used to transmit data and messages between computer systems used in intelligent transport systems. NTCIP offers both the rules for communication and the vocabulary needed to allow electronic traffic control equipment from different manufacturers to operate with each other as a system. The NTCIP is the first set of standards for the transportation industry that allows traffic control systems to be built using a "mix and match" approach with equipment from different manufacturers. Therefore, NTCIP standards reduce the need for dependence on specific equipment vendors and software of a custom class.

- *Transmodel version 5.0* (http://www.transmodel.org/en/cadre1.html): TRANSMODEL is a reference data model for Public Transport operations developed in several European projects. It deals mainly with the needs of urban bus operators, trolleybuses, tram, and light rail.

- *TMDD Traffic Management Data Dictionary for Center-to-Center Communications* (http://www.ite.org/standards/tmdd/): Standard used by the transportation industry to define and support center-to-center interface communications as part of the regional deployment of an Intelligent Transport System (ITS). TMDD provides dialogs, message sets, data frames, and data elements to manage the sharing of these devices and the regional distribution of data and responsibility for incident management. TMDD is defined in the standards family of National Transportation Communications for Intelligent Transportation System Protocol (NTCIP).

- *SIRI—Server Interface for Real-Time Information* (Daly et al., 2013): It is the European interface that allows real-time information exchange on public transport operations.

5.3.3 Ontologies

Gruber originally defined the notion of ontology as an "explicit specification of a conceptualization." In 1997, Borst defined ontology as a "formal specification of a shared conceptualization." For these reasons we think that is the best option to represent a domain like traffic and mobility. But to the best of our knowledge there are scarce studies into this area.

To this day we can find only a few different references for ontologies in the field of Smart Cities and traffic, namely, with the most remarkable being: Ontology of Traffic Networks (OTN) (REWERSE, 2017), which extends GDF formalizing in OWL, Ontology of Urban Planning Process (OUPP); Kaza and Hopkins (2007) based on the CityGML model on survey data; LinkedGeoData (OSM) (LinkedGeoData, 2017); and the Townontology COST initiative (TOWNONTOLOGY, 2008), among others.

Although there have been some successful results (Metral et al., 2005; Malgundkar et al., 2012), after a deeper analysis, we can conclude that these solutions are not compatible with each other, and cover only partially the needs in the context of Smart Mobility, so an important effort should be necessary to extend and harmonize these initiatives. On the other hand, the potential of ontologies and semantics, in general, is maximized in scenarios with unstructured, textual, and multimedia data or is intended to address issues such as dynamic service composition and semantic reasoning, In some cases prioritizing the efficiency of information acquisition and access processes, it is recommended to use conventional data model structures and implement specific adapters when needed.

Following the most important ontologies about traffic, mobility and logistic domains are described:

- *SCRIBE—IBM.* This IBM-funded project is exploring how to build a set of modular ontologies, compiling standards, to model the infrastructure and processes of a smart city.

- *Rewerse—Ontology on transportation network.* The OTN ontology is an initiative of the European project rewerse, which attempts to model an ontology in (as well as expanding) the GDF standard.
- *Datex II Ontology.* At the University of Valencia there is a research group that has tried to create an ontology from DATEX II.
- *Townontology* (http://www.towntology.net/). Set of diverse ontologies that model the layout of cities.
- *iCargo/e-Freight Ontology.* Ontologies of the European projects iCargo and e-Freight, which focus on logistics and modeling the transport and business events of this. The iCargo ontology is superior and more improved than that of e-Freight. Both come from the Common Logistics Framework (http://www.its.sintef9013.com/CF/v01/).
- *LinkedGeoData.* These datasets are in RDF format and follows the principles of LinkedData, collecting the information collected by OpenStreetMap.

5.4 SUPERVISED LEARNING

Supervised learning is a big branch of Machine Learning (ML) methodologies. The main characteristic of a whole lot of techniques is that during the learning of the model we can feed the learning with the right answer for each training instance. Therefore, methodologies in this subset are basically feedbacking from their responses' distance to the expected response iteratively until they can.

In this ML area many different classifications can be made depending on:

- The purpose of the trained model (Predictive or Descriptive).
- The output of the trained model (Classification or Regression).
- The kind of algorithms used as learners (Decision Trees, Artificial Neural Networks, Support Vector Machines, Bayes-based and Ensembles).
- The Real Time application or the time-frame of the collection of the data used for the training the model (Off-Line vs On-Line)
- Finally within On-Line we may have Incremental versus Batch learning, and within Incremental, we have those considering Concept Drift and those assuming Concept Stability.

We will discuss a little bit about each of the above categories, providing some examples from the Intelligent Transportation arena.

5.4.1 Predictive Versus Descriptive

In general, all methods in Knowledge Discovery and Datamining (KDD) can be used to look at the past or the future, if they are used for observing and understanding aspects hidden within a particular dataset. In this set of methods you basically try to extract useful knowledge. It may range from data visualizations or summaries (Al-Dohuki et al., 2017), dataset dimensionality reduction (Bleha and Obaidat, 1991) or frequent patterns mining like in Giannotti et al. (2007), or anomalies detection, like in Xiao and Liu (2012), all of them enabling us humans to better understand a particular process.

When the purpose of the KDD process is to produce models that can predict future instances, that is the so-called Predictive. In intelligent transportation it may be useful for predicting traffic and emissions (Groot et al., 2013), pedestrian intentions (Keller and Gavrila, 2014), to name a few. Also, there are many works in transportation where the approach picked is mixed descriptive and predictive, such as in the famous Theory of Planned Behavior (Forward, 2009).

5.4.2 Classification Versus Regression

Supervised learning pursues building models that, eventually will accept unseen instances to produce an output. That output generally comes in two fashions: it may be a class within a set of previously set classes, or it may be a continuous value, typically a real number. The first kind is called classification and it implies "labeling" every newly coming instance with the previously trained classifier, meaning that a new instance is believed to belong to one specific class. There are lots of classification examples regarding transportation. To pick some, in Bolbol et al. (2012) the pedestrian intention about crossing or not the street is classified, in Kirak and Rendell (1992) they train a classifier to detect the transportation mode by using GPS samples, and in Shin et al. (2006) vehicle types are classified.

In the second kind of supervised learning, the model is trained to produce a desired numeric value after a set of instances are presented to it and the used learning has yielded a so-called regression model. The goal is that the model is able to produce sensible values when unseen instances are input. In a more mathematical way, regression supervised learning aims at building a function (nonlinear in most of the cases) that will be able to produce images corresponding to new input values. Celikoglu and Cigizoglu (2007) provide a good example of the regression model.

5.4.3 Learners

This section describes some of the most popular learners used in supervised learning, in particular for intelligent transportation applications. Decision Trees (DT) has become one of the most popular methodologies for tackling supervised learning problems. The reasons are several: first of all, they are easily understood and implemented. Their algorithmic foundations are quite easy to follow and to comprehend. In a few words, a decision tree is a set of nodes organized hierarchically from up to down, from a root node to leaves nodes. Each node includes a test that every new instance will pass by in the form of logic questions done to one or more of its attributes each time. In the end every new observation will end up in one of the tree leaves and each leaf is assigned a category.

After the training process, the resulting model is human readable, meaning useful knowledge can be grabbed from the model structure itself. There are plenty of examples of DTs applied to Intelligent Transportation. One of the most popular kinds of trees is C4.5, which was used in Shin et al. (2006) in combination with other machine learning techniques.

The Artificial Neural Networks (ANN) family of methodologies are also one of the most popular, maybe because they have an immense body of literature behind them, as possibly the most popular Machine Learning technique in the last decades. Therefore, its both theoretical and experimental foundations are more than solid. From the single layer perceptron to Deep Learning or Recurrent Neural Networks, there are quite a lot of kinds of branches coming from the same simple, bio-inspired principles. Basically, after a stimulus (a new observation), that stimulus is propagated across a network of nodes or neurons. The propagation of such stimulus is weighted at each link joining two nodes. Also, each neuron usually implements an activation function, which is conceptually a smooth threshold before propagating that signal forward. By means of the comparison of the expected and actual output of the whole network, usually the link weights, and sometimes other parameters, are incrementally adjusted, instance by instance. There are many examples of applications of ANNs in Intelligent Transportation, e.g., Celikoglu and Cigizoglu (2007), where public transportation flow is modeled using a generalized regression neural network.

Support Vector Machine (SVM) is also quite common as a supervised learning technique. The aim of this technique is to create the best possible separating border between the classes to be classified, maximizing the distance between instances from each side of that border, by using a highly nonlinear model. Because of that, the resulting borders are rather smooth in comparison to many other learning approaches. They are most frequently used for classification, but can also be applied in regression problems. A good example of the use of this kind of learning strategy is published in Bolbol et al. (2012) and Xiao and Liu (2012). Bayesian and other probabilistic approaches are quite popular too. An example of this learning method type is in Keller and Gavrila (2014).

We end up this selection of common methodologies with the so-called ensembles, which are combinations of learners. Sometimes they are all of the same kind of learner (bagging or boosting), and sometimes they may be combinations of different kinds (staking) of learners. The idea behind this is to mutually complement the strengths of several learners, ideally with little overlapping and correlation between the models. A good example is in Wu (2013), where a SVM ensemble is used to detect traffic incidents.

5.4.4 Real Time Application

A clear split can be observed when looking at the need of getting models in real time (or not). Some applications do not need to be run in real time (Off-Line KDD). Some applications need to produce usable models whilst the process is happening (On-Line KDD). Within Real Time applications one may approach it as doing iteratively off-line KDD, applying it to chunks of data. That is called batch learning.

On the other hand, there are also applications where the models are constructed incrementally across time, as new data is approaching. That is called Incremental Learning, Data Stream Mining, Data Flow Mining, and others. That is possibly the newest branch in KDD and, as a consequence, it poses bigger challenges than those described in the following sections.

5.4.5 Concept Drift Handling

Finally, these Incremental Learning approaches can be classified in two categories depending on whether they assume that the process they are trying to model presents statistical stability or not. In other words, if it is a stationary stochastic process or not. That can be measured, for example, by inspecting its average, variance, and covariance to be constant across time.

Although challenging enough, that is not the case for many processes and applications in transportation. In many of them their statistics present a drift, so-called Concept Drift. We can split supervised on-line learning depending on whether their models are able to cope and follow that concept drift or not. Of course, the number of methodologies with that super power, the ability of following concept drifts and discriminating between when it is an actual concept drift and when it is just noise, and the model does not need to move too much tracking it, is rather scarce. A good example of Concept Drift handling in a public transportation situation is in Moreira–Matias et al. (2014).

5.5 NONSUPERVISED LEARNING

From a general perspective, unsupervised learning refers to all such techniques stemming from Machine Learning and Computational Intelligence aimed at inferring knowledge from datasets without an explicit knowledge of the labels assigned to the input data. As opposed to its supervised counterpart, unsupervised learning seeks hidden structures in datasets such that data samples can be arranged and classified in groups (patterns) without even knowing the number of such patterns underlying the data at hand. Although the so-called clustering approaches are by far the most representative techniques, many other schemes can be classified as elements within the portfolio of unsupervised learning methods depending on their intended purpose (Hastie et al., 2009).

We next provide a general taxonomy of unsupervised learning models, with an emphasis on how they have been recently utilized in the transport and mobility domain:

1. Clustering, by which objects (i.e., unlabeled data instances or samples) are grouped in a number of categories or groups depending on a measure of similarity that numerically quantifies to what extent a sample is similar to another. In this first category the number of algorithms capable of clustering data is huge, depending on their notion of similarity and how groups are algorithmically discovered, but they all share the principle that instances within the same cluster should be strongly similar to each other, or at least more similar that samples lying in different clusters. There is thus a trade-off between the density of samples within each of the clusters (intracluster distance) and their separability (intercluster distance) whose discovery and balance is what characterize different clustering approaches.

At first, clustering can be partitional (hard) or fuzzy (soft) depending on whether instances are allowed to belong to more than one cluster. From this seminal discrimination diverse criteria can be adopted to classify clustering techniques: to begin with, the core idea on which the technique relies to find clusters can be based on the concept of centroids (i.e., central virtual samples not necessarily belonging to the dataset, as in, e.g., K-Means, K-Medoids, or CLARANS), the notion of connectivity among clusters (yielding agglomerative or divisive hierarchical clustering algorithms such as BIRCH), the definition of density and neighborhood among samples (as in, for instance, DBSCAN, OPTICS, MeanSHIFT, and SUBCLU), the assumption of mixtures of standard statistical models underneath the dataset at hand (namely, Expectation-Maximization algorithms) and other assorted strategies and principles such as message passing between pair of samples (e.g., Affinity Propagation) and the decomposition and dimensionality reduction of similarity graphs prior to its segmentation (correspondingly, spectral clustering).

In all the above techniques it is indeed the application itself that determines the measure of similarity between samples, deferring the quality assessment of the set of discovered clusters to internal and external (supervised) evaluation metrics. The former aims at evaluating the structural quality of the cluster arrangement found by the algorithm without requiring any knowledge on the true distribution of clusters in the data, while the latter requires known labels for their computation. Therefore, internal measures may serve as the fitness function that permits to formulate clustering as an optimization problem, for which heuristics have been extensively explored in the literature.

Thorough surveys on generic clustering approaches incorporating recently proposed algorithmic alternatives can be found in Xu and Wunsch (2005). As for the transportation and mobility domain the use of clustering techniques is common practice, as evinced recently in several use cases such that the detection of frequent routes and points of interest (Gong et al., 2015; Hong et al., 2016; Kim and Mahmassani, 2015), the identification of congestion points (Galba et al., 2013), the analysis of travel time statistics and its variability with weather conditions (Hans et al., 2014; Kwon et al., 2016), traffic identification and planning (Rao et al., 2016; Hu and Yan, 2004; Higgs and Abbas, 2013), transportation system design (Salavati et al., 2016), the understanding of demand models (Anand et al., 2014; Davis et al., 2016), and the recognition of driving styles (Hodge and Austin, 2004), among others.

2. Outlier detection refers to all such data mining techniques aimed at discriminating samples from an unlabeled dataset that do not fit to the distribution of the remainder of the dataset. There is no well-established definition of an outlying sample with respect to a given set of data, as the rareness of the sample can be featured in different domains such as amplitude, time, likelihood with respect to a previously assumed statistical distribution fitted to the rest of the dataset, or any other domain alike (Hodge and Austin, 2004).

Supervised learning algorithms such as neural networks (Munoz and Muruzábal, 1998) and decision trees (John, 1995) can be utilized for detecting outliers by learning the relationships between the feature space, normal samples, and positive (confirmed) outliers in the training set. On the contrary, unsupervised (unlabeled) algorithms detect outlying samples by assuming that all training data are labeled as normal. As in the case of clustering, these techniques can be roughly classified depending on the distance, statistics, or criterion adopted for declaring a sample as abnormal with respect to the training set: among them it is worth mentioning those hinging on statistics, from naive approximations based on the assumption of a statistical distribution generating the data samples at hand to those gravitating on the concepts of density (e.g., Local Outlier Factor), deepness, and deviation in the feature space. Clustering-based methods are also thoroughly utilized as an intuitive, simple yet effective approach for the detection of clusters, with a dominance of density-based and centroid-based schemes.

Outlier detection algorithms have been recently applied within the ITS domain mainly for preprocessing and cleansing data before traffic forecasting models (Abdullatif et al., 2016; Wang et al., 2016), but also for the detection of accidents and traffic jams (Sun et al., 2016; Wang et al., 2013), the assessment and characterization of driving behavior (Zheng and Hansen, 2016; Zhou et al., 2016), and the detection of abnormal routes from GPS traces (Chen et al., 2016).

3. Matrix decomposition techniques, which refers to all those techniques aimed at decomposing the dataset at hand in another alternative dataset where statistical relationships between the variables involved in the dataset can be explained in a more insightful fashion (Liu, 2016). Several techniques fall within this broad definition, such as the

extraction of principal components using linear projections under a maximum variance criteria (i.e., PCA and all its variants) to the decomposition of multivariate data as a sum of additive, independent components (namely, ICA) and other techniques such as nonnegative matrix factorization, factor analysis, latent dirichlet allocation, and singular value decomposition, among others. Despite their simplicity, this family of techniques has been proven to be effective as auxiliary processing steps towards unveiling hidden patterns (Yang et al., 2013; Li et al., 2013) and imputing missed entries (Asif et al., 2014) in spatio-temporal data, with an excellent review in Demšar et al. (2013).

4. Dimensionality reduction: this last category covers all methods and tools aimed at transforming highly-dimensional datasets so that the similarities between samples can be visually inspected in the transformed dataset (Huo, 2007). Also referred to as manifold learning in the literature, differences between the techniques that can be included in this subset of unsupervised learning techniques stem again from the myriad of criteria adopted for transforming the input data: from the preservation of the pairwise distances between samples in the original and the transformed space featured by Multi-Dimensional Scaling (MDS) to the extension of the concept of distance preservation to local neighborhoods (as in Locally Linear Embedding, LLE). The literature dealing with this type of technique is not as rich as in the previous cases, yet interesting applications have been recently contributed as nontraditional methods for anomaly detection in logistics (Agovic et al., 2009) and the inference of movement patterns in crowds (Yand and Zhou, 2011).

5.6 PROCESSING ARCHITECTURES

Emerging new technologies have been deployed under the term of massively parallel-processing (MPP) to address some of the issues to be tackled in this context.

• *Storage Technologies.* We need repositories capable of storing a huge volume of data and with distribution, scalability, and performance characteristics. NoSQL databases present a simple, lightweight mechanism for storage and retrieval of data that provides higher scalability/availability, distribution and high performance in load and query time. We find different structures based on column, graphs, etc.

- *Batch-processing/analysis.* Programming models are needed that are able to parallelize and manage the distribution of large volumes of data. Map Reduce programming model is a programming model for processing large data sets, and the name of an implementation of the model by Google. MapReduce (Hadoop) is typically used to do distributed computing on clusters of computers.
- *Real Time Analytics.* Here, we can find solutions able to deal with timing constraints of data analysis process and storage. Processing must be done in real-time. Three main options: Complex Event Processing (CEP), In-Memory-Data Grids (IMDG), and RT-specific data analysis platforms. CEPs capture information from message streams, databases, or applications in real time, also define taxonomies of interrelated events, supporting their identification and automatic actions or warnings (e.g., alarms). Several solutions are currently in the market (e.g., Esper, Drools Fusion, and others), with different processing definition richness. Also, in the recent times, we can find a new concept In-Memory Data Grid (IMDG), able to process huge datasets in real time, using the system main memory as storage. This is specially indicated for local and volatile data, being distributed as plugins or libraries for reference applications servers.

Finally, also there exist several trend platforms specific for real-time data analysis, some reference samples are the following: Teradata, a parallel massive processing system, based on the concept of "shared nothing"; Storm & Kafka, characterized by high fault tolerance and scalability, identifies several streaming steps: collection, transportation and process; and, finally, Spark & Shark (and Spark streaming model) with similar capabilities but implements a functional development approach.

On the other hand, when we analyze platforms to provide mobility services in real time in the context of a city, there are some requirements that point to cloud-based deployments:

- *Data volume*: real-time systems and especially ITS and mobility need to handle and store a huge amount of data. Hybrid solutions combining both NoSQL, with deep scalability and data acquisition capabilities, and relational, whose programming and querying is easier.
- *Application load fluctuation*: this will be solved by profiting from scalability at infrastructure, application, and data level. Load balancing is another key issue.
- *Real-time requirements on information processing*: even though there is the perception of an unique engine per city or company, the reality is

that they will all share the same resources. This feature is called multitenancy.

- *High availability*: one of the major advantages of cloud computing is the high availability. The cloud generally provides a high availability rate by node replication and load balancing on demand.

Finally, it should be mentioned briefly that in the context of mobility analysis we find commercial products from CISCO, Siemens, Indra, among other system integrators, and also relevant European projects, such as SUPERHUB (2016), which integrates multimodal travel planning, journey resourcing, and ticket purchasing; Co-Cities (2016), which addresses the feedback for final users; and MoveUs (2016), which integrates the just mentioned capabilities with the energy efficiency aspects: measurement, prediction, and savings; as well as several other platforms defined in R&D European projects.

5.7 APPLICATIONS

The availability of massive mobility and transport massive data allows the deployment of a smart mobility management and the provision of advanced services. Here, some specific real problems and existing data-based approaches are briefly introduced.

5.7.1 Transport Demand Modeling

To identify, represent, and predict behavior in travel demand are mandatory at the three levels of decision taking: strategic, evaluating long-term infrastructure improvement strategies, definition of Sustainable Urban Mobility Plans (SUMP); tactical, including the management and planning; and finally, operational, including shorter-term congestion management policies. Transport demand models are mathematical models that predict long-term mobility demand based on current conditions and future projections. They have specific analytical capabilities, such as the prediction of mobility demand, choice of route day, selection of the route itself, and representation of traffic flow in the road network. These systems are limited to accurately estimating changes in parameters (such as speed, delays, and queues) (Anda et al., 2016) resulting from the implementation of data acquisition processes, uncertainty on model structure, and assumptions.

Hereby, focusing on approaches that allow the generation of O/D matrices, the first attempts to exploit mobile phone CDR data were based on the correlation between movements of anonymous mobile phones and vehicle displacements. The concept transient O/D matrix, presupposing that for segments without observed CDR data due to inactivity, could be enough to deduce usage patterns by simply counting trips for each pair of consecutive calls made within a time frame and most of the O/D data identified by a routing algorithm (Devillaine et al., 2012). Such transient O/D matrices must be scaled up to match the real traffic counts by applying optimization and finally, apply correction factors based on smartphone penetration rates.

The exploitation of transactional data starts with the estimation of alighting stops, usually based on the Trip-Chaining algorithm assuming dependences between round trips and stations. The next step is the identification of multistage trips (e.g., transfers). Public transport information provides also semantic meaning to itinerary (Williams and Hoel, 2003). In any case, the value of this data varies depending on the city and its fare policies (e.g., tapping onboarding and at destination).

GPS consists of a spatiotemporal track, being used for location-based mobile apps, guiding and routing, etc. One of the most promising uses is the extraction of semantic meaning from GPS data (Huang et al., 2010), where different approaches and methods can be used: distance-based, probabilistic-based, generative based on HMMs, etc.

5.7.2 Short-Term Traffic State Prediction

Most of the work in the literature related to traffic state prediction is usually made for the near-term future; precise predictions that use intensively the latest values obtained and have a short validity. They also can be long-term predictions, which are not so precise but are valid for all times. Probably the most powerful way for short-term prediction of the state of the traffic is by means of performing a battery of runs with a traffic microsimulator (e.g., Aimsun, SUMO).

On the other hand it is very expensive because all the microdynamics are solved and in order to have a statistically relevant solution multiple Montecarlo solutions need to be obtained. Moreover, for these solutions to be accurate, a complete description of the road network, including turning probabilities at every intersection and boundary conditions at the edge of the network, is required. Other techniques also applicable in forecasting the state of the traffic are specific time series prediction algorithms. For this analysis the most commonly predicted feature is the traffic flow, measured in vehicles per hour; other researchers, such as Mai et al. (2012), consider other features, such as the average speed or the occupancy of the road, etc. The first approaches to a forecasting model based on historical average algorithms were the autoregressive integrated moving average (ARIMA) and Kalman filtering models, assuming the premise that traffic patterns are seasonal. Knowing the typical traffic conditions for a particular day of the week will allow predicting the conditions on any other day as long as it is the same day of the week. These algorithms perform reasonably well during normal operating conditions, but they do not respond well to external system changes such as weather, special events, or modified traffic control strategies. Seasonal ARIMA (SARIMA) (Vashanta Kumar and Vanajakshi, 2015) models can be applied to handle these specific aspects of the time series.

The nonparametric (e.g., ANNs) and machine learning (ML) models are part of a completely different and trending approach (eCompass, 2017); although in general time series analysis techniques are the most precise methods, they are usually highly computationally intensive and faster methods are required to provide real-time estimations.

In most of the literature, the predictions are made for a near future and do not take into account external variables like events, weather, or traffic incidences or alterations. The main problem of these methods (Zarei et al., 2013) is that the predictions are computationally expensive and are only valid for a few minutes after they have been computed. The relaxation time for traffic dynamics, the time after which the state of the traffic does not depend on the previous values, depends on the size of the network, its density, and the values of the typical flow of vehicles.

5.7.3 Planning/Routing

Determining optimal routes in road networks from a given source to a given target location is a problem frequently addressed in everyday life. There are also tools and applications like logistic planners and traffic simulators that need to solve a huge number of such route queries. Route planning techniques have evolved quickly in terms of efficiency and accuracy during the last years. Two of the first routing algorithms proposed consider the distance (Dijkstra, Bellman-Ford, using predefined metrics) between one node to each other in a graph.

The Classic Algorithm A* extends these calculations with supplementary information and a greedy strategy to improve its results. Metaheuristic methods have been also used successfully in the computation of the shortest path. Nevertheless, these solutions yield very slow query times when we deal with realistic road networks, and this fact hampers their usage in real-time or interactive applications (Geisberger, 2011; Schultes, 2008). On the other hand, applying aggressive heuristics do not always provide accurate results. Road networks present structural properties (e.g., networks are sparse, layered, almost planar, and present hierarchical structures) that support different speed-up techniques. These techniques are based on a preprocessing step, where auxiliary information about the network is obtained and annotated, and then used to accelerate subsequent queries. Speed-up techniques can be classified into: Goal-Directed Search techniques, comprising different algorithms like geometric A* search, heuristic A* search based on distance estimations, landmark-based A* search, signposts, geometric containers, or edge flags, managing different graph annotation schemes and exploration strategies; and Hierarchical Approaches, taking advantage of the topological properties of the network (e.g., the separator-based multilevel method or the popular contraction hierarchies).

On the other hand, the structure of public transportation networks is a bit different from roads. The key property that determines the difference with road networks is the inclusion of the time-dependency aspect, each public transport service is subject to a timetable.

The eCOMPASS project (2017) presents an updated deep analysis of current problems, algorithms, and underlying approaches. Basically, there are two different approaches to cover timetable requirements: time-expended and time-dependent, both prioritizes flexibility under changing constraints or memory needs and response time (Schultes, 2008). Adaptations of successful speed-up techniques from road routing are complex and without a clear extrapolation of results. In multimodal routing planning, multicriteria methods are a suitable approach.

Additionally, various uncertainties must be considered when dealing with public transportation, such as unexpected interruption or cancelation of the service or delays, all of them caused by different reasons: weather, breakdowns, mass events, demand increase, strikes, among others. The propagation delay can be catastrophic and avoidance of mistrust by the users, through delay propagation robustness and real-time travel replanning capabilities, is a must. At design time, this is being partially addressed by stochastic sensibility analysis; but the model adjustments, when available, must be done by hand by experts. The European EU project

PETRA (2017) has developed a multimodal network, while taking into account uncertainty, providing robust trip schedules, while predicting sensible robust future states by simulation.

REFERENCES

Abdullatif, A., Rovetta, S., Masulli, F., 2016. Layered ensemble model for short-term traffic flow forecasting with outlier detection. In: 2016 IEEE 2nd International Forum on Research and Technologies for Society and Industry Leveraging a Better Tomorrow (RTSI). IEEE, pp. 1−6.

Agovic, A., Banerjee, A., Ganguly, A., Protopopescu, V., 2009. Anomaly detection using manifold embedding and its applications in transportation corridors. Intell. Data Anal. 13 (3), 435−455.

Al-Dohuki, S., Wu, Y., Kamw, F., Yang, J.L., Zhao, Y., Wang, F., 2017. SemanticTraj: a new approach to interacting with massive taxi trajectories. IEEE. Trans. Vis. Comput. Graph. 23 (1), 11−20.

Anand, S., Padmanabham, P., Govardhan, A., 2014. Application of factor analysis to k-means clustering algorithm on transportation data. Int. J. Comput. Appl. 95, 15.

Anda, C., Fourie, P., Erath, A., 2016. Transport modelling in the age of big data. Work Report. (FCL) Future Cities Laboratory. Subgapore-ETH Centre (SEC).

Asif, M.T., Dauwels, J., Goh, C.Y., Oran, A., Fathi, E., Xu, M., et al., 2014. Spatiotemporal patterns in large-scale traffic speed prediction. IEEE Trans. Intell. Transp. Syst. 15 (2), 794−804.

Beeri, C., Bernstein, P.A., Goodman, N., 1978. A sophisticate's introduction to database normalization theory. Proceedings of the Fourth International Conference on Very Large Data Bases (vol. 4, pp. 113−124), VLDB Endowment.

Bleha, S.A., Obaidat, M.S., 1991. Dimensionality reduction and feature extraction applications in identifying computer users. IEEE. Trans. Syst. Man. Cybern. 21 (2), 452−456.

Blum, A.L., Langley, P., 1997. Selection of relevant features and examples in machine learning. Artif. Intell. 97 (1), 245−271.

Bolbol, A., Cheng, T., Tsapakis, I., Haworth, J., 2012. Inferring hybrid transportation modes from sparse GPS data using a moving window SVM classification. Comput. Environ. Urban. Syst. 36 (6), 526−537.

Celikoglu, H.B., Cigizoglu, H.K., 2007. Public transportation trip flow modeling with generalized regression neural networks. Adv. Eng. Softw. 38 (2), 71−79.

Chen, X., Cui, T., Fu, J., Peng, J., Shan, J., 2016. Trend-residual dual modeling for detection of outliers in low-cost GPS trajectories. Sensors 16 (12), 2036.

Co-Cities, 2016. "Co-Cities project. Cooperative Cities." [Online]. Available: <www.co-cities.eu/> (01.02.17).

Daly, E.M., Lecue, F., Bicer, V., 2013. Westland row why so slow?: fusing social media and linked data sources for understanding real-time traffic conditions. In: Proceedings of the 2013 International Conference on Intelligent user interfaces. ACM, pp. 203−212.

Davis, N., Raina, G., Jagannathan, K., 2016. A multi-level clustering approach for forecasting taxi travel demand. In: 2016 IEEE 19th International Conference on Intelligent Transportation Systems (ITSC). IEEE, pp. 223−228.

Demšar, U., Harris, P., Brunsdon, C., Fotheringham, A.S., McLoone, S., 2013. Principal component analysis on spatial data: an overview. Ann. Assoc. Am. Geogr. 103 (1), 106−128.

Devillaine, F., Munizaga, M., Trépanier, M., 2012. Detection of activities of public transport users by analyzing smart card data. Transp. Res. Rec.: J. Transp. Res. Board 2276, 48−55 December 2012, ISSN 0361 -1981.

eCompass, 2017. eCompass- eco-fiendly urban multi-modal route planning services for mobile users project [Online]. Available: <http://www.ecompass-project.eu/> (01.02.17).

Feng, Z., Zhu, Y., 2016. Survey on trajectory data mining: techniques and application. IEEE Access Digital Object Identifier 10.1109/ACCESS.2016.2553681.

Forward, S.E., 2009. The theory of planned behaviour: The role of descriptive norms and past behaviour in the prediction of drivers' intentions to violate. Transp. Res. Part F Traffic Psychol. Behav. 12 (3), 198−207.

Galba, T., Balkić, Z., Martinović, G., 2013. Public transportation BigData clustering. Int. J. Electr. Comput. Eng. Syst. 4 (1), 21−26.

Geisberger, R., 2011. "Advanced route planning in transportation networks," Ph.D. dissertation, Karlsruher Instituts für Technologie.

Giannotti, F., Nanni, M., Pinelli, F., Pedreschi, D., 2007. Trajectory pattern mining. In: Proceedings of the 13th ACM SIGKDD International Conference on Knowledge Discovery and Data Mining. ACM, pp. 330-339.

Gong, L., Sato, H., Yamamoto, T., Miwa, T., Morikawa, T., 2015. Identification of activity stop locations in GPS trajectories by density-based clustering method combined with support vector machines. J. Mod. Transp. 23 (3), 202−213.

Groot, N., De Schutter, B., Hellendoorn, H., 2013. Integrated model predictive traffic and emission control using a piecewise-affine approach. IEEE Trans. Intell. Transp. Syst. 14 (2), 587−598.

Hans, E., Chiabaut, N., Leclercq, L., 2014. Clustering approach for assessing the travel time variability of arterials. Transp. Res. Rec.: J. Transp. Res. Board 2422, 42−49.

Hastie, T., Tibshirani, R., Friedman, J., 2009. Unsupervised learning. The Elements of Statistical Learning. Springer, New York, pp. 485−585.

Higgs, B., Abbas, M., 2013. A two-step segmentation algorithm for behavioral clustering of naturalistic driving styles. In: 16th International IEEE Conference on Intelligent Transportation Systems (ITSC 2013). IEEE, pp. 857−862.

Hodge, V.J., Austin, J., 2004. A survey of outlier detection methodologies. Artif. Intell. Rev. 22 (2), 85−126.

Hong, Z., Chen, Y., Mahmassani, H.S., Xu, S., 2016. Spatial trajectory clustering for potential route identification and participation analysis for carpool commuters. In: Transportation Research Board 95th Annual Meeting (No. 16-7013).

Hu, D.W., Yan, G.H., 2004. Application of clustering analysis in macroscopic planning of highway transportation hub. J. Highway Transp. Res. Dev. 9, 035.

Huang, L., Li, Q., Yue, Y., 2010. Activity identification from GPS trajectories using spatial temporal POIs' attractiveness. In: Proceedings of the 2nd ACM SIGSPATIAL International Workshop on Location Based Social Networks, pp. 27−30.

Huang, T., Sethu, H., Kandasamy, N., 2016. A new approach to dimensionality reduction for anomaly detection in data traffic. IEEE Trans. Netw. Serv. Manage. 13 (3), 651−665.

Huo, X., Ni, X.S., Smith, A.K., 2007. A survey of manifold-based learning methods. Recent Adv. Data Mining Enterprise Data 691−745.

IBM, 2017. IBM traffic standards <http://www.ibm.com/support/knowledgecenter/es/SSTMV4_1.6.1/transport/ref_itsstandards.html> (13.01.17).

John, G.H., 1995. Robust decision trees: removing outliers from databases. KDD 174−179.

Kaza, N., Hopkins, L.D., 2007. Ontology for land development decisions and plans. Stud. Comput. Intell. (SCI) 61, 47−59.

Kdnuggets, 2017. Seven Techniques for Data Dimensionality Reduction, <http://www. kdnuggets.com/2015/05/7-methods-data-dimensionality-reduction.html> by Rosaria Silipo (13.01.17).

Keller, C.G., Gavrila, D.M., 2014. Will the pedestrian cross? A study on pedestrian path prediction. IEEE Trans. Intell. Transp. Syst. 15 (2), 494–506.

Kim, J., Mahmassani, H.S., 2015. Trajectory clustering for discovering spatial traffic flow patterns in road networks. In: Transportation Research Board 94th Annual Meeting (No. 15-5443).

Kirak, K., Rendell, L.A., 1992. A practical approach to feature selection. In: Proceedings of the Ninth International Workshop on Machine Learning, pp. 249–256.

Kohavi, R., John, G.H., 1997. Wrappers for feature subset selection. Artif. Intell. 97, 273–324.

Kwon, O.H., Park, S.H., 2016. Identification of influential weather factors on traffic safety using K-means clustering and random forest. Advanced Multimedia and Ubiquitous Engineering. Springer, Singapore, pp. 593–599.

Lakhina, A., Crovella, M., Diot, C., 2005. Mining anomalies using traffic feature distributions. ACM SIGCOMM Comput. Commun. Rev. 35 (4), 217–228, ACM.

Leather, H., Bonilla, E., O'Boyle, M., 2009. Automatic feature generation for machine learning based optimizing compilation. In: International Symposium on Code Generation and Optimization. CGO 2009. IEEE, pp. 81–91.

Lévy-Leduc, C., Roueff, F., 2009. Detection and localization of change-points in high-dimensional network traffic data. Ann. Appl. Stat. 637–662.

Li, L., Li, Y., Li, Z., 2013. Efficient missing data imputing for traffic flow by considering temporal and spatial dependence. Transp. Res. Part C: Emerg. Technol. 34, 108–120.

LinkedGeoData, 2017. <www.linkedgeodata.org> (01.01.17).

Liu, X., Liu, X., Wang, Y., Pu, J., Zhang, X., 2016. Detecting anomaly in traffic flow from road similarity analysis. International Conference on Web-Age Information Management. Springer International Publishing, pp. 92–104.

Lord, P., Macdonald, A., Lyon, L., Giaretta, D., 2004. From data deluge to data curation. In: Proceedings of the UK e-science All Hands Meeting, pp. 371–375.

Mai, T., Ghosh, B., Wilson, S., 2012. Miltivariate Short-term traffic flow forecasting using Bayesian vector autoregressive moving average model. In: Proceedings of the 91st transportation Research Board Annual Meeting.

Maier, A., Schriegel, S., Niggemann, O., 2017. Big data and machine learning for the smart factory—solutions for condition monitoring, diagnosis and optimization. Industrial Internet of Things. Springer International Publishing, pp. 473–485.

Malgundkar, T., Rao, M., Mantha, S.S., 2012. GIS Driven urban traffic analysis based on ontology. Int. J. Managing Inform. Technol. (IJMIT) 4 (1), February.

Metral, G., Falquet, G., Karatzas, K., 2005. Ontologies for the integration of Air Quality Models and 3D City Models. Institut d'Architecture. University of Geneva.

Morais, C.D., 2015. Where is the Phrase "80% of Data is Geographic" From? Retrieved May 2.

Moreira-Matias, L., Gama, J., Mendes-Moreira, J., de Sousa, J.F., 2014. An incremental probabilistic model to predict bus bunching in real-time. International Symposium on Intelligent Data Analysis. Springer International Publishing, pp. 227–238.

MoveUs, 2016. MoveUs project. ICT Cloud-based Platform and Mobility Services available, Universal and Safe for all Users. [Online]. Available: <www.superhub-project. eu/> (01.02.17).

Munoz, A., Muruzábal, J., 1998. Self-organizing maps for outlier detection. Neurocomputing 18 (1), 33–60.

Petra, 2017. "Petra project. personal transport advisor: an integrated platform of mobility patterns for smart cities to enable demand-adaptive transportation system." [Online]. Available: <http://petraproject.eu/> (01.02.17).

Rao, W., Xia, J., Lyu, W., Lu, Z., 2016. A K-means clustering method to urban intersection traffic state identification using interval data. In: Transportation Research Board 95th Annual Meeting (No. 16-4769).

REWERSE, 2017. REWERSE Project. Reasoning on the web. A1.D1. Ontology of Transportation Networks. <http://rewerse.net/deliverables/m18/a1-d4.pdf> (01.01.17).

Salavati, A., Haghshenas, H., Ghadirifaraz, B., Laghaei, J., Eftekhari, G., 2016. Applying AHP and clustering approaches for public transportation decision making: a case study of Isfahan City. J. Public Transp. 19 (4), 3.

Schadt, E.E., Li, C., Ellis, B., Wong, W.H., 2001. Feature extraction and normalization algorithms for high-density oligonucleotide gene expression array data. J. Cell. Biochem. 84 (S37), 120—125.

Schultes, D., 2008. Route planning in road networks, Ph.D. dissertation, Karlsruher Instituts für Technologie.

Shin, W.S., Song, D.H., Lee, C.H., 2006. Vehicle classification by road lane detection and model fitting using a surveillance camera. JIPS 2 (1), 52—57.

Snášel, V., Nowaková, J., Xhafa, F., Barolli, L., 2017. Geometrical and topological approaches to big data. Future Gener. Comput. Syst. 67, 286—296.

Sun, C., Hao, J., Pei, X., Zhang, Z., Zhang, Y., 2016. A data-driven approach for duration evaluation of accident impacts on urban intersection traffic flow. In: 2016 IEEE 19th International Conference on Intelligent Transportation Systems (ITSC). IEEE, pp. 1354—1359.

Sun, D.J., Liu, X., Ni, A., Peng, C., 2016. Traffic congestion evaluation method for urban traffic congestion evaluation method for urban arterials. Transportation Research Record 2461, 9—15.

TOWNONTOLOGY, 2008. COST ACTION C21 — TOWNONTOLOGY, 2008. Urban Ontologies for an Improved Communication in Urban Civil Engineering Projects.

Uguz, H., 2011. A two-stage feature selection method for text categorization by using information gain, principal component analysis and genetic algorithm. Knowledge-Based Syst. 24 (7), 1024—1032.

Vashanta Kumar, S., Vanajakshi, L., 2015. Short-Term Traffic Flow Prediction Using Season ARIMA Model With Limited Input Data. Springer.

Wang, Y., Xu, J., Xu, M., Zheng, N., Jiang, J., Kong, K., 2016. A feature-based method for traffic anomaly detection. In: Proceedings of the 2nd ACM SIGSPATIAL Workshop on Smart Cities and Urban Analytics. ACM, p. 5.

Wang, Z., Lu, M., Yuan, X., Zhang, J., Van De Wetering, H., 2013. Visual traffic jam analysis based on trajectory data. IEEE. Trans. Vis. Comput. Graph 19 (12), 2159—2168.

Widhalm, P., Yang, Y., Ulm, M., Athavale, S., González, M.C., 2015. Discovering urban activity patterns in cell phone data. Transportation 42 (4), 597—623.

Williams, B.M., Hoel, L.A., 2003. Modeling and forecasting vehicular traffic FLor as a seasonal ARIMA process: theoretical basis and empirical result. J. Transp. Eng. 664—672.

Wu, S., 2013. A review on coarse warranty data and analysis. Reliab. Eng. Syst. 114, 1—11. Available from: http://dx.doi.org/10.1016/j.ress.2012.12.021.

Xiao, J., Liu, Y., 2012. Traffic incident detection by multiple kernel support vector machine ensemble. In: 2012 15th International IEEE Conference on Intelligent Transportation Systems. IEEE, pp. 1669—1673.

Xu, R., Wunsch, D., 2005. Survey of clustering algorithms. IEEE Transactions Neural Netw. 16 (3), 645—678.

Xu, Z., Li, W., Niu, L., Xu, D., 2014. Exploiting lowrank structure from latent domains for domain generalization. In: Proceedings of European Conference on Computer Vision (ECCV), pp. 628–643.

Yang S., Kalpakis K., Biem A., 2013. Spatio-temporal coupled bayesian robust principal component analysis for road traffic event detection. In: 16th International IEEE Conference on Intelligent Transportation Systems (ITSC 2013). IEEE, pp. 392–398.

Yang, S., Zhou, W., 2011. Anomaly detection on collective moving patterns: Manifold learning based analysis of traffic streams. In: 2011 IEEE Third International Conference on Privacy, Security, Risk and Trust (PASSAT) and 2011 IEEE Third Inernational Conference on Social Computing (SocialCom). IEEE, pp. 704–707.

Yang, Y., Pedersen, J.O., 1997. A comparative study on feature selection in text categorization. ICML 412–420.

Zarei, N., Ali Ghayour, M., Hashemi, S., 2013. Road traffic prediction using context-aware random forest based on volatility nature of traffic flows. ACIIDS 196–205.

Zheng, Y., 2015. Trajectory Data Mining: An Overview. ACM Transactions on Intelligent Systems and Technology, Vol. 6, No. 3, Article 29, Publication date: May 2015. doi: <http://dx.doi.org/10.1145/2743025>.

Zheng, Y., Hansen, J.H., 2016. Unsupervised driving performance assessment using free-positioned smartphones in vehicles. In: 2016 IEEE 19th International Conference on Intelligent Transportation Systems (ITSC). IEEE, pp. 1598–1603.

Zhou, Z., Dou, W., Jia, G., Hu, C., Xu, X., Wu, X., et al., 2016. A method for real-time trajectory monitoring to improve taxi service using GPS big data. Inform. Manage. 53 (8), 964–977.

FURTHER READING

Aggarwal, C.C., Reddy, C.K. (Eds.), 2013. Data Clustering: Algorithms and Applications. CRC Press.

Brodesser, M., 2013. Multi-modal route planning, Ph.D. dissertation, niversity of Freiburg Faculty of Engineering Department for Computer-Scien.

Wang, P., Hunter, T., Bayen, A.M., Schechtner, K., González, M.C., 2012. Understanding Road Usage Patterns in Urban Areas, Scientific Reports, 2, December 2012, ISSN 2045-2322.

Superhub, 2017. SupeHub project. SUstainable and PERsuasive Human Users moBility in future cities. [Online]. Available: <www.superhub-project.eu/> (01.02.17).

Wu, X., Zhu, X., Wu, G.Q., Ding, W., 2014. Data mining with big data. IEEE Trans. Knowl. Data Eng. 26 (1), 97–107.

PART II

Applications

CHAPTER 6

Driver Assistance Systems and Safety Systems

Felipe Jiménez
Universidad Politécnica de Madrid, Madrid, Spain

Contents

6.1 INTEGRATED SAFETY MODEL

For many years, the "Vision Zero" has been designed for the long term to eliminate the deaths and serious injuries of traffic accidents (Corben et al., 2009). In addition, zero emissions and zero traffic jams are also being pursued. This new approach means that responsibility ceases to be completely user-centered, but other actors that directly or indirectly intervene in traffic should be included, such as government, infrastructure managers, vehicles manufacturers, police, etc. New technologies play an important role in achieving this aim (Organisation for Economic Co-operation and Development, 2003).

In relation to the vehicle, safety systems have traditionally been divided according to their main orientation: avoidance of accidents (active or primary safety) or reduction of their consequences (passive or secondary safety) (e.g., Elvik and Vaa, 2004).

At present, the concept of integrated safety is quite widespread (Aparicio et al., 2008), and it aims to provide optimum protection of

Intelligent Vehicles
DOI: http://dx.doi.org/10.1016/B978-0-12-812800-8.00006-0
209

vehicle occupants, pedestrians, and properties by reducing the likelihood of an accident (traditionally understood as active safety) or by the reduction of the effects in case it occurs (traditionally understood as passive safety). In addition, such a safety model encompasses even more so that it extends to tertiary safety systems (beyond the accident and its immediate consequences). In this way, all safety models, despite their differences, coincide in showing the interrelationship between safety systems, eliminating the classical differentiation of primary, secondary, and tertiary safety systems, and presenting overlaps that are achieved through sharing of information between systems.

In this way, they can be distinguished and defined:

- *Driver assistance systems*: systems developed with the fundamental aim of avoiding errors in the driver due to tiredness or inattention, or unsatisfactory decisions during the driving activity.
- *Primary safety systems*: devices that help to avoid or minimize unsafe acts and behaviors of the driver and the vehicle itself that are liable to cause accidents, by acting on the vehicle during the pre-accident phase.
- *Precollision systems*: elements that allow, using information provided by sensors, to coordinate actions by vehicle control systems and occupant protection during the precollision and collision phases, in order to reduce or eliminate occupant or pedestrians injuries, caused by the collision.
- *Secondary safety systems*: devices intended to prevent or minimize damage to persons or things transported in the vehicle or with which it may interact, in the event of an accident.
- *Tertiary safety systems*: devices whose aim is to reduce the consequences of an accident after it has occurred.

Fig. 6.1 shows this differentiation but also the overlap between phases. It is noted that, sufficiently far from an accident situation, the driver requires assistance to be better informed and have a more pleasant and rested driving. As the accident situation is approaching, the driver requires to be alerted and, when the time to collision falls below a certain threshold, the driver may need automatic actions that improve their own performance or reduce the reaction times. After the accident, consequences should be reduced as much as possible, which implies taking into account variables such as the degree of severity of the accident or characteristics of the occupants and opponents. Thus, at each stage, the demands on the safety systems are different (Table 6.1).

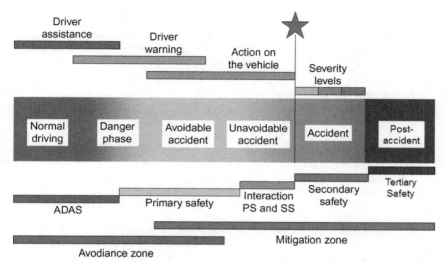

Figure 6.1 Integrated safety model (Aparicio et al., 2008).

Table 6.1 States in the integrated safety model

State	Aim
Normal driving	To increase comfort to reduce errors
Warning	To detect danger conditions
Avoidable collision	To act to avoid collision
Unavoidable collision	To reduce effects of collision
Postcollision	To reduce consequences after collision

Driver assistance systems as well as primary safety systems have a great potential in the coming years. On the other hand, the opposite situation is the case with secondary safety systems, which emerged earlier, had a notable impact, and are more developed today, but great progress is not expected except in three specific fields: actions for specific groups of people (elderly, children, disabled), systems adaptation to accident conditions, and measures to protect vulnerable users (pedestrians, cyclists, etc.). These trends can be observed in Fig. 6.2 where the next steps towards autonomous and connected driving are marked.

The high integration of the functions leads to improvements in the safety superior to the sum of the contributions of each one of them since the individual functionality is increased. Thus, the development of new measures is motivated simply by the large amount of information

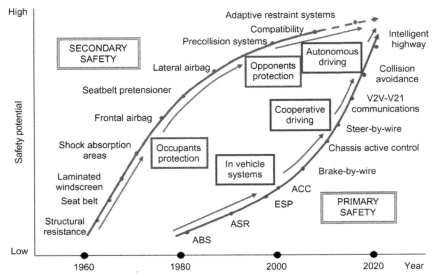

Figure 6.2 Evolution of primary and secondary safety systems potential.

available. In this context, all information is integrated to be evaluated as a whole. Then, the greater the density and quality of this information is, the greater the probability that the hypotheses taken by the assistance system will approach reality. In addition, the efficiency in the treatment of the information will be greater. This is of great importance if one takes into account that driving in a nonfully structured environment requires hypotheses of possible events in the next few seconds in order to anticipate them. At present, research is being carried out to evaluate the potential of this information fusion, specifically addressing the following aspects:

- Increased reliability by confirming the same information more than one signal.
- Consistently interpreting danger situations.
- Reducing ambiguity by integrating information and reducing hypotheses in the interpretation of signals.
- Improving detection when using different systems.
- Increasing robustness, ensuring that, at all times, some sensor is able to transmit useful information.
- Avoid giving false alarms that lead to incorrect and disturbing actions for traffic that cause the driver to lose confidence in the system.
- Increasing spatial and temporal coverage.

- Reducing costs by the use of low-performance sensors, and compensate the deficiencies with the support of the others.
- Systems must be able to interact with users.

6.2 SYSTEMS FOR IMPROVING DRIVING TASK

6.2.1 Assistance Systems Aim

Although the task of driving is a relatively easy activity to perform by a human in most situations, some capabilities are required. In addition, there are situations where the demand of the system may exceed the capabilities of the driver or the driver does not perform the driving function correctly. Thus, more than 90% of accidents have as one of their causes the human factor and solely the human factor is responsible for two-thirds of accidents (Hobbs, 1989). This is due to the adoption of risky behaviors in a more or less conscious way or to improper performance in a critical situation. These failures may be motivated by a lack of information or lack of capacity to process and react conveniently within the time available. On the other hand, incorrect driver reactions can imply that different vehicle systems work close to their physical limit or even overcome it.

The definition of "improving driving task" is quite ambiguous and involves concepts such as comfort, information, safety, and efficiency, but they are not completely independent.

The Advanced Driver Assistance Systems (ADAS) are aimed at alleviating these deficiencies, reducing the mental load on the driver, increasing and improving the information available to promote safe driving without errors, and achieving better responses in the control of the vehicle. The operation of ADAS systems and their interaction with the driver must be parallel and the driver must understand the actions taken by the system to encourage their acceptance and not end up compromising safety if the driver is not advised to react properly.

Primary safety systems are aimed at reducing the number of traffic accidents. At times, the border with the ADAS systems is diffuse. As a differentiating aspect it can be taken that the assistance systems are focused on the normal driving conditions and, with them, it is intended to unload from the drivers some tasks or to help them in order that they can focus on more important issues. On the contrary, primary safety systems affect risk situations, which do not occur in normal traffic, such as emergency braking, stability control of the vehicle during cornering in conditions close to the physical limits, etc.

An example of this differentiation can be found in speed control systems. Thus, the cruise control system which maintains a predefined speed introduced by the driver is encompassed within the group of assistance systems, since it discharges the driver of a task but does not have a direct impact on safety. On the other hand, adaptive cruise control (ACC) systems and intelligent ISA systems have a clear impact on safety by controlling the distance with the preceding vehicle in the first case and by controlling the speed with respect to a limit (legal or technical, preset or calculated).

Furthermore, some assistance systems are simply oriented to provide information (McQueen et al., 2002) or make easier the driving task, for example, helping the driver in the parking maneuver with sensors or even with automatic actions. In most cases, risk is not present, but, in others, the same system could be essential. This is the case, for example, of rear cameras in large vehicles in which visibility of the surroundings is limited.

Within primary safety systems, a distinction is made between those that warn the driver and those acting on the vehicle. The latter can achieve good results given that the response times are very short, lower than the driver's, although an adequate perception of driving conditions and the environment must be ensured for correct decision making.

On the other hand, in addition to the systems oriented towards improving safety, others have also been incorporated in vehicles aimed at improving efficiency to reduce fuel consumption. In this sense, there are systems that are based on the improvement of operation conditions of the vehicle while other more advanced ones also take into account the orography of the road to take advantage of the positive and negative slopes. Finally, knowledge of traffic, traffic lights, etc., also allows better adjusting of the system recommendations.

Several systems are oriented just to make the driving task more comfortable. This fact has direct impact on safety because the driver could be in better condition and pay attention to what is really important in the driving task while the assistance system takes care of those repetitive actions. In this sense, it should be noted that nowadays vehicles provide several options for personalizing their performance to make them more friendly for drivers, both in operation functions such as driving sensitivity in pedals, steering wheel, and powertrain response, and in comfort functions such as regulation of seats, rear-view mirrors, headrests, and steering wheel position.

Alongside the positive points that these systems contribute, such as reducing the driver's mental load, fostering attention to the most important elements, and improving decision and driver performance, there are a number of uncertainties that need to be analyzed to evaluate the true effectiveness of these systems, as well as their acceptance among drivers. The main potential limitations are related to the difficulty of the users predicting how the system will perform automatically, effects of the complete nonadaptation to the systems or loss of capacities caused by their use, and the effects of information overload or excessive discharge that entails the partial loss of a driver's attention. On the way to overcoming systems limitations, the integration of assistance systems and primary safety systems is the basis of the first steps of vehicle automation (SAE, 2016).

6.2.2 Classification

In general, a common way of understanding the assistance and primary safety systems is to correlate possible errors or problems in the task of driving and systems that can alleviate them (avoid them, promote a behavior change, reduce their effect, etc.). Although it is not easy to classify systems regarding the final objective (comfort, information, safety or efficiency), Table 6.2 shows a list of systems grouped in homogeneous groups and, if possible, correlated to common driver errors that could be minimized by implementing these systems.

General information systems are designed to provide the driver with information that may be useful during the trip. Thus, from the classic navigators that only worked with static digital maps that were updated periodically, they tend to incorporate new dynamic information such as weather and traffic conditions that allow alerting the driver of unforeseen situations and recalculating routes to optimize criteria of time, distance, consumption, etc., in each situation. In addition, using computer vision systems, it is possible to identify traffic signals so that information can be provided at all times to the driver even if the driver has not perceived or has forgotten them (McQueen et al., 2002).

In terms of dynamic control, the assistance systems aim to optimize the vehicle's behavior in traction, braking, lateral and vertical dynamics, in order to take full advantage of the physical limits to which the dynamics of the vehicle is subject (e.g., Burton et al., 2004; Paine, 2005). Thus, attempts are made to avoid wheel-locking, skidding, loss of directional control, or excessive reduction of vertical load on one of the wheels (E-IMPACT European project).

Table 6.2 Systems classification

System group	Examples	Comfort	Information	Safety	Efficiency	Driving problem or error
General information	Navigation system (relevant places, events, ...)		X		X	Driving along the incorrect lane. Route choice mistake.
	Traffic information system		X		X	Sudden maneuver due to lack of prevision near a crossing. Improper action due to unusual traffic, road, or weather conditions
	Road surface conditions		X	X		Improper action due to unusual traffic, road or weather conditions
	Traffic signs automatic recognition		X			Misperception of signals
Dynamic response control	Antilock braking system ABS			X		Excessive barking force for the road conditions
	Emergency braking assistance system BAS			X		Inadequate braking maneuver
	Automatic braking			X		Inadequate braking maneuver
	Electronic stability programme ESP			X		Loss of vehicle control
	Roll over warning			X		Excessive speed along a curve or sudden maneuver
	Active suspension control			X		Unbalanced vertical dynamic forces
	Traction control			X		Excessive traction force

Category	System				Issue
Visibility	Lights automatic activation	X		X	Incorrect obstacles perception due to poor visibility conditions. Sudden change in visibility conditions. Inadequate use of lights.
	Adaptive lights for curves	X		X	Blind spots in curves when driving at night
	Lights control	X		X	Inadequate use of lights. Blinding drivers in the opposite direction
	Wipers automatic activation	X		X	Incorrect obstacles perception due to poor visibility conditions. Inadequate use of wipers
Speed control	Head-up display		X		Excessive time looking at the dashboard
	Speed limiters	X		X	Excessive speed
	Cruise control			X	Repetitive task
	Intelligent speed adaptation system			X	Unsafe speed
	Adaptive cruise control	X		X	Unsafe speed and safe distance from vehicles ahead
	Congestion assist	X		X	Repetitive task
	Ecodriving assist		X	X	Excessive fuel consumption due to incorrect driver behavior
Lane keeping	Lane departure warning	X		X	Dangerous scenarios when departing lane
	Automatic lane keeping assist	X		X	Repetitive task
Collision prevention	Obstacles detection in front of the vehicle			X	Incorrect actions before obstacles, vehicles, or pedestrians
	Blind spot monitoring			X	Misperception of some obstacles around the vehicle
	Intersection assist			X	Misperception of some obstacles around the vehicle
	Obstacles detection with night vision			X	Poor visibility in night conditions

(Continued)

Table 6.2 (Continued)

System group	Examples	Comfort	Information	Safety	Efficiency	Driving problem or error
Parking aids	Distance sensors		X			Misperception of some obstacles around the vehicle
	Rear camera		X	X		Misperception of some obstacles around the vehicle
	Automatic parking assist	X				Repetitive task
Driver monitoring	Tiredness or drowsiness detection		X	X		Inadequate driver psycho-physical state
	Alcohol-lock system		X	X		Driver in bad conditions for driving after drinking alcohol
Vehicle state monitoring	Remote diagnosis		X	X		Damaged vehicle systems
	Tire pressure monitoring		X	X		Inadequate tire pressure
Connectivity	Fuel level, distance traveled, …		X			
	Maintenance program		X			
	Location		X			
	Comfort systems remote control (climate control)	X				
	Connection to smartphone apps		X			
	Fleet management applications		X			
Comfort	Climate control	X				
	Seats heating	X				
	Seats and headrest position configuration	X				
Value added services	WiFi access	X	X			
	Multimedia	X	X			
	Voice recognition function	X				

Improving the visibility is essential so that the driver can obtain complete and reliable information about the environment. Although many of the systems in this group are very focused on the automation of repetitive and simple tasks to discharge the driver, the most advanced systems act on the system of illumination of the vehicle to adapt the beam to the characteristics of the road, mainly in curves and intersections where the requirements are not equal to those required for straight stretches. In the same way, systems are introduced that detect vehicles running in the opposite direction to modify the light beams and to avoid blinding other users.

In terms of the tasks of driving under normal conditions regarding speed and trajectory control, the new systems try an automation of the most repetitive functions. Thus, starting from the simple speed limiters, these systems have evolved to the cruise control (CC) that only maintains a target speed, the adaptive cruise control (ACC) that also maintains a safety distance with the vehicle in front of him, or the intelligent speed adaptation system (ISA) that evaluates the best speed regarding the conditions of the road, its geometry, traffic, type of vehicle, etc. (Aparicio et al., 2005; Carsten and Tate, 2005). Taking advantage of the information available in digital maps, some systems recommend the speed that provides optimum fuel consumption for a predefined travel time. Other systems on the same line simply assess whether driver behavior is efficient and provide warnings to follow ecodriving patters (Hellström et al., 2009). In relation to the task of lane keeping, the first systems alerted the driver when they exceeded or were about to cross the line that delimits the lane if they had not activated the turn signal (Visvikis et al., 2008). At present, many new vehicles incorporate systems that automatically correct the direction to keep the vehicle in the lane.

The early obstacles detection is a key aspect of autonomous driving but is already provided in the assistance systems in order to alert the driver of the presence of other vehicles, pedestrians, etc. in their trajectory, acting on the controls of the vehicle if the driver does not respond (Jiménez et al., 2012; Jiménez et al., 2015). This detection of obstacles extends to detection in front of the vehicle, in the visual field of the driver, but blind spots are also monitored. To improve the perception in night conditions, systems are also being implemented specifically designed for these situations where the conventional perception or the own vision of the driver are limited.

The assistance systems are also present in simple maneuvers, such as parking maneuvers, from warnings of the proximity of obstacles to the execution of the maneuver fully automatically (Endo et al., 2003).

The assistance systems also include monitoring of the state of the driver, trying to detect situations in which the driver is not in the proper conditions to drive, such as fatigue, cognitive or visual distraction, or drowsiness. To do this, the most advanced systems merge information from various sources, such as driving parameters, head and eyes tracking with computer vision, etc. (Bergasa et al., 2006).

On the other hand, the incorporation of all the electronic systems allows the periodic verification of their proper operation. In addition, this task is essential in order that the vehicle knows at all times that its powertrain, safety, etc. systems work satisfactorily. This diagnosis can be performed on several levels. The most basic one comprises detecting the malfunction of a system or sensor and providing the warning to the driver. For a deeper diagnosis, special equipment used in the workshop is needed. The most advanced systems perform this deep diagnosis remotely so that the vehicle sends the operating information to a central unit (Mazo et al., 2005).

Finally, the current vehicles come equipped with various functions focused specifically on improving comfort and connectivity applications, from which the driver can connect with the vehicle and exchange information, give orders, know its location or its maintenance program, etc. A special case is the fleet management system that can be implemented on this architecture for information exchange (Bieli et al., 2011).

It should be noted that the systems and services derived from vehicle-to-vehicle and vehicle-to-infrastructure communications have not been addressed here (they are described in Chapter 7: Cooperative Systems), but they undoubtedly extend the potential for new applications and support to those already implemented with onboard sensors (Jiménez et al., 2013; Anaya et al., 2015). In addition, only systems with the lowest levels of automation have been addressed, since autonomous driving will be dealt with in more detail in Chapter 8, Cooperative Systems (European Technology Platform on Smart Systems Integration, 2015).

6.3 ELECTRONIC AIDS FOR REDUCING ACCIDENTS CONSEQUENCES

6.3.1 Secondary Safety Systems

Passive safety systems, such as the seat belt and the airbag, are really effective when used correctly and the occupant is in the right position. However, there are problems in situations when the occupant is out of position or when generalizing the action of such systems to all types of occupants, regardless of their weight, height, etc. In relation to the occupants, their age and anthropometric characteristics pose different protection requirements. Therefore, the main work in secondary safety is focused on specific groups of people (vulnerable users, elderly, children, etc.), as well as providing a more accurate response to each specific situation according to the occupants, type of collision, severity, etc. (Chute, 2001).

Adaptive restraint systems are different from conventional ones in the fact that they try to assess by means of sensors the collision circumstances so that, in the event of an accident, they act in the most favorable way according to those conditions. Therefore, for proper operation, it is necessary to monitor certain variables using appropriate sensors.

In this way, an adaptive retention system is one that is able to adapt its characteristics and behavior to the type of collision and the occupant (MacLennan et al., 2008). Their goal is to reduce injuries under any kind of conditions. Examples of such an adaptation in the case of airbags are the control of the power of deployment depending on the severity of the collision and the height of the occupant, as well as the disconnection if the occupant is too close to the device at the time of the deployment.

Thus, these adaptive systems must be controlled by electronic systems that have to perform the characterization functions of the occupant and the position, characterization of the impact, decision of the activation sequence of the systems, and deployment during and after the impact.

6.3.2 Interaction Between Primary and Secondary Safety Systems

The current trend towards integration of primary and secondary safety measures is due to the flow of information between vehicle systems, which may be available to all if necessary (EEVC−WG 19, 2004). The high integration of functions leads to improvements in safety superior to the sum of the contributions of each of them, since it increases the individual functionality of the systems, as well as reducing the false alarms. Detection by a single sensor has limitations inherent in the characteristics of that sensor. However, the combination of other information sources reinforces redundant data and provides new ones.

The interaction between primary and secondary safety systems is based on the fact that the time between the detection of the risk and the collision is usually higher than the reaction time of most protection systems, so they can be prepared in advance. If they have the appropriate information, they can change from a "comfort mode" to a "safe mode."

Table 6.3 Classification of precollision systems

Actions oriented to accident severity reduction	Reduce impact speed
	Change impact zone
	Change impact orientation
Geometric and structural modifications for increasing compatibility	Extensible front bumper
	Vehicle height modification
	Hood elevation
Optimization of performance of restraint system	Seat belt pretensions
	Airbag deployment adapted to collision and occupants
	Seat position adjustment
Other systems	Retractable steering column
	Sunroof automatic closure
	Doors automatic unlock system
	External airbags

Table 6.3 shows the precollision systems grouped according to the final orientation of the system, distinguishing those that are aimed at reducing the severity of the collision, increasing the compatibility (mainly in the geometric or structural issues adapting the height and stiffness, since the incompatibility of masses is complicated to be solved), or improving the performance of retention systems (with early activation).

6.3.3 Tertiary Safety Systems

Finally, tertiary safety systems are aimed at reducing the consequences after the accident. The most commonly known example is the emergency call or eCall, which seeks the best care of victims, both in first aid and evacuation and immediate and specialized assistance. It is a system that can be activated by the occupants of the vehicle or do it automatically from the signals of sensors onboard. Thus, the vehicle may contact the assistance services (by activation of the driver or by the vehicle itself) in case of an accident and provide information useful in the rescue phase. Its importance lies in the fact that, at present, infrastructure operators and authorities rely on the notification of accidents through the telephone (fixed and mobile) or through data from video cameras, radar, or other similar equipment. In the special case of telephone calls, the location of the accident or incident cannot be accurately determined in many cases. Although the basic configuration focuses on transmitting the position primarily, in order to provide faster and more efficient attention, the vehicle involved in the accident should be able to emit the following information, evidently conditioned by the sensors available in the vehicle:

- Location of the accident (by means of satellite positioning).
- Direction of the vehicle at the time of the accident (critical aspect on motorways and highways).
- Time of accident.
- Vehicle data such as manufacturer, model, version.
- Vehicle condition: fire, fuel leaks, submerged.
- Call activation mode specification.
- Status of occupants: number of occupants, characteristics, condition of the occupants.
- Monitoring of vital signs through advanced biometric sensors.
- Severity of injuries: collision type and severity, impact speed, direction and trajectory followed, use of the seat belts.

Thus, if this information is collected by specific centers (note the need to filter false alarms) and by emergency services, shortening of the response time of services can be achieved and the necessary resources in each case depending on the accident could be mobilized. In this way, the severity of the injuries is reduced, the probability of survival is increased and the accident management is improved through this rapid response.

Another system encompassed within tertiary safety is data logging, information that is useful for manufacturers, accident reconstruction, and emergency services. In that sense, systems that store only vehicle information are considered, while others also consider video recording, for example.

In addition, there are other systems aimed at reducing the negative effects of the accident, reducing successive accidents or avoiding situations of greater risk derived from the state of the vehicles. For example, fuel cut-off, battery power cut-off, automatic door unlocking, or emergency signals automatic activation can be highlighted.

6.4 FUTURE EVOLUTION OF ASSISTANCE AND SAFETY SYSTEMS

The evolution of vehicles tends to result in more and more information and this information has higher quality and precision coming from onboard sensors, positioning, wireless communications, etc. This fact increases the systems' ability to inform the driver and even act automatically.

In this way, the evolution of these systems is leading to driving automation (Meyer and Beiker, 2014; Kala, 2016), although there are clear leaps as presented in the next chapters. However, it is understood that the introduction of autonomous driving passes through the successive implementation of assistance systems that are supported by one another. Thus, the synergy between all the information channels and the processing and decision units of the vehicles results in a more precise understanding of the environment and a more reliable decision making. This reliability can be fostered to face dynamic situations and more complex scenarios. Specifically, advancing concepts from the following chapters, systems called assistance systems would be classified in levels 1 and 2 of automation (Driver Assistance and Partial Automation, respectively), while the more advanced automated driving functions are associated with levels 3, 4, or 5 (Conditional, High, and Full Automation, respectively) (SAE, 2016).

Cooperative systems will also play an important role in considering the interaction of the different vehicles with each other and with the road, which leads to a much higher flow of information, extending the "visual horizon" to a larger "electronic horizon" (Jiménez and Aparicio, 2008). Each vehicle receives information, processes it, and sends it. The advantages can be appreciated in the fields of safety, environmental protection, and transport efficiency. In this way, the vehicle, in addition to possessing its own data and perceiving its surroundings by means of onboard sensors, can receive information from other vehicles, infrastructure, or traffic centers, as well as having accurate positioning. In addition, this vehicle is a source of information, which can be transmitted abroad (infrastructure, other vehicles, control centers, etc.). The communication between vehicles allows each user to anticipate certain situations of risk, offering greater times to the driver to adapt to the new conditions. Common cases that can be addressed through these communications are collision avoidance systems, intelligent intersections, crashes with stopped vehicles on the roadway, etc. Note that these cooperative systems allow responding to demands at all stages of the integrated safety model presented above. They are also considered as catalysts for the implementation of autonomous driving (Gill et al., 2015), since it has been shown that autonomous vehicles without external communications and only using the information provided by their onboard sensors are ineffective (and sometimes incapable) of addressing complex driving scenarios.

Related to secondary safety systems, current trends look for the optimization of systems responses considering impact, occupants, and opponents characteristics, so new sensors are to be considered in the vehicle, and active safety systems are also used for information acquisition.

REFERENCES

Anaya, J.J., Talavera, E., Jiménez, F., Serradilla, F., Naranjo, J.E., 2015. Vehicle to vehicle geonetworking using wireless sensor networks. Ad Hoc Netw. 27, 133—146.

Aparicio, F., Arenas, B., Gómez, A., Jiménez, F., López, J.M., Martínez, L., et al., 2008. Ingeniería del Transporte. Ed: Dossat. Madrid (in Spanish).

Aparicio, F., Páez, J., Moreno, F., Jiménez, F., López, A., 2005. Discussion of a new adaptive speed control system incorporating the geometric characteristics of the roadway. Int. J. Veh. Autonom. Syst. 3 (1), 47—64.

Bergasa, L.M., Nuevo, J., Sotelo, M.A., Barea, R., Lopez, M.E., 2006. Real-time system for monitoring driver vigilance. IEEE Trans. Intell. Transp. Syst. 7 (1), 63—77.

Bielli, M., Bielli, A., Rossi, R., 2011. Trends in models and algorithms for fleet management. Proc. Soc. Behav. Sci. 20, 4—18.

Burton, D., Delaney, A., Newstead, S., Logan, D., Fildes, B., 2004. Evaluation of antilock braking systems effectiveness. 04/01. Noble Park North, Victoria: Royal Automobile Club of Victoria (RACV) Ltd.

Carsten, O., Tate, F., 2005. Intelligent speed adaptation: accident savings and costbenefit analysis. Acc. Anal. Prevent. 37 (3), 407—416.

Clute, G., 2001. Potentials of adaptive load limitation presentation and system validation of the adaptive load limiter. In: Proceedings of the 17th International Technical Conference on the Enhanced Safety of Vehicles, Amsterdam, Holland.

Corben, B., Logan, D., Fanciulli, L., Farley, R., Cameron, I., 2009. Strengthening road safety strategy development "Towards Zero" 2008-2020 - Western Australia's experience scientific research on road safety management SWOV workshop 16 and 17 November 2009. Saf. Sci. 48, 2010, 1085—1097.

EEVC—WG 19, 2004. Primary and secondary interaction. Final report.

Elvik, R., Vaa, T., 2004. The Handbook of Road Safety Measures. Elsevier.

Endo, T., Iwazaki, K., Tanaka, Y., 2003. Development of reverse parking assist with automatic steering. In Proceedings of 10th World Congress and Exhibition on Intelligent Transport Systems and Services, Madrid, Spain, 16—20 November 2003.

European Technology Platform on Smart Systems Integration — EPoSS. 2015. European Roadmap Smart Systems for Automated Driving. Berlin: EPoSS.

Gill, V., Kirk, B., Godsmark, P., Flemming, B., 2015. Automated Vehicles: The Coming of the Next Disruptive Technology. The Conference Board of Canada, Ottawa.

Hellström, E., Ivarsson, M., Åslund, J., Nielsen, L., 2009. Look-ahead control for heavy trucks to minimize trip time and fuel consumption. Control Eng. Pract. 17, 245—254.

Hobbs, F.D., 1989. Traffic Planning and Engineering. Pergamon Press, Oxford.

Jiménez, F., 2013. Libro verde de los Sistemas Cooperativos, ITS España- Universidad Politécnica de Madrid (in Spanish).

Jiménez, F., Aparicio, F., 2008. Aportación de los ITS a la sostenibilidad y mejora del transporte por carretera. Dyna Ingeniería e Industria 83 (7), 434—439.

Jiménez, F., Naranjo, J.E., Gómez, O., 2012. Autonomous manoeuvrings for collision avoidance on single carriageway roads. Sensors 12 (12), 16498—16521.

Jiménez, F., Naranjo, J.E., Gómez, O., 2015. Autonomous collision avoidance system based on an accurate knowledge of the vehicle surroundings. IET Intell. Transp. Syst. 9 (1), 105—117.

Kala, R., 2016. On-Road Intelligent Vehicles. Motion Planning for Intelligent Transportation Systems. Elsevier.

MacLennan, P.A., Ashwander, W.S., Griffin, R., McGwin Jr., G., Loring, L.W., 2008. Injury risks between first- and second-generation airbags in frontal motor vehicle collisions. Acc. Anal. Prevent. 40, 1371—1374.

Mazo, M., Espinosa, F., Awawdeh, A.M.H., Gardel, A., 2005. Diagnosis electrónica del automóvil. Estado actual y tendencias futuras. Fundación Instituto Tecnológico para la seguridad del automóvil (FITSA), Madrid, in Spanish.

McQueen, B., Schuman, R., Chen, K., 2002. Advanced Traveller Information Systems. Artech House.

Meyer, G., Beiker, S., 2014. Road Vehicle Automation. Springer, Switzerland.

Organisation for Economic Co-operation and Development, 2003. Road safety. Impact of new technologies. OECD.

Paine, M., 2005. Electronic Stability Control: Review of Research and Regulations. Vehicle Design and Research Pty Limited, Beacon Hill, New South Wales, p. G248.

SAE, 2016. Taxonomy and Definitions for Terms Related to On-Road Motor Vehicle Automated Driving Systems, s.l. SAE International.

Visvikis, C., Smith, T.L., Pitcher, M., Smith, R., 2008. Study of lane departure warning and lane change assistant systems. Transp. Res. Lab.

CHAPTER 7

Cooperative Systems

Ana Paúl[1], José A. Fernández[1], Felipe Jiménez[2]
and Francisco Sánchez[1]
[1]CTAG – Centro Tecnológico de Automoción de Galicia, Porriño, Spain
[2]Universidad Politécnica de Madrid, Madrid, Spain

Contents

7.1 INTRODUCTION

The use of cooperative systems able to exchange information among vehicles (V2V) and infrastructure (V2I) appears to be one of the most promising solutions to improve significantly road traffic safety and efficiency.

Intelligent Vehicles
DOI: http://dx.doi.org/10.1016/B978-0-12-812800-8.00007-2
227

In the late 1980s and early 1990s, researchers from the PROMETHEUS European project already visualized and started to work on this type of technology within the COPDRIVE, PRO-COM, and PRO-NET subprojects. However, the enormous limitations and technical constraints in terms of positioning, communication, computing, embedded security systems, vehicular networks, or standardization processes did not allow the significant advances, which were necessary at that time, to develop the basic technologies (Fig. 7.1, Fig. 7.2).

Nowadays, more than 20 years later, all these fields have experienced an exponential change that allows the complete overcoming of the old technological barriers. The concept of cooperative systems has been successfully evaluated in many European projects, pilots, and field operational tests (FOTs) such as CVIS, SafeSpot, Coopers, DRIVE-C2X, FOTsis, COMPASS4D, CO-GISTICS, or SCOOP@F. Additionally, the European Union has assigned a frequency band from 5.875 to 5.905 GHz

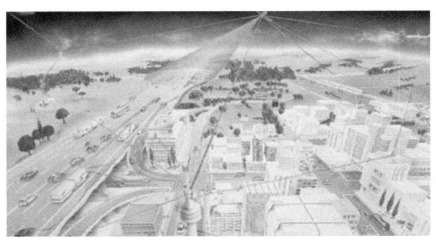

Figure 7.1 Cooperative communication vision of PROMETHEUS European Project.

Figure 7.2 Cooperative communication tests in PROMETHEUS European project.

devoted to cooperative applications (IEEE 802.11p). Different initiatives and bodies like COMeSafety, Car2Car Communication Consortium, C-ITS Platform, and C-ROADS have established a common framework for the development and implementation of cooperative systems.

The evolution of mobile communication technologies in the path towards 5G allows nowadays the development of the LTE/V technology as a potential alternative candidate communication solution for cooperative services. In this sense, at this very moment, there are different projects and activities analyzing the possibilities, limitations, and capacities of each technology and exploring as well hybrid communication approaches.

Nevertheless, independently of the technological approach selected (IEEE 802.11p, 3G/4G, LTE-V2X, 5G, or hybrid), the proper time for the deployment of C2X systems seems to have arrived. In this sense, at this moment several cooperative mobility pilots in urban and interurban environments are strongly contributing to this deployment and are summarized in Sections 7.5.1 and 7.5.2. Additionally, the most relevant collaborative initiatives and platforms supporting the deployment of cooperative systems are introduced in Section 7.5.3.

Finally, it is important to mention the relevance, in the near future, of C2X communication services to extensively support and improve the behavior and capabilities of upcoming automated vehicles.

7.2 C-ITS FRAMEWORK

7.2.1 General Architecture

An ITS station is a functional entity that provides an Intelligent Transport System (ITS) with communication capability dedicated to transportation scenarios.

Cooperation among road traffic participants and between road traffic participants and roadside infrastructure is based on different communication technologies, and information can be communicated between vehicles, from a vehicle to a roadside infrastructure element, and between vehicles and backend applications. This is what is conventionally known as C-ITS.

Based on ETSI considerations on ITS communication architecture, a C-ITS system architecture should comprise the following basic components:

- Vehicle ITS Station (VIS) (AKA On Board Unit (OBU))
  A VIS is equipped with communication hardware for information exchange with other vehicles or with roadside infrastructure.

The communication hardware is connected to the vehicle onboard network to collect data within the vehicle. As a result, vehicle data can be exchanged between vehicles. The communication hardware may also support wireless Internet access in order to communicate with services running on top of a central component, thus allowing information from the vehicle to be sent immediately to the central system.

- Roadside ITS Station (RIS) (AKA Road Side Unit (RSU))
  Some typical examples of RIS are the Variable Message Signs or traffic lights equipped with communication hardware. This way, the RIS can communicate with vehicles, e.g., to send information to vehicles or to act as relay stations for (multihop) communication between vehicles. Additionally, an RIS may be connected to the Internet. This enables an RIS to communicate with central components, and to forward information received from vehicles to central components.

- Central ITS Station (CIS)
  A CIS is typically an organizational entity, where centrally managed applications and services are operated. This can be, e.g., a traffic management center controlling roadside infrastructure, a fleet management center, or an advertisement company that distributes location-based advertisement (via the roadside components) to the vehicles. Vehicles or roadside infrastructure components may send information to the central component, and the central component may, in turn, send information to vehicles or roadside infrastructure components.

- Personal ITS Station (PIS)
  A PIS represents mobile consumer devices, such as mobile phones or navigation devices, which may provide numerous ITS applications. Typically these devices are personal and use appropriate communication hardware. The devices can also support cooperative ITS applications based on communications with other road users or roadside infrastructure. A wide variety of ITS applications may arise also from communications with backend applications. From an architectural perspective a PIS device, which is integrated or connected to a vehicle should be considered as part of the vehicle equipment. In this case the personal device can provide additional information, generated internally (e.g., navigation data), or received from other sources via its communication facilities (e.g., HMI device). Furthermore, a vehicle may directly use the communication facilities of a connected personal device and may also make use of its output devices to the vehicle owner.

In general, a cooperative system does not necessarily include all the components but may include a subset of the components (depending on the deployment scenario and the functions). These four components are able to communicate with each other using several communication networks. According to the requirements of the different use cases, communication can be performed either directly within the same communication network, or indirectly across several communication networks, providing in any case an overall support for all the implemented functions related to improving traffic safety and efficiency levels.

7.2.2 Support Technologies

Assuming wireless communication as a key enabler for C-ITS deployment, there are a number of access technologies based on this concept that will support C-ITS applications.

Also, anticipating that new access technologies will appear on a regular basis, any of them should be able to:

- Operate in a very dynamic environment with high relative speeds between transmitters and receivers.
- Support extremely low latency in the safety-related applications.
- Tolerate the high load generated by the periodic transmission of multiple messages by multiple actors, and the high vehicle density typical of congested traffic scenarios.
- Consider that part of V2x messages (especially V2V) are local in nature, meaning they are most important to nearby receivers.

Respecting these principles, two are the ones nowadays forecasted to cover C-ITS needs:

- Ad hoc broadcast V2X technology (ITS G5 short range WLAN).
- Mobile broadband (3G/4G/5G cellular networks).

These two technologies are in line with the ISO and ETSI common approach for ITS station reference architecture. Described in their respective standards ISO/IEC 21217 and ETSI EN 302 665, this description becomes the core of the set of Cooperative ITS standards and divides the Intelligent Transport Systems into the four different types of ITS subsystems previously mentioned.

The ITS station reference architecture accommodates a diversity of ITS stations and provides a diversity of communications means and technologies. The architecture unifies the technologies so that it can be of benefit to a diversity of applications (road safety, traffic efficiency, and comfort/mobility) that are agnostic to the communications technologies.

The architecture is defined in such a way that Cooperative ITS standards can be deployed using any existing and forthcoming access technologies so they can be added without changing the existing Cooperative ITS standards as long as an interface complying with the ITS station standards is developed. ITS applications in turn would benefit from the newly available access technologies as communication profiles are defined according to the communication requirements (expressed in a technology agnostic way) of the ITS applications and the current capabilities of the ITS station.

Here the overall descriptions of the two mentioned technologies are provided, depicting the most relevant considerations concerning their final exchange usage.

7.2.3 Public Land Mobile Networks (Cellular Networks)

• Overview and considerations for C-ITS support
  The mobile telephony systems have evolved from voice/SMS/GPRS to mobile broadband supported by modern smartphones. 3G WCDMA is the first generation of mobile broadband and has good coverage in populated areas. It has low latency and can be used to exchange messages between vehicles for many C-ITS use cases.

  From an ITS perspective, the improved latency of WCDMA and LTE is of particular interest. Research projects have shown that car-to-car messages and warnings can be transported via the mobile network in 350–500 ms using WCDMA and below 50–150 ms using LTE. The existing infrastructure can be reused, no changes in the networks are needed, and as additional requirements appear new application servers can be added in the mobile network and/or on the Internet level to enable geographical addressing (geo-casting) and to optimize message streams for the cellular distribution channel. Also the upcoming LTE broadcast system can be used for distribution.

  4G LTE is a much improved mobile broadband being deployed now. The coverage of mobile broadband will expand rapidly as the new spectrum with good propagation characteristics will be deployed and existing spectrum licenses will become "technology neutral." This means that the existing 2G GSM towers and spectrum can be reused and this will greatly speed up the network transformation. 4G LTE Advanced is a further developed version meeting the global 4G requirements set by ITU-R (International Telecommunication Union—Radio communication).

5G is the next generation of mobile network technology. It is under definition now with an expected deployment in around 2020. The performance target is to make it "ten times better" than 4G in many categories including latency and reliability and it is designed to support all aspects of the "Internet of Things."

As far asWith regards to cars and trucks, increasing numbers are getting connected via mobile networks for many different services; from a C-ITS aftermarket perspective that means that many C-ITS use cases can be supported by smart apps (at least to receive messages and warnings). There is a great potential for C-ITS to use the infrastructure investments in the mobile networks and the mass market of smartphones.

The core part of the mobile networks is now being evolved from circuit switched to packet switched technology using Internet Protocol. This enables a whole range of new possibilities in multimedia communication, secure communication and encryption, setting Quality of Service, and differentiated billing.

However, it is not only base stations and antennas that make up a mobile network. There are many interacting layers that perform the services. From an ITS perspective, an efficient multiservice platform is needed where all stakeholders, service providers, and those needing information from the traffic system have access via simple interfaces that provides a common technology platform with economy of scale and efficiency. The networks have functions to handle the connectivity ("SIM cards"), the software in the mobile device, the services and information, and the interfaces to more or less integrated services and providers. Correct handling of the rights to data and media must be in place and a cost-efficient payment and clearing system is required as well.

- Deployment and cost structure
Depending on the choices of cellular operators, technology, country, and región, the coverage will vary. In almost every car, the driver has a mobile phone today that can have navigation and ITS applications. There is no backend server system for ITS applications deployed yet but the rest of the infrastructure for V2X is under deployment in certain countries.

For mobile communication, a standardization of the cooperative functions is needed. Backend architecture via different backend providers, different traffic management centers, different OEM backends for safety is missing until now.

The market development of connected car services is likely to result in such attractive services for users and the stakeholders of the value chain, that the penetration rate of built-in cellular connectivity (or closely integrated smart phones) in new vehicles, will reach close to 100% in a few years. These vehicles can then support V2X services based on the DENM-messages.

The cost for spectrum, investment in radio network HW/SW, operational cost, and maintenance cost is paid by subscription fees and transaction-based fees or flat fees for the general (smart phone) usage of the mobile systems. For new additional V2V, V2I, and I2V applications there will be subscription costs and communication costs similar to smartphones to cover investment and operations. But note that many different business models are possible, enabled by the systems capabilities to flexibly bundle the communication cost into the complete safety or efficiency service. Data rates needed for V2X (max. 1MB/s) are relatively small compared to common applications used now on smartphones.

7.2.4 ITS G5 (Vehicular Wi-Fi)

- Overview and considerations for C-ITS support
  ITS G5, also known as IEEE 802.11p (and DSRC in the US) is a short range WLAN standard developed for ad hoc broadcast communication between vehicles and to the road side infrastructure. The specification is derived from the well-known Wi-Fi specification (IEEE802.11), specially adapted to the vehicular environment, supporting high driving speeds and low latency requirements. It operates in a dedicated spectrum on 5.9 GHz. In the US this technology is called Wireless Access in Vehicular Environments (WAVE) and in Europe ITS-G5.

  Radio communication systems in the 5 GHz range can today offer communications with a high data rate, ranges typically 300–500 m and up to 1000 m in direct LOS, low weather-dependence, and global compatibility and interoperability for ITS communication.
- The connectivity required by the applications can be summarized as:
  - Inter-Vehicles Communications (V2V) (this includes multihop routing involving several vehicles).

- Vehicle to Roadside (uplink) V2I and Roadside to Vehicle I2V (downlink).

ITS-G5 is designed to support time critical road safety applications where fast and reliable information exchange is necessary. ITS-G5 communication is the kind of communication technology which currently fulfils real-time requirements in the best way, because of direct communication with very low latency in a highly dynamic ad hoc network (15–100 ms) as it has been tested in the different FOTs that have been carried out during the last 10 years both in urban and interurban environments (e.g., COMPASS4D, DRIVE C2X, SCOOP@F,...)

No roaming or translation between different providers is necessary. Additionally, a physical "prioritization" is built into any direct communication, because the closer the communications partner, the faster and smaller is the probability of errors in the communication. This schema fits very well with the intended safety applications.

The short-range ad hoc network has no "server" or similar that keeps track of the vehicles' positions in order for the communication to be routed to the recipients in the vicinity in the V2V and V2I mode. But for traffic management centers to reach out a backend is needed and cross-border and cross-road operator coordination will be needed in many places.

- Deployment and cost structure

Different European automotive OEMs have stated that they will equip their cars with this technology. The aim is to approach 100% penetration in motor vehicles within the coming 15–20 years.

Up-front costs are foreseen for the vehicle buyer to provide ITS G5 onboard equipment, but no costs for data transfer or spectrum. There will be small lifetime costs in the context of vehicle software/data provisioning and update and periodic technical inspection.

Investment in roadside ITS stations (HW + SW) are necessary, the integration into existing infrastructure systems is possible and reduces the investments into civil works, electrical power, internet connection, backend servers as well as operation and maintenance cost. It is assumed that the road operators will make the investment.

The number of R-ITS-S depends heavily on the use case. For a roadworks warning only the warning trailers need R-ITS-S equipment, power needs to be available, and mobile networks are used for back-haul communication with traffic management centers. It is expected that existing traffic centers can handle the incoming data so no new backend infrastructure is needed. Classical traffic center tasks are supported by the probe vehicle data, so with a rising equipment rate the need for classical traffic flow detectors will go down, so that a counter funding is possible if the equipment rate reaches >30%.

Overall fixed costs are foreseen per road side unit for Wi-Fi capacity as well as annual maintenance costs. The average lifetime of the roadside units is expected to be around 10 years.

7.2.5 Standardization Level

The European Commission, through Mandate M/453, invited the European Standardization Organizations (ESOs, e.g., CEN, ISO, ETSI, . . .) to prepare a coherent set of standards, specifications, and guidelines to support European Community-wide implementation and deployment of Cooperative ITS (C-ITS) to warrant both interoperability by means of defining a common policy of data definition and a common framework for security and privacy aspects.

This set of standards fills in the ISO/OSI reference model where four horizontal protocol layers and two vertical layers depict a C–ITS station architecture from a protocol stack point of view.

Horizontal layers:
- *ITS Access Technologies* cover various communication media and related protocols for the physical and data link layers. The access technologies are not restricted to any particular type of media, though most of the access technologies are based on wireless communication (e.g., *ITS G5, cellular network*, . . .)
- *ITS Network and Transport* comprises protocols for data delivery among ITS Stations, and from ITS Stations to other network nodes, such as in the Internet. ITS network protocols particularly include routing of data from source to destination through intermediate nodes, and efficient dissemination of data in geographical areas (e.g., *Geonetworking*). ITS transport protocols provide end-to-end delivery of data, and depending on requirements of ITS facilities and applications, additional services, such as reliable data transfer, flow control, and congestion avoidance. A particular task of ITS Network and Transport is the usage of Internet protocol IP version 6. This includes the integration of IPv6 with specific network protocols for communication among ITS Stations, dynamic selection of ITS access technologies and handover between them, as well as interoperability issues of IPv6 and IPv4.

- *ITS Facilities* are a collection of functions to support applications for various tasks. The facilities provide data structures to store, aggregate, and maintain data of different types and sources (such as from various vehicle sensors and from received data from communication, e.g., *CAM, DENM, MAP, SPAT, . . .*). As for communication, ITS facilities enable various types of addressing modes to applications, provide ITS specific message handling, and support establishment and maintenance of communication sessions. An important facility is the management of services, including discovery and download of services as software modules from the ITS Application Service System and their management in the ITS Station.
- *ITS Applications* refer to the different applications and functions.

Vertical layers:
- *ITS Management* is responsible for configuration of an ITS Station and for cross-layer information exchange among different layers.
- *ITS Security* provides security and privacy services, including secure message formats at different layers of the communication stack, management of identities and security credentials, and aspects for secure platforms (firewalls, security gateway, and tamper proof hardware).

From a deployment perspective, the general needs regarding the topic of C-ITS standardization goes through:
- The need for elaboration of test standards.
- The need for profiling of standards to ensure interoperability in implementation.
- The need for proper maintenance of standards due to implementation needs.

In order to meet the abovementioned needs a first step was taken in the C-ITS platform launched by EC: to collect an overview of the standards that are being used within C-ITS deployments initiatives in Europe. This exercise was only focused on C-ITS deployments that are currently being implemented or are going to be deployed in the next 2—3 years so they become relevant for interoperability in Europe for this period of time, not in 5 years or even longer periods of time. The next step should go through laying the necessary foundations to further discuss how profiles can and have to be defined for EU-wide interoperable C-ITS deployment in the future by means of analyzing actual profiling of standards and the elaboration of test standards, which have not been the main targets of this exercise yet.

7.3 SERVICES

7.3.1 Introduction

The purpose of cooperative systems is to implement a wide range of applications and services to improve safety, efficiency, comfort, and reduce the environmental impact of transport through the exchange of information between vehicles, infrastructure, and control centers. This information allows a broadening of the horizon that it is possible to "see" in order to adopt actions. However, a distinction must be made between connected driving, where data can be used to make decisions, and cooperative driving where vehicles communicate to make decisions together. Undoubtedly, the first step is essential for the second one and many of the systems that will be presented later are framed in the first group.

These systems can bring significant improvements over the current situation. In recent years, several services and systems have been developed based on the mentioned communications, some at prototype level and others already have been implemented in real situations, which have proven their effectiveness. However, a global approach to the problem is difficult to consider, and only partial approaches to the feasibility of the services implementation have generally been undertaken.

The main qualitative leap between the conventional vehicle, which includes its own systems, and the vehicle connected in a cooperative environment is that, in addition to possessing its own data and perceiving its surroundings by means of onboard sensors, it can receive information from other vehicles, infrastructure, or traffic centers.

Thus, in cooperative systems between vehicles or between vehicles and infrastructure three levels of cooperation can be established:

- The vehicle receives information from the road or other vehicles with which it is possible to anticipate situations (retentions, works zones, adverse weather, etc.), that the driver cannot perceive.
- The infrastructure receives information from the vehicle, which allows it to know traffic variables and to manage the network more efficiently.
- Information is exchanged in both directions, since vehicle and infrastructure are sources of information, and this fact increases information and its quality, benefiting both parties.

Cooperative systems (C-ITS) are considered the next qualitative leap in vehicle technology and are a priority. In order to promote the use of these C-ITS, the European Commission (DG-MOVE) created in

Table 7.1 C-ITS Day-1

Service	Communications type	Application type
Emergency electronic brake light	V2V	Safety
Emergency vehicle approaching	V2V	Safety
Slow or stationary vehicle(s)	V2V	Safety
Traffic jam ahead warning	V2V	Safety
Hazardous location notification	V2I	Motorway
Roadworks warning	V2I	Motorway
Weather conditions	V2I	Motorway
In-vehicle signage	V2I	Motorway
In-vehicle speed limits	V2I	Motorway
Probe vehicle data	V2I	Motorway
Shockwave damping	V2I	Motorway
GLOSA/Time to Green (TTG)	V2I	Urban
Signal violation/intersection safety	V2I	Urban
Traffic signal priority request by designated vehicles	V2I	Urban

November 2014 the Platform for the deployment of Cooperative Intelligent Transport Systems (C-ITS Platform) with the intention of giving a unified vision to the services derived from this technology at the European level. Thus, in the January 2016 report (C-ITS Platform, 2016), the C-ITS Platform defined the implementation roadmap for Cooperative Systems in Europe, starting with a series of basic services called C-ITS Day-1 (Table 7.1), followed by Day-1.5 services (Table 7.2), ending with Cooperative Autonomous Driving for 2030 (OECD, 2015). In this way, the C-ITS are considered by the European Commission as catalysts for the deployment of autonomous vehicles and the first step towards autonomous cooperative driving, promoting the integration with automatic control systems.

In the next subsections, the most representative services, both included in Day-1 or Day-1.5, and systems not referred to in those lists, are briefly described. A more detailed description can be found in Jiménez et al. (2012).

7.3.2 Systems Oriented to Information Provision

Services within this category can result in a number of benefits such as indirect improvement of safety or efficiency, as they seek to warn the driver of special situations on the road so that the driver is alerted and informed, and can act accordingly. In addition, some of them can

Table 7.2 C-ITS Day-1.5

Service	Communications type	Application type
Off-street parking information	V2I	Parking
On-street parking information and management	V2I	Parking
Park & Ride Information	V2I	Parking
Information on AFV fueling & charging stations	V2I	Smart routing
Traffic information and smart routing	V2I	Smart routing
Zone access control for urban areas	V2I	Smart routing
Loading zone management	V2I	Freight
Vulnerable road user protection (pedestrians and cyclists)	V2X	VRU
Cooperative collision risk warning	V2V	Collision
Motorcycle approaching indication	V2V	Collision
Wrong way driving	V2I	Wrong way

be considered as value-added services that make driving a more comfortable activity. Some of these services are described below.

Information to the driver before the trip
The information to the driver before the trip allows the driver to plan in a more efficient way. Within this planning some issues can be considered such as route selection, schedule selection, stops planning, etc. For a correct choice, the information must be accurate and up to date and this information must be complemented by information during the trip, as there may be changing conditions (traffic conditions, weather, etc.) that condition the initial selections.

In addition to this information, a wide variety of information on nearby places of interest, hotels, restaurants, petrol stations, parking lots, connection possibilities with public transport, etc. can be provided.

Information to the driver during the trip
The information along the route allows the driver to improve the information available before the trip and that can change over time. It requires to be updated, and its rapid transmission is critical.

These information systems, although they are considered as systems to promote informed driving, have clear connotations with safe and efficient driving. For example, it is intended to warn drivers about particular conditions such as traffic jams (to prevent accidents when reaching those stretches), adverse weather conditions (to slow down speed), accidents, roadworks, etc. Several studies show estimates in the reduction of accidents with the deployment of these systems in adverse weather conditions, for example. Other studies show that the information systems achieve modifications in the routes followed by the users in a significant percentage of occasions.

The variable information panels in the infrastructure are a means of disseminating information that has been carried out for several years. The variable information provided directly in the vehicle allows the customizing of the information to each vehicle according to the circumstances.

The following applications are particular cases of special relevance within this group of services.

- Signaling information: although there are systems for collecting traffic signaling through onboard sensors, communications also provide a means of providing this information to the user with the added value that they can adapt to the dynamic conditions of the environment.
- Information on road areas with special characteristics: the system alerts the driver when he approaches a road section where some special circumstance happens. In this way, the driver can be prepared before perceiving that situation and can accommodate his driving behavior to the conditions, also increasing his alert level. Notices are provided to all vehicles approaching the section. This information can be disseminated from the vehicles that are in those sections or from control centers or the infrastructure.
- Warning of sliding road and weather conditions: this case is a particular situation of the previous system. In the basic operation, the vehicle or sensors in the infrastructure can detect the characteristics of the road surface or weather conditions. If abnormal conditions are observed that could affect safety, they transmit that information to the other vehicles that approach that road section. In addition, this information is transmitted to units in the infrastructure, either to variable information panels or to control centers. Based on this information, new speed limits according to the type of vehicle could be suggested.
- Warning of heavy traffic. this system, together with a real-time recalculation of the route, allows the reduction of travel time. These systems also include a prediction of short-term traffic conditions. Thus, dynamic navigation refers to that type of navigation that recalculates the route to reach a certain predetermined target as a function of transient incidents that are detected before reaching them.

Floating vehicles
The new dynamic applications imply a continuous data updating. These transient effects must be captured and processed in real time. In traffic management, the location of incidents is vital, and delays in the transmission of such information can cause the loss of much of its value. An alternative to sensors in the infrastructure are vehicles ("floating vehicles") that transmit information such as position, speed, activation of some vehicle systems, etc., to a control center that is responsible, after proper processing, for its distribution to the rest of the users, although it is also possible for the transmission of punctual data between vehicles as they approach a certain road section. The use of these vehicles is an efficient solution to determine travel times when compared to other solutions in the infrastructure, although there are problems in terms of information delay and possible low penetration level of vehicles on a stretch of road.

Applications or multimedia content download while driving
During the driving it is possible to download information, applications, or multimedia contents.

Digital map updates download
This service considers the downloading of digital maps updates, as its degree of precision and detail is increasingly critical, as well as its degree of updating, since a very common problem arises when there are modifications in the infrastructure that are not included in the map. This service allows access to information more frequently.

Social networks
The main idea of this type of system is to encourage the collaboration of users to report incidents, errors in signaling, problems in infrastructure, or errors in digital maps. In this way, it would create a structure that would allow much more agile actions.

Vehicle positioning
This service has applications in different areas such as fleet management, stolen vehicle tracking, goods tracking, etc.

Remote diagnosis and assistance
Diagnostic techniques have evolved in parallel with the advances of the vehicle itself and there is specific equipment in the workshops capable of identifying faults in the vehicle from the connection to the internal communications bus of the vehicles. In addition to this conventional diagnosis, wireless communications allow the vehicle to exchange information with control centers or workshops remotely (Mazo et al., 2005).

The concept of remote diagnosis involves the wireless connection between the vehicle and an assistance center that will process the error codes, and will provide the actions to be performed, as well as indicate to the workshop the tasks to be undertaken, if necessary. Undoubtedly, this will result in a faster response to a breakdown.

7.3.3 Systems Oriented to Improve Safety

Cooperative services and systems that can provide increased safety can be present in almost every phase of the integrated safety model. Thus, although the main set is oriented towards assistance and primary safety, there are also applications in primary—secondary interaction systems and in tertiary safety, such as emergency call.

Some cooperative systems within this block are being implemented also using onboard sensors. However, by increasing the perception horizon using communications, it is possible to increase the potential of these systems.

Speed control
The fundamental idea of the system is to be able to provide information on the most appropriate traffic speed depending on the characteristics of the road, traffic, weather, and the vehicle itself. The improvement over conventional systems is the ability to customize information and adapt it in a flexible and fast way to dynamic situations that the vehicle comes to know thanks to communications from the infrastructure or control centers.

Warning of abnormal driving conditions
When a vehicle performs an unusual maneuver, such as a sudden steering maneuver or emergency braking, it is indicative that there may be a road situation that could lead to an accident or even the maneuver itself may imply a threat to other road users. In this way, when the vehicle detects this maneuver, it sends a warning to all the vehicles that approach it so that they are alerted.

A similar situation occurs if a vehicle is traveling at an abnormally low speed or is stopped on the road; it sends a signal to vehicles approaching that road section where it is located so that they are alerted even before they can see it. In this way, they can adapt driving conditions, avoiding potential danger situations and improving the capacity of the road since traffic flow could be reorganized before that stretch of road.

Detection of vehicles or other obstacles

The detection of obstacles, vehicles, pedestrians, or other objects can be performed by onboard sensors. Cooperative systems also provide a solution for avoiding vehicle collisions. To do this, the vehicle must be positioned on a detailed digital map, where the theoretically free driving areas are known.If the position and the kinematics of the other vehicles are transmitted, it can be determined if any of them imply a risk.

These systems based on onboard sensors have the advantage that their operation is independent of the systems that are mounted in the other vehicles and allows the detection (and identification) of obstacles different from vehicles. These advantages are not shared by systems based on communications between vehicles, although they provide a greater scope in the monitoring of the environment, being able to advance the alerts and decisions, but a certain necessary level of penetration of these systems in the market is required. It could also extend to the detection of pedestrians or cyclists if they emit their position just like a vehicle would. Another disadvantage is the precision required in positioning since in certain environments the safety margin may be very small, but false alarms should be avoided as far as possible.

Particular cases of systems involving two vehicles are as follows:

• *Safety distance warning*: this system controls that the safety distance between two vehicles traveling in the same lane and direction is not less than a certain value, considered as safe. Thus, by transmitting the position and speed of the vehicles, it is possible to determine the safety distance between the vehicles and, if detected to be reduced below a limit, a warning would be issued to the driver or actions could be performed automatically to maintain the safe distance.

• *Blind spot monitoring*: this system involves determining whether there is any vehicle, road user, or obstacle where the driver's field of vision is limited, before performing a lane change maneuver. By means of cooperative systems, vehicles emit their position and speed with which the others can evaluate if they are in that zone that is not perceived by the driver or not. Such systems can be particularly useful for vulnerable users, such as bicycles or motorcycles, who, because of their size, may be more difficult to perceive by drivers of larger vehicles. Thus, this system is useful in lane change maneuvers or when the larger vehicle wants to turn right or left to take another street.

• *Cooperative lights control*: this system allows a vehicle to switch from headlights to dipped beam automatically in the presence of another vehicle that could be blinded. For this, the vehicles must transmit at least their position in order to determine the moment at which the switching should be done. Once the vehicle has passed, the system can return to the initial situation of activating the headlights.

- *Vehicle warning in the opposite direction*: in the event that a vehicle is driving in the opposite direction, this system provides the information to this vehicle, to vehicles approaching it in the right direction, so that they are aware of the danger that this vehicle implies for them before they see it, and to the control center of the infrastructure, so that it adopts the appropriate actions.

Other applications also based on vehicle detection but which may involve a more coordinated management because a greater number of users are involved are the following ones:

- *Safe overtaking assistance.* The system analyzes the relative speeds of the vehicles and their positions, evaluating if safe overtaking is possible. Thus, the system can observe if, given the performance of the vehicle and the kinematic characteristics of the others, it is possible to carry out an overtaking in safe conditions by having enough space to complete the maneuver. This overtaking maneuver should not only take into account the vehicles traveling in the opposite direction, but also those that circulate in the same direction. The communications-based system allows a more complete knowledge of the vehicles that are circulating on a stretch of road and are not limited to the scope of the onboard sensors or to possible occlusions by the orography or vehicles of large dimensions.

- *Collision avoidance at intersections.* This system evaluates the positions of the vehicles at or near the intersection and estimates the possible trajectories. Some applications to make the intersections safer are:
  - Warning of an incident at the intersection, which allows to better inform the driver who can take the appropriate actions.
  - Warning of the approach of an emergency vehicle and its trajectory, providing suggestions to other drivers on how they should react to allow their route.
  - Warning of red traffic lights violation or stop signs.
  - Warning about the fact that a vehicle that circulates on a route that does not have priority will not respect the priority of others given its speed and position.
  - Warning of the need to stop or slow down before an intersection.

- Assistance for merging a high capacity road. This system looks for free spaces where the vehicle can merge into the existing flow. It can also negotiate the introduction of the vehicle into the flow cooperatively, alerting other vehicles that they should facilitate this maneuver.

Red light and time-to-green warning
By communicating between the traffic light and the vehicle, it is possible to inform the driver of the remaining time in the red or green phase. In this way, driving is much smoother, avoiding unnecessary accelerations when the traffic light is red and encouraging more progressive decelerations, with the consequent positive effect on consumption and pollutant emissions. Likewise, in the event that a vehicle is going to pass through the red light, the system can alert the other vehicles at the intersection.

7.3.4 Systems Oriented to Improve Efficiency

The improvement of traffic efficiency can be addressed from two approaches: global traffic management and driving behavior of the individual vehicle.

The efficiency of interurban traffic includes the detection and management of accidents, the management of variable message signals, management of the use of the lanes, speed management, etc. Information on road conditions and navigation systems (for example, systems that use traffic and weather information in real time, adapting the itineraries of the vehicles) are also included in this category.

The efficiency of urban traffic is responsible for the optimization of traffic flow, for example through the management of traffic lights, detection of accidents, minimization of traffic delays, priority for specific vehicles, and management of speed and lanes. Parking availability services are also included.

Efficiency from the point of view of the vehicle itself focuses on improving driving behavior to minimize environmental impact and consumption, although these behaviors also often lead to improvements in safety and comfort. In a scenario where a large amount of information is available, its appropriate use can encourage more efficient driving behavior.

Systems to support traffic management
Traffic management means achieving an efficient balance between travelers' needs and network capacity, bearing in mind that many users of diverse means of transport share the limited space of the infrastructure.

Traffic management is a very broad area that collects different services, being able to distinguish those from urban and interurban roads, although many of the applications are common from a conceptual point of view. Here are some of the most representative services:

- *Traffic management*: among the ITS services, one of the most well-known is traffic management based on detectors, traffic lights and signals, and the system of processing and sending of information, which can respond in real time to the specific characteristics of traffic at intersections. From the traffic monitoring, algorithms are established to optimize the intersections controlled by traffic lights in order to minimize variables, i.e., the total delay in the network, the number of stops, the travel time, etc. To do this, it starts from a model of the road network and simulates traffic behavior. This management can also be extended to prioritize public transport vehicles or emergency vehicles.

- *Access management*: vehicles are notified that they are approaching areas of restricted traffic where vehicles cannot circulate or only some types of vehicles can. This is especially useful in the case of dangerous goods vehicles. It can also be applied to limit entry to city centers of vehicles that exceed certain levels of pollution. Access control can also be used to regulate traffic in some areas and reduce congestion levels, limiting vehicles entering a certain road. In this sense, the efficiency of access control in high capacity roads has been verified (U.S. Department of Transportation, 2003).

- *Lanes management*: detailed knowledge of the traffic in each of the lanes and the location of each vehicle in each lane allows a more efficient management of such lanes as the use of the road shoulder, the use of bus lanes, or other types of dedicated lanes or reversible lanes. On the other hand, the location of the vehicles in the lanes allows that particular warnings are given to them in the event that in the lane through which they circulate there is an incident.

- *Incident management*: in this area, different levels are covered, such as the detection of the incident, the management of the resources to solve it so that the action is as fast and efficient as possible, and the distribution of the information of the incidence to other users for safety and efficiency issues.

Vehicle priority

Vehicle priority is focused on two types of vehicles: emergency vehicles and public transport.

In the first case, emergency vehicles warn those in the vicinity that they are approaching and that they should give way, indicating to them where they should move to foster this action. These warnings that enable V2V communications reduce uncertainties and indecision, and allow other drivers to begin their maneuvers with greater anticipation and safety. The communication of these vehicles with the infrastructure to generate "green corridors" where the traffic lights turn green or maintain in green phase to facilitate their way has also been discussed.

The same idea of the green corridors can be applied with the public transport, mainly applied when one of these vehicles accumulates delay with respect to the schedule initially anticipated (Furth, 2005). However, it is necessary to study how these measures affect the traffic flow. This makes it necessary, beyond local control, for a management of larger areas.

Information, management, and reservation of parking lots
The search for a place to park a vehicle means extending the actual time of a journey. In addition, in certain areas, this search leads to higher levels of congestion. The cooperative systems propose several levels of action:
- Information on parking lots available in a particular area or car park in the area.
- Directing to free parking lots via navigation systems.
- Reservation of the parking lot once the area and the time of arrival are selected.

The last of the states is the most advanced one and involves the reservation of a parking lot in a temporal window and supplying the information to the driver about the reservation, guiding him, if necessary, to the place. In the event that, due to traffic, delays occur in the time of arrival at the reserved place, this reservation could be modified by the system, informing the driver of the changes of time slot and place.

Similarly, it is possible to offer services aimed at delivery vehicles, reserving loading and unloading areas, optimizing the movements of these vehicles, reducing the stress of drivers, and avoiding nonlegal behaviors such as the realization of these tasks outside the designated areas, obstructing traffic.

Systems to encourage eco-driving
It is estimated that improvements in vehicle efficiency can lead to a 40% reduction in CO_2 emissions from passenger cars and a 10% reduction in commercial vehicles. Eco-driving techniques offer reductions in emissions between 10% and 25% (Hiraoka et al., 2009), while the potential for improvements in infrastructure, the use of the most efficient modes of

transport, and the implementation of information technology systems provide savings of 10%–20%.

Within this scope, the aim is to provide drivers with information to adapt their driving style to the conditions of the road. Thus, although many systems are based on static information such as that contained in digital maps (adaptation of the speed to the orography of the road), the use of communications allows the adaptation of the speed to the traffic conditions or the red–green cycles of traffic lights in urban environments.

Based on the fact that the way of driving significantly conditions consumption, the main idea for the minimization of consumption lies in the use of an "electronic horizon" more extensive than the "visual horizon," and taking advantage of information from those stretches of road that are not visible to the driver in order to adjust the speed. In order to make decisions, the system has to identify and classify the different situations that occur in traffic. These situations can be fixed-positioned (signals) or variable (retentions), and involve constant limits (conventional fixed signals) or variables (variable signals).

7.4 CHALLENGES TOWARD DEPLOYMENT

Taking as reference the report elaborated in January 2016 by the C-ITS Platform launched by the European Commission, below is a summary of the main issues to be faced in order to effectively deploy C-ITS systems.

7.4.1 Technical Issues

- Access to in vehicle data
  The increasing connectivity and digitalization of vehicles is currently changing the automotive industry landscape. Specific data that were previously accessed via a physical connection in the vehicle (e.g., FMS standard, OBD connector, ...) are now more and more accessible remotely (vehicle manufacturers servers, C-ITS, eCall, ICT platforms, ...). Independently of the model/solution retained to give access to in-vehicle data and resources, the main objective should be to allow customers the freedom to choose which service they desire, meeting their specific needs, in order to ensure open choice for customers.

- Decentralized Congestion Control

Cooperative intelligent transport systems (C-ITS) operating at 5.9 GHz for short-range communication use an ad hoc network topology. This implies that there is no central coordinator, such as a base station or an access point, granting access to the wireless channel. All network participants are peers and share the wireless channel whenever they have something to transmit. However, when many network participants simultaneously want to access the channel, the performance of applications can be severely degraded due to saturation. To overcome this, a decentralized congestion control (DCC) scheme must be implemented. DCC specifications for day one applications on a single channel are already in place and will function satisfactorily for low to moderate densities of ITS stations (ITS-Ss), but might not be sufficient for multichannel and day two applications demanding higher data throughput and enhanced spectrum efficiency (e.g., automated driving use-cases or Infrastructural and Vulnerable Road User use-cases).

- Hybrid communication

Considering the different access technologies and recognizing ITS G5 as a technology already capable of covering most of the relevant requirements of C-ITS services, it can't be denied that cellular technologies are on the way to adding to their current catalog of features those services needed to also efficiently cover C-ITS requirements (e.g., D2D).

Under this assumption, focusing on technical aspects, currently and in the mid-term, both technologies should be considered complementary and therefore a hybrid communication concept approach is recommended to take advantage of benefits from the usage of both of them (e.g., operation in the absence of network/network coverage; support of critical safety use-cases, . . .).

It is therefore essential to ensure that C-ITS messages can be transmitted independently from the underlying communications technology (access-layer agnostic) wherever possible. At the same time, the possible simultaneous provision of similar information through different information providers (including onboard sensors) could lead to validation problems at the receiving side, which might find it difficult to judge which information is better (more accurate, more timely, . . .) to use. Thus, the use of different communication channels when sending/receiving C-ITS messages and the corresponding question on how to validate which message is most recent/relevant/accurate should be considered.

In any case, connected, cooperative, and automated ITS services and applications have highly varying functional and technical communication requirements. Depending on the business cases and requirements, it should be possible to select the most effective access and communication technologies for each purpose. A hybrid communication approach including multiple technologies (including future new developments) and radios is the only way to support continued deployment of ITS services and applications today and in the future.

- Other technical issues

  Not only the increasing number of C–ITS in vehicle equipment but also the introduction of C–ITS for vulnerable road users will increase the number of C–ITS devices dramatically and will lead to much higher channel load (especially in urban environments). A better protection of VRUs is clearly an important safety goal. Therefore, there are some hurdles to be overcome: to ensure reliable and more accurate positioning technology for pedestrian detection (e.g., to distinguish if the pedestrian is walking on the sidewalk, wants to cross the street, or is already crossing the street) or to face a potential designation of additional frequency spectrum for this and other (future) C–ITS services.

7.4.2 Implementation Issues

- Human machine interaction

  The deployment of C–ITS services will result in a more complex interaction between the driver and the vehicle that could affect negatively road safety. In fact, the vehicle interfaces are becoming more complicated, and as a result, driver distraction may be increased. In order to avoid this risk, specific measures have to be taken into account and a revision of the European Statement of Principles on Human–machine Interface might be needed to cover C–ITS.

- Nonequipped users

  For a long time, users equipped with C–ITS will coexist with other users not equipped with these technologies, resulting in potential problems for the road safety. For the different C–ITS services, the situation could have different effects on the user behavior, whether they do or don't have an equipped vehicle. Nevertheless, the same problems are appearing for other kinds of advanced driver assistance technologies, and the question is whether C–ITS should be addressed differently. On the other hand, the administration could play an important role in trying to avoid future problems in road safety by introducing new legislation to shorten the deployment period. Questions are also arising about the role that the retrofit devices would play in a near future.

• Training and awareness
With the deployment of C-ITS, users will access more information and functionalities within the cars, but in some cases, they will miss the required knowledge to benefit from them without affecting road safety. Potential risks are: false perception and overreliance on the system, ignorance on the noncontinuity of the service due to the vehicle or to the infrastructure, etc. Specific training on driver assistance or information systems may be required in order to avoid dangerous situations and let the users understand the possibilities and limitations of these new technologies.

• Other implementation issues
Although C-ITS deployment is considered a crucial factor to improve road safety and traffic efficiency in the long-term, it is still unclear how the costs and business models will be approached. There is a need for a good business case, showing benefits in terms of cost-effectiveness and performance over existing systems. In this way, it is important to integrate C-ITS decisions for their deployment within the long-term investment plans, and to analyze how C-ITS can build on existing investments. Some legal and liability issues have also to be address, for example, who is liable if the technology failure leads to an accident.

Ensuring interoperability is another issue that has to be addressed, as well as identifying the missing standards and deciding on standard profiling. A hybrid communication approach, with the coexistence of several communication technologies, seems to be the most realistic scenario for C-ITS in the near future.

7.5 MAIN RELATED INITIATIVES AT EUROPEAN LEVEL

In the last decades, the European Commission has supported several initiatives oriented to improve road safety and traffic efficiency through the development and demonstration of Cooperative ITS. In fact, the first steps towards cooperative vehicle-to-vehicle and vehicle-to-infrastructure systems were initially covered by a number of projects launched under the EU's sixth Framework Program. This is the case of COMeSAFETY, CVIS, SAFESPOT, and COOPERS, all of them collaborative projects that laid the foundation for C-ITS services, based on different approaches.

The COMeSAFETY (eSafety Cooperative Systems for Road Transport) project is aimed at providing an integrating platform for both the exchange of information and the public and private awareness, seeking worldwide harmonization among the relevant stakeholders. The project actively supported the process of spectrum allocation for cooperative ITS systems in Europe by participating in relevant standardization bodies.

The role of the infrastructure operators was thoroughly analyzed in Cooperative Systems for Intelligent Road Safety (COOPERS). This project focused on the provision of real-time local situation-based, safety-related traffic and infrastructure status information distributed via dedicated I2V communication links. The project results were demonstrated on four important road segments with high traffic density (Fig. 7.3).

Cooperative Vehicle-Infrastructure Systems (CVIS) contributed significantly to the development of the basic set of Communication Standards (CALM) for Cooperative Systems. The CVIS objective was to create a unified technical solution allowing all vehicles and infrastructure elements to communicate with each other in a continuous and transparent way. An open architecture was proposed and, for the first time, the challenge of interoperability was considered, as it required different car manufacturers to agree on a common system specification to ensure full functionality of their applications.

Finally, SAFESPOT (Cooperative systems for road safety "Smart vehicles on Smart Roads"), focused on onboard equipment and enabling technologies for cooperative systems preserving highly safety critical tasks. This project proposed a "Safety Margin Assistant" to increase the amount of information available to drivers, where the communication of warnings and advice to approaching vehicles would provide extra reaction time in order to help prevent an accident.

The three aforementioned projects were broadly disseminated in several pilot sites and events, in order to raise awareness of the relevance of cooperative services and to set the bases for future developments. In this sense, during the ITS World Congress in Stockholm (2009), a common demonstration on how cooperative systems could contribute to transport safety and efficiency was offered, with the demonstration of several applications.

From those pioneer initiatives, the European Commission has been supporting new projects and initiatives, covering both urban and

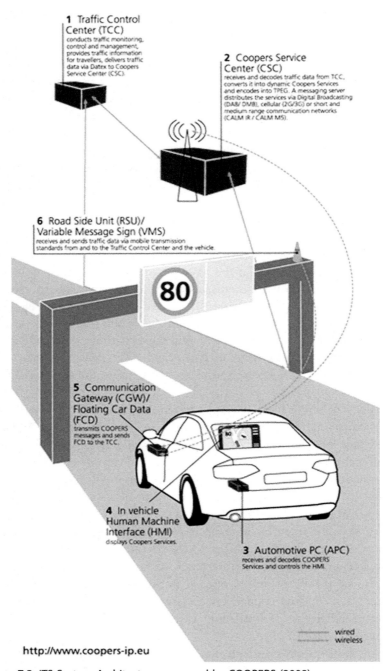

Figure 7.3 ITS System Architecture proposed by COOPERS (2009).

interurban scenarios, standardization issues, and predeployment pilots, among other aspects. Many of these projects have been focused on the test of C-ITS in real environments by launching dedicated pilots in urban or interurban scenarios. Although the distinction between urban and nonurban areas regarding C-ITS can be considered artificial, since a large number of issues are not specific of a concrete geographical environment, in order to better understand the evolution and progress achieved in the field of C-ITS, the following sections present some relevant projects focusing on both of these scenarios. Some relevant European initiatives supporting the introduction and deployments of C-ITS are also described.

7.5.1 Interurban Mobility Pilots

7.5.1.1 DRIVE C2X—DRIVing Implementation and Evaluation of C2X Communication Technology in Europe

DRIVE C2X was a project launched under the Seventh Framework Programme, with the objective of assessing cooperative systems by creating a harmonized Europe-wide testing environment that could help future deployment of those systems. The project focused on communication among vehicles (C2C) and between vehicles, a roadside and backend infrastructure system (C2I).

Previous projects such as PReVENT, CVIS, SAFESPOT, COOPERS, and PRE-DRIVE C2X have proven the feasibility of safety and traffic efficiency applications based on C2X communication. DRIVE C2X went beyond the proof of concept and addressed large-scale field trials under real-world conditions at multiple national test sites across Europe (Sánchez et al., 2011).

The systems tested were built according to the common European architecture for cooperative driving systems defined by COMeSafety, thus guaranteeing the compliance with the upcoming European ITS standards.

DRIVE C2X investigated the impacts of cooperative driving functions in field conditions with the help of ordinary drivers. In principle, two different application areas featured cooperative driving:

- C2C where vehicles exchanged information on safety-relevant or other useful information.
- C2I where vehicles were exchanging information with backend systems like traffic lights or road operators.

The project wanted to know how drivers react to different services that cooperative systems provide. The opinions, attitudes, and behavior of drivers were of paramount importance when considering the later market introduction of cooperative systems.

The system architecture of DRIVE C2X was based on (Fig. 7.4):

- The DRIVE C2X Vehicle, equipped with radio hardware based on IEEE 802.11p and UMTS for data exchange with other vehicles or with roadside infrastructure. The protocol stack supported ad hoc communication based on GeoNetworking, which enabled a rapid and efficient message exchange among vehicles using single-hop and multihop communication. The system was connected to the vehicle onboard network (CAN bus) to collect data within the vehicles, so that vehicle data could be exchanged between vehicles.
- The roadside infrastructure (DRIVE C2X roadside unit), fully integrated into the ad hoc communication network, e.g., to send information to the vehicles or to act as relay stations for (multihop) communication. Additionally, a roadside infrastructure component might be connected to the Internet, to communicate with the central component.
- The central component, for example a traffic management center controlling roadside infrastructure, a fleet management center, etc. Vehicles or roadside infrastructure might send information to the central component, and the latter may, in turn, send information to vehicles or roadside infrastructure.

The large-scale tests in DRIVEC2X were conducted following the FESTA methodology, which was extended to cover cooperative systems (Sawade et al., 2012), including issues such as data collection, handling, and data processing. Interoperability was also addressed in the project through several interoperability and conformance test sessions with different vendor systems.

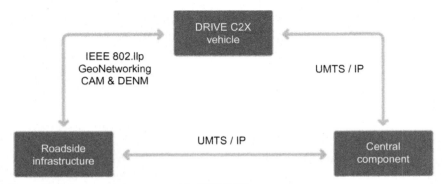

Figure 7.4 Components proposed in DRIVEC2X.

Figure 7.5 Example of DRIVEC2X cooperative service in Vigo Test-site (SISCOGA).

In summary, more than 750 drivers participated in the field operational tests, with 262 vehicles over 1.8 million kilometers in seven different test-sites around Helmond (The Netherlands), Gothenburg (Sweden), Tampere (Finland), Frankfurt (Germany), Brennero-Trento (Italy), Yvelines (France), and Vigo (Spain) (Fig. 7.5).

7.5.1.2 FOTsis—European Field Operational Test on Safe, Intelligent, and Sustainable Road Operation

FOTsis was a large-scale field testing of the road infrastructure management systems needed for the operation of seven close-to-market cooperative I2V, V2I, and I2I technologies, in order to assess in detail both their effectiveness and their potential for a full-scale deployment on European roads. It aimed to contribute to the safety, mobility, and sustainability challenges faced by the European road transport system.

This project, that was also funded by the Seventh Framework Program, followed a different approach to DRIVEC2X, focusing on the infrastructure's capability to provide the different cooperative services (Alfonso et al., 2013). Thus, aspects such as infrastructure management tools for safety and mobility, interaction between infrastructure and users, communication networks, or regulatory framework, were covered. FOTsis proposed an integrated architecture following the open architecture established in COMeSAFETY, so that additional services could be also supported.

The communications network of the proposed architecture is able to interconnect all fixed and mobile devices (RSUs, vehicles, etc.). The infrastructure communications (I2I) are mainly based on existing

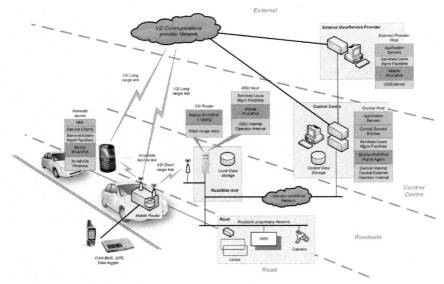

Figure 7.6 FOTSIS Integrated Architecture.

technologies: IP-networks over Ethernet or using advanced protocols such as MultiProtocol Label Switching (MPLS) which are able to guarantee the desired functionality, QoS, and security. These channels are implemented on the physical level using fiber optic. Additionally, the use of Digital Media Broadcasting (DMB) and Digital Video Broadcasting (DVB-H) provide broadcast and multicast tools used basically for the dissemination of information or control commands to infrastructure elements such as Variable Message Signs (VMS) or sensors (Fig. 7.6).

FOTsis included nine test sites in four different countries (Spain, Portugal, Germany and Greece), where the following services were tested and demonstrated:

- Emergency Management.
- Safety Incident Management.
- Intelligent Congestion Control.
- Dynamic Route Planning.
- Special Vehicle Tracking.
- Advanced Enforcement.
- Infrastructure Safety Assessment.

As a result, the FOTsis project demonstrated the feasibility of deploying C-ITS services from an infrastructure perspective, thus complementing the already significant work that had been done from the vehicle side. From a political perspective, FOTsis demonstrated the possibility of deploying C-ITS solutions to road users irrespective of vehicle category or level of intelligence embedded in the car, which will be crucial in the transition period towards the full connected and automated driving. In this sense, the degree of penetration of C-ITS on Europe's road will depend on the ability of authorities to offer hybrid solutions that take into account the needs of all drivers. The contribution of FOTsis was significant as it helped road authorities on their way to offering C-ITS services to users of conventional vehicles by counting on communication between the traffic control centers and users deploying nomadic devices.

7.5.2 Urban Mobility Pilots

7.5.2.1 COMPASS4D—Cooperative Mobility Pilot on Safety and Sustainability

The COMPASS4D project was launched in 2013, under the Seventh Framework Program, with the aim of promoting C-ITS services close to deployment in urban scenarios that could increase drivers' safety and comfort, by reducing road accidents as well as avoiding traffic jams, with a positive impact on the local environment. Three main objectives were established:

- Specification of a methodology for the evaluation of the COMPASS4D services.
- Development of measurement and assessment tools for safety, efficiency, sustainability, maintenance, traffic management, and driver-specific metrics.
- Evaluation of services' contribution to improved journey time reliability, reduced accident rates, improved energy efficiency, support for reductions in carbon emissions, and user acceptance/experience.

The project followed a methodology based on testing three different C-ITS services in seven cities across Europe: Bordeaux (France), Copenhagen (Denmark), Helmond (The Netherlands), Newcastle (UK), Thessaloniki (Greece), Verona (Italy), and Vigo (Spain).

The three cooperative services selected were the following:

- *Red Light Violation Warning (RLVW)*. This service consists of sending messages that will increase drivers' alertness at signaled intersections in order to reduce the number of collisions or the severity of them. This service also addresses exceptional situations such as alerting other vehicles that an emergency vehicle is approaching or violating a red light.

- *Road Hazard Warning (RHW)*, aiming at reducing the number and the severity of road collisions by sending warning messages to drivers approaching a hazard (obstacles, road accident, etc.). The messages sent will raise drivers' attention level and inform them about appropriate behavior in specific situations such as queues after a blind spot.
- *The Energy Efficient Intersection (EEI)*. This service was proposed to reduce energy consumption and vehicle emissions at signaled intersections. Selected vehicles (Heavy Goods Vehicles, Emergency Vehicles, Public Transport) would be granted a green light when approaching the intersection, thus avoiding stops and delays. This service will also provide information to other drivers to anticipate current and upcoming traffic light phases and adapt their speed accordingly (GLOSA) (Fig. 7.7).

These services were implemented through a combination of established technologies and available precommercial equipment. Dedicated short-range communications (ITS-G5) and cellular networks (3G/LTE) were deployed, following European (ETSI TC ITS) standards. In addition, COMPASS4D identified solutions to deployment barriers and has elaborated business models to make the services self-sustainable for a wide commercialization. This work included cooperation with standardization organizations and global partners to achieve interoperability and harmonization of services.

More than 650 vehicles participated in the field operational test, involving 1,215 drivers and the installation of 134 RSUs. After analyzing the collected data, the following conclusions were outlined:

- Light vehicles show a benefit from the cooperative system and services in terms of time saved, but the energy efficiency savings are typically small in absolute terms.
- Heavy-goods vehicles show a much greater saving in energy efficiency (up to 5%–10%).

Figure 7.7 COMPAS4D services demonstrated at Vigo test site (SISCOGA).

- Buses show large savings (up to 10%) depending on the route and, especially, on the position of bus stops relative to the nearest equipped, signalized intersection ahead.
- The priority service has a greater positive impact than the speed advice for a wider range of vehicle types.

COMPASS4D demonstrated the great potential of the introduction of cooperative services in the urban environment, both in terms of safety, and reduction of vehicles' CO_2 emissions and fuel consumption, and has served as example for many cities that are already deploying some cooperative applications.

7.5.2.2 CO-GISTICS—Cooperative Logistics for Sustainable Mobility of Goods

CO-GISTICS is the first European project fully dedicated to the deployment of cooperative intelligent transport systems (C-ITS) focused on logistics. The project consortium includes relevant logistics stakeholders, among other entities, in order to promote the introduction of a set of cooperative logistics services combining cooperative mobility services and intelligent cargo with real-life logistical aspects.

The project considers the following cooperative services:
- *Intelligent Truck Parking and Delivery Areas Management*: optimize traffic activities on the route and reduction of stops.
- *Cargo Transport Optimization*: optimize and increase the efficiency of cargo transport operations.
- *CO_2 Footprint Monitoring and Estimation*: use of GPS data or CANBUS related data to measure the fuel consumption.
- *Priority and Speed Advice*: use of C-ITS at intersections to indicate the speed to reduce number of stops and accelerations.
- *Eco-Drive Support*: use of C-ITS to provide time to red/green light at intersections.

These services have been deployed in seven European logistic hubs: Arad (Romania), Bordeaux (France), Bilbao (Spain), Frankfurt (Germany), Thessaloniki (Greece), Trieste (Italy), and Vigo (Spain). A total of 315 vehicles, including tracks and vans, are participating in the different tests.

CO-GISTICS final results will be presented in a final event that will take place by mid-2017. The main achievements expected at the end of the project will cover the following aspects:
- Pilot and deployment of C-ITS in European logistics hubs.

- Reduction of fuel consumption and the equivalent CO_2 emissions.
- Improvement of logistics efficiency in urban áreas.
- Harmonization of testing.
- Cooperation with logistics and freight public/private bodies.

7.5.3 Collaborative Platforms and Supporting Initiatives

7.5.3.1 Car2Car Communication Consortium

The Car2Car Communication Consortium (C2C-CC) is a nonprofit, industry driven organization originally founded in 2001 by six European vehicle manufacturers with the aim of creating and establishing an open European industry standard for C2C communications and enabling the development of active safety applications by specifying, prototyping and demonstrating the C2C applications. From its origin, this initiative promoted the allocation of a royalty-free European-wide exclusive frequency band for this kind of applications, which finally resulted in the 5.9 GHz spectrum, aligned with a similar spectrum allocation in USA, Canada, Mexico, and Australia (Fig. 7.8, Fig. 7.9).

This initiative is nowadays supported also by equipment suppliers, research organizations, and other partners, and continues to promote the creation of European standards in close cooperation with the European

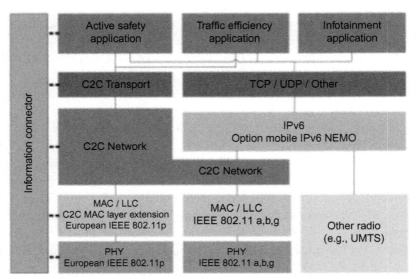

Figure 7.8 Protocol architecture of the C2C Communication System (C2C-CC manifesto, 2007).

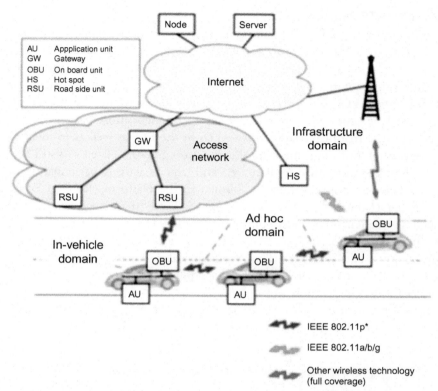

Figure 7.9 Reference architecture proposed in C2C-CC manifesto in 2007.

Figure 7.10 Examples of use cases proposed in C2C-CC (Emergency vehicle-to-car communication, Avoidance traffic jams, Warning of roadworks).

and international standardization organizations. Other activities carried out by the C2C–CC include the analysis of business cases, the preparation of roadmaps, and the proposal of strategies for the systems' deployment (Fig. 7.10).

7.5.3.2 C-ITS Platform

The Cooperative ITS Platform (C-ITS) is a cooperative framework promoted by DG MOVE (European Commission) in November 2014, that includes national authorities, C-ITS stakeholders, and the European Commission, with the objective of promoting the deployment of Cooperative ITS services in an interoperable way in the European Union. It is expected to provide policy recommendations for the development of a roadmap strategy for C-ITS in the EU, as well as to identify potential solutions to some critical cross-cutting issues related to C-ITS.

There are more than 60 members and experts actively participating in the different working groups established in the platform (see Fig. 7.11). In general, the C-ITS platform addresses both, main technical issues, such as frequencies, hybrid communications, (cyber-)security, and access to in-vehicle data and resources, and legal issues, such as liability, data protection, and privacy. The Platform also covers standardization, cost—benefit analysis, business models, public acceptance, road safety, international cooperation, etc.

As a result of the work performed by the different working groups, a Final Report was generated in January 2016 with the main results and

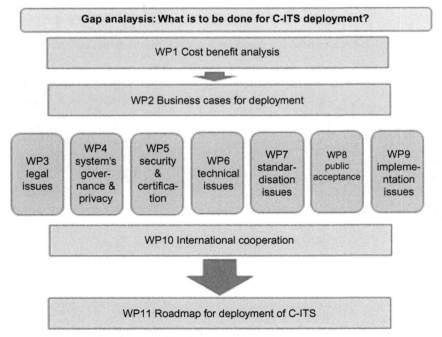

Figure 7.11 C-ITS Platform working groups.

conclusions towards a common vision on the interoperable deployment of Cooperative Intelligent Transport Systems in the European Union. It has a remarkable final agreement on a list of Day 1 and Day 1.5 Services (see Section 7.3.1), after having considered their expected societal benefits and the maturity of technologies. Other conclusions included in this report are related to:

- Security and certification.
- Radio frequency and hybrid communication.
- Decentralized congestion control.
- Access to in-vehicle data and resources.
- Liability.
- Data protection and privacy issues.
- Road safety issues.
- Acceptance and readiness to invest.
- Costs and benefits.

The C-ITS Platform is now evolving the results, focusing on the links between connectivity and automation, in particular in relation to infrastructure and road safety issues, and in the definition of the next steps regarding policy recommendations for all relevant actors in the C-ITS value chain.

7.5.3.3 The Amsterdam Group
The Amsterdam Group is a strategic alliance of committed key stakeholders with the objective of facilitating common deployment of cooperative ITS in Europe. A first approach to the deployment of C-ITS in Europe was proposed, based on a number of phases where complexity would increase with the augmentation of vehicles equipped with ITS systems and the more important infrastructure coverage (see Fig. 7.12).

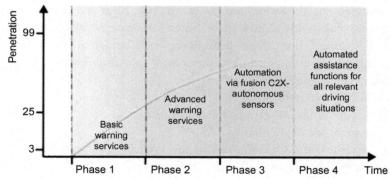

Figure 7.12 Phases for C-ITS deployment proposed by The Amsterdam Group.

Some key players are included in this Group, such as Conference of European Directors of Roads (CEDR), Association Européene des Concessionaires d'Autoroutesetd'ouvrages à Péage (ASECAP), POLIS, and the C2C-CC.

The Amsterdam Group is active in facilitating information exchange, discussion, and creation of solutions between the involved stakeholders in the context of C-ITS. The objectives of the Amsterdam Group include:

- Facilitate dialog between actors, e.g., on corridors results and needs.
- Integrated communication interface towards the individual members of the Amsterdam group.
- Exchange of experience between corridors, front runners, etc.
- Communicate with EC and other bodies (e.g., ETSI/CEN).
- Functional specifications for C-ITS services as input for standardization.
- Taking away barriers for deployment.

As part of the existing activities, the Amsterdam Group has defined a roadmap between industry and infrastructure organizations on the initial deployment of Cooperative ITS in Europe. In this roadmap, the needed steps for a Joint Deployment Strategy are addressed, as well as the common open issues and possible timeline when they are to be solved.

7.5.3.4 CODECS—COoperative ITS DEployment Coordination Support

CODECS is a Coordination and Support Action funded under the Horizon 2020 Programme, aiming at sustaining the initial deployment of cooperative systems and services in Europe. The project has established a stakeholder network for stimulating a transparent information flow and exchange of lessons-learned from initial deployment. Through workshops, webinars, and personal consultation, CODECS takes inventory on the status and implementation approaches in early deployment activities (technologies, specifications, and functions), the roles and responsibilities of different stakeholders, as well as issues for strategic decision making. CODECS consolidates these procedures, stakeholder interests, preferences, and requirements and plays them back into the network. Through interactive discussion, it develops

- A V2I/I2V standards profile;
- White papers closing gaps in standardization; and
- A blueprint for deployment.

These results of CODECS sustain the interoperability of systems and services across hot spots of early deployment, enabling end-users to witness the benefits of cooperative road traffic first-hand, in turn impacting on the penetration rates. CODECS promotes the idea of cooperative road traffic to a broad target audience to support this effect.

To give guidance for a future concerted C-ITS rollout also for later innovation phases with corresponding research, testing, and standardization, CODECS transforms the fused stakeholder preferences in

- An aligned use case road map; and
- Recommendations for strategic decision making.

With this goal setting, CODECS supports the Amsterdam Group, the C-ITS Deployment Platform of the European Commission, Standards Setting Organizations, and other key deployment players to come to a concerted C-ITS rollout across Europe.

7.5.3.5 C-Roads Platform

The C-Roads Platform is an initiative cofunded through the Connecting Europe Facility Programme (CEF), based on the cooperation between Member States and road operators working on the deployment of harmonized and interoperable C-ITS services in Europe. It follows a bottom-up approach that includes sharing experiences and knowledge from the national pilots, in order to identify barriers and common problems that have to be solved to facilitate C-ITS deployment at the European scale.

The main goals of C-Roads Platform can be summarized as follows:

- To link C-ITS pilot deployment projects in EU Member States.
- To develop, share, and publish common technical specifications (including the common communication profiles).
- To verify interoperability through cross-site testing.
- To develop system tests based on the common communication profiles by focusing on the hybrid communication mix, which is a combination of ETSI ITS-G5, and existing cellular networks.

Although this initiative was officially launched at the end of 2016, many C-ITS corridors and national pilots participating in the platform were already in operation. Thus, the platform established very soon the Terms of Reference for all platform members, so that all of them share the vision towards an interoperable deployment of C-ITS services at European level.

The C-Roads platform structure is based on a Steering Committee, and several working groups, acting in a synergistic way with the local pilots established at local or national level (see Fig. 7.13).

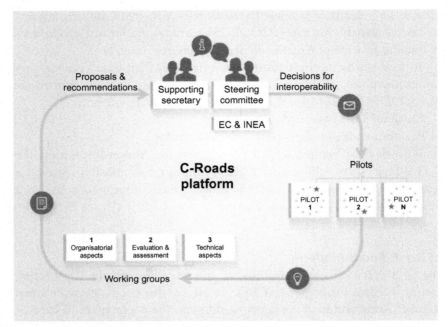

Figure 7.13 C-Roads Platform approach.

From the beginning, the following working groups and task forces have been proposed:

- WG1—Organizational Aspects, dealing with the identifications of legal barriers and obstacles for the C-ITS deployment, the involvement of end users, and the collection and exchange of business models.
- WG2—Technical Aspects, that address the recommendations on driver information through C-ITS services, the harmonization of current and future C-ITS services, and the definition of a harmonized communication profile. Three task forces have been organized within this WT to cover specifically security aspects, service harmonization, and infrastructure communication.
- WG3—Evaluation and Assessment, for the definition of a common methodology to evaluate and assess the impact of the C-ITS services in terms of interoperabililty, sustainability, safety, efficiency, and environmental aspects.

There are currently eight countries participating as core members in the platform, but new members are entering in order to join efforts and to share experiences from the activities carried out in their local corridors. In this sense, some of the most relevant active cooperative corridors, whose results have been monitored in the platform are the following:

7.5.3.6 Cooperative ITS Corridor

This is a cooperation between the German, Dutch, and Austrian road operators to start the gradual introduction of cooperative systems, sharing the view of an intelligent and accident-free-mobility thanks to the networking of vehicle and infrastructure. It follows a cross-border approach and involves industrial partners to help bring vehicles and telematics infrastructure onto the market. The corridor covers several roads in the section Rotterdam−Frankfurt−Vienna. The first cooperative services planned in this corridor are:

- Roadworks Warning, from the traffic control centers via the roadside infrastructure to the drivers.
- Vehicle Data for improved traffic management, vehicles transmitting data about the current situation on the road to the roadside infrastructure and the traffic control centers.

This initiative is expected to pave the way for the future public−private partnerships, needed to achieve the full deployment of C-ITS.

7.5.3.7 Intercor—North Sea−Mediterranean Corridor

Four member states participate in this corridor covering approximately 968 km across the Netherlands, Belgium, UK, and France. InterCor will focus on the deployment of Day-1 services as recommended by the European C-ITS platform, such as Roadworks warning, Green Light Optimized Speed Advisory, In-vehicle signage, and Probe vehicle data. The final goal is to search a harmonized strategic rollout and common specification for C-ITS, taking advantage of the results obtained in the cross-border tests.

7.5.3.8 SCOOP@F (France)

SCOOP@F was proposed in France as a main initiative to test and predeploy C-ITS on approximately 2000 km of roads, including intercity roads and highways in Ile-de-France and Bretagne, the Paris−Strasbourg highway, Bordeaux and its bypass, and county roads in the Isère. A total of 3000 vehicles will participate in the test that will exchange information about their position, speed, road obstacles, and other details with the infrastructures.

In the project, two different phases or waves were proposed: the first one to cover priority services through ITS-G5 communications, and the second focused on new services and a hybrid approach (cellular and ITS-G5 communications). Additionally, an extension of the project included also other local corridors in Spain, Portugal, and Austria, in order to perform cross-tests.

7.5.3.9 SISCOGA Corridor (Spain)

Since 2011, when the SISCOGA corridor was created in the region of Galicia, several research, development, and deployment activities have been carried over this corridor, at national and international level. The corridor covers more than 120 km of interurban roads and 10 km of urban road in the city of Vigo. More than 80 RSUs (ITS-G5 technolgoy) are currently active, enabling cooperative services, such as accident or traffic jam, adverse weather information, roadworks information, etc. (see Fig. 7.14). This corridor is now being extended and upgraded with a hybrid architecture approach that combines ITS-G5, LTE, and LTE/V communication technologies.

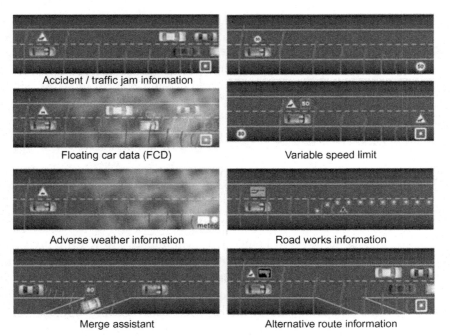

Figure 7.14 Example of cooperative services available at SISCOGA corridor.

7.6 NEXT STEPS

As explained above, various different initiatives are pushing the large-scale deployment of Day-1 and Day-1.5 cooperative services. In this sense, the different initiatives explained in Section 7.5 will strongly support the real deployment of cooperative systems in different regions. Users have already experienced the benefits of cooperative traffic safety and efficiency services in pilot projects such as DRIVEC2X or COMPASS4D. However, the main challenge is now how to achieve a full interoperability and harmonized approach at the European level, dealing with upcoming technical aspects (new communication technologies to be combined with existing ones), business models behind infrastructure investments, and upcoming legal and security issues.

Another currently expanding area in the road transport sector is the domain of automated and autonomous driving. Connectivity services could bring a major benefit for this kind of vehicles, allowing them to understand much better the present and future situations that vehicles will have to cope with. It could be also very interesting for road operators and traffic administrations to receive information from individual connected vehicles to develop, in the end, a safer and more efficient road transport holistic approach.

Examples of some interesting specific use cases of connected services for automated vehicles are the following:

- Connected services to support automated vehicles in Highway Entrances.
- Connected services to support automated vehicles in Highway Exits.
- Electronic Horizon services to inform about future hazards affecting the route of the automated vehicle.
- Connected services to support automated vehicles at intersections in urban scenarios to enhance safety and efficiency (see Fig. 7.15).
- Cooperative sensing services to extend the detection range of ego-sensors through the information coming from other vehicles, from the infrastructure elements or from the cloud (see Fig. 7.16).
- Enriched map data services.
- Valet parking services.
- Software Over The Air (SOTA) services.

At this moment, there are several research projects and initiatives dealing with this specific challenge. One very interesting example is the European R&D project Automated Driving Progressed by Internet of

Figure 7.15 C2C Cooperative intersection assistant.

Figure 7.16 Example of tests to evaluate a cooperative sensing use case. The traffic light is informing the automated vehicle about GLOSA and about the presence of a pedestrian, concealed by a bus, crossing at an intersection.

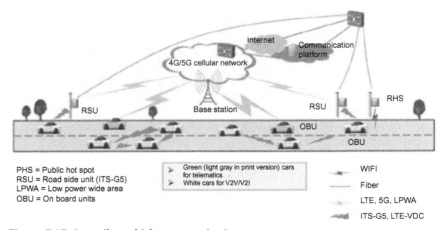

Figure 7.17 Autopilot vehicle communication concept.

Things (AUTOPILOT). This project, started in January 2017, intends to merge automotive and IoT technologies to move automated driving towards a new dimension. In this sense, cars are becoming moving "objects" in an IoT ecosystem that involves vehicles, road infrastructure, and surrounding objects. AUTOPILOT aims to develop new services on top of IoT to involve autonomous driving vehicles, like autonomous car sharing, automated parking, or enhanced digital dynamic maps to allow fully autonomous driving. AUTOPILOT will be tested in real conditions, at five permanent pilot sites in Finland, France, The Netherlands, Italy, and Spain (Fig. 7.17).

REFERENCES

Alfonso, J., Sánchez, N. Menéndez, J.M., Cacheiro, E., 2013. Cooperative ITS communications architecture: the FOTsis project approach and beyond. IET Intelligent Transport Systems, ISSN 1751-956X, vol. 9, No. 6.
Car 2 Car Communication ConsortiumManifesto, 2007. Available at <https://www.car-2-car.org/>.
C-ITS Platform, 2016. Final report. Available at <http://ec.europa.eu/transport/sites/transport/files/themes/its/doc/c-its-platform-final-report-january-2016.pdf>.
Furth, P.G., 2005. Public Transport Priority for Brussels: Lessons from Zurich, Eindhoven, and Dublin. Report Completed Under Sponsorship of the Brussels Capital Region Program "Research in Brussels".
Hiraoka, T., Terakado, Y., Matsumoto, S., Yamabe, S., 2009. Quantitative evaluation of eco-driving on fuel consumption based on driving simulator experiments. Proceedings of the 16th ITS World Congress. Stockholm, Sweden.
Jiménez, F., et al., 2012. Libro Verde de los Sistemas Cooperativos. ITS Spain and TechnicalUniversity of Madrid.
Mazo, M., Espinosa, F., Awawdeh, A.M.H., Gardel, A., 2005. Diagnosis electrónica del automóvil. Estado actual y tendencias futuras. Fundación Instituto Tecnológico para la seguridad del automóvil (FITSA), Madrid, in Spanish.
OECD, 2015. Automated and Autonomous Driving Regulation under uncertainty. Available at <http://www.itf-oecd.org/sites/default/files/docs/15cpb_autonomous-driving.pdf>.
Sánchez, F. Paúl, A., Sánchez, D., Sáez, M., 2011. SISCOGA − a Field Operational Test in the North West of Spain on future C2X Applications. In: Proceedings of the 13th EAEC European Congress.Valencia.
Sawade, O., Sánchez, D., Radusch, I., 2012. Applying the FESTA Methodology to Cooperative system FOTs. In: Proceedings of the 19th ITS World Congress, Vienna, Austria.
U.S. Department of Transportation, 2003. Intelligent Transportation Systems Benefits and Costs US DOT Washington DC.

FURTHER READING

Amsterdam Group, 2016. Available at <https://amsterdamgroup.mett.nl/>.
C-ITS Corridor, 2015. Available at <http://c-its-korridor.de/?menuId=1&sp=en>.

CODECS, 2016. Available at <http://www.codecs-project.eu/>.

COGISTICS, 2016. Available at <http://cogistics.eu/>.

Commission Recommendation, 26 May 2008 on safe and efficient in-vehicle information and communication systems: update of the European Statement of Principles on human machine interface, OJ L 216, 12.8.2008. Available at <http://eur-lex.europa. eu/legal-content/EN/TXT/?uri=CELEX%3A32008H0653>.

COMPASS4D, 2015. Available at <http://www.compass4d.eu/>.

C-Roads Platform, 2016. Available at <https://www.c-roads.eu/platform.html>.

C-Roads Terms of Reference, 2016. Available at <https://www.c-roads.eu/platform/ documents.html>.

DRIVEC2X, 2014. Available at <http://www.drive-c2x.eu/project>.

FOTsis, 2013. Available at <http://www.fotsis.com/>.

InterCor, 2016. Available at <https://ec.europa.eu/inea/sites/inea/files/fiche_2015-eu- tm-0159-s_final.pdf>.

Konstantinopoulou, L., Han, Z., Fuchs, S., Bankosegger, D., 2010. Deployment Challenges for Cooperative Systems. CVIS-SafeSpot-COOPERS White Paper. Available at <http://ertico.assetbank-server.com/assetbank-ertico/action/viewHome>.

SCOOP@F, 2014. Available at <https://ec.europa.eu/inea/sites/inea/files/fichenew_2013- fr-92004-s_final.pdf>.

CHAPTER 8

Automated Driving

Jorge Villagra[1], Leopoldo Acosta[2], Antonio Artuñedo[1],
Rosa Blanco[3], Miguel Clavijo[4], Carlos Fernández[5], Jorge Godoy[1],
Rodolfo Haber[1], Felipe Jiménez[4], Carlos Martínez[4], José E.
Naranjo[4], Pedro J. Navarro[5], Ana Paúl[3] and Francisco Sánchez[3]

[1]CSIC, Madrid, Spain
[2]Universidad de La Laguna, Santa Cruz de Tenerife, Spain
[3]CTAG - Centro Tecnológico de Automoción de Galicia, Porriño, Spain
[4]Universidad Politécnica de Madrid, Madrid, Spain
[5]Universidad Politécnica de Cartagena, Cartagena, Spain

Contents

8.1 FUNDAMENTALS

Following the Webster definition, an autonomous vehicle "has the right or power of self-government," "exists or act separately from other things

Intelligent Vehicles
DOI: http://dx.doi.org/10.1016/B978-0-12-812800-8.00008-4

or people," "is undertaken or carried on without outside control," and "responds, reacts, or develops independently of the whole."

If we take these words at face value, it is therefore clear that there is a common mistake confusing this definition with the kind of systems currently appearing on mass media, where the driver has very little intervention. As a matter of fact, it is doubtful that autonomous driving, as it is defined above, would bring the commonly accepted benefits (capacity, efficiency, cleanness). New mobility paradigms, where autonomous on-demand vehicles are at the heart, would rather need to be connected and automated vehicles. Only in this context would cars be able to "drive at close range and increase the infrastructure's capacity, prevent a big deal of accidents by communicating with other vehicles and with the infrastructure, save or eliminate parking space by driving one passenger after other" (Holguín, 2016). But what are the differences then between autonomous and automated and connected vehicles?.

Automated vehicles can be defined as those in which at least some safety-critical aspects occur without direct driver input. Or in other words, an automated vehicle is one that can, at least partly, perform a driving task independently of a human driver. When these vehicles, with different levels of automation can communicate among them and with the infrastructure/cloud, a very relevant socioeconomic impact can be obtained, namely safety, congestion and pollution reduction, capacity increase, etc. By contrast, autonomous cars have theoretically the ability to operate independently and without the intervention of a driver in a dynamic traffic environment, relying on the vehicle's own systems and without communicating with other vehicles or the infrastructure.

Original equipment manufacturers (OEMs), Tier ones, and new big players are not developing the same product in this area, but it is not always easy to differentiate their unique selling points because all of them are using erroneously the term "autonomous" cars/driving.

To cope with this problem, the Society of Automotive Engineers issued in 2014 the international norm J3016 (SAE, 2016), bringing order to different prior proposals of standardization from NHTSA and the SAE. It serves as general guidelines for how technologically advanced an automated vehicle is, providing a common taxonomy for automated driving in order to simplify communication and facilitate collaboration within technical and policy domains.

There exist six levels of driving automation spanning from no automation to full automation (see Fig. 8.1). These levels are descriptive and

SAE level	Name	Narrative Definition	Execution of Steering and Acceleration/ Deceleration	Monitoring of Driving Environment	Fallback Performance of Dynamic Driving Task	System Capability (Driving Modes)
Human driver monitors the driving environment						
0	No Automation	the full-time performance by the *human driver* of all aspects of the *dynamic driving task*, even when enhanced by warning or intervention systems	Human driver	Human driver	Human driver	n/a
1	Driver Assistance	the *driving mode*-specific execution by a driver assistance system of either steering or acceleration/deceleration using information about the driving environment and with the expectation that the *human driver* perform all remaining aspects of the *dynamic driving task*	Human driver and system	Human driver	Human driver	Some driving modes
2	Partial Automation	the *driving mode*-specific execution by one or more driver assistance systems of both steering and acceleration/ deceleration using information about the driving environment and with the expectation that the *human driver* perform all remaining aspects of the *dynamic driving task*	System	Human driver	Human driver	Some driving modes
Automated driving system ("system") monitors the driving environment						
3	Conditional Automation	the *driving mode*-specific performance by an *automated driving system* of all aspects of the dynamic driving task with the expectation that the *human driver* will respond appropriately to a *request to intervene*	System	System	Human driver	Some driving modes
4	High Automation	the *driving mode*-specific performance by an automated driving system of all aspects of the *dynamic driving task*, even if a *human driver* does not respond appropriately to a *request to intervene*	System	System	System	Some driving modes
5	Full Automation	the full-time performance by an *automated driving system* of all aspects of the *dynamic driving task* under all roadway and environmental conditions that can be managed by a *human driver*	System	System	System	All driving modes

Figure 8.1 SAE J3016 automation levels for driving (SAE, 2016).

technically-oriented, indicating minimum rather than maximum system capabilities for each level.

A key distinction appears between level 2, where the human driver performs part of the dynamic driving task, and level 3, where the automated driving system performs the entire dynamic driving task. It is worth noting that in no way does it propose a particular order of market introduction.

Fig. 8.1 classes the six automation levels following different classification aspects, relevant to understand the implications of each level. The execution, monitoring, and fall-back can be performed either by the human driver or the system, being the differentiator between levels 1−4. Driving modes are an additional aspect, which allows to talk about full automation, when all of them (e.g., expressway merging, high speed cruising, low speed traffic jam, closed-campus operations...) can be handled by a system.

The classical ADAS belong to Levels 0−1 and many examples are already on the market. Some additional solutions of Level 2 have begun to appear in the last couple of years. However, the market emergence of products in Levels 3−5 is quite controversial.

Although the new players sell the vision that these systems will be on the market in a short period of time (before 2020), the European OEMs, affiliated and represented under the European Road Transport Research Advisory Council (ERTRAC) have a more conservative roadmap (see Fig. 8.2). They envision different pathways for urban environments (high automation in areas with low speed and/or dedicated infrastructure) and elsewhere (building on Level 0 use of ADAS to full automation of Level 5 for trucks and cars).

This chapter aims to shed some light into this fascinating debate, providing an overview of the current state of the technology in automated driving, focusing on the potential of the current technology and the

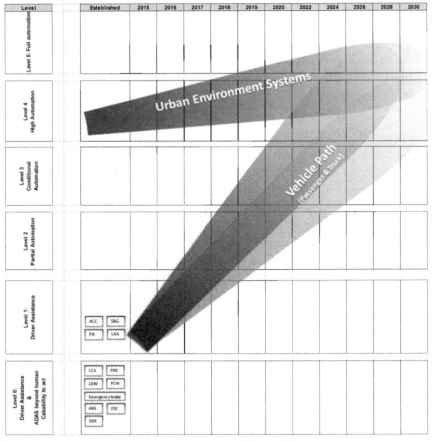

Figure 8.2 Pathways for automated driving (ERTRAC, 2015).

different socioeconomic aspects that will condition the deployment of these complex systems.

To that end, the most relevant technology bricks needed for the automation of a car will be introduced in Section 8.2. In Section 8.3 some relevant aspects of cooperative automated vehicles will be presented. It will be followed by a chapter describing one of the most challenging barriers for automated driving, Verification and Validation (Section 8.4). A brief introduction to the most relevant projects and prototypes through the world is presented in Section 8.5. Finally, a brief description of the socioregulatory aspects will be introduced in Section 8.6.

8.2 TECHNOLOGY BRICKS

This section is devoted to briefly summarizing the key technologies needed for a vehicle to incorporate some degree of automation. The enabling technologies, presented in Part A, are complemented with additional subsystems to conform a closed-loop control architecture, as detailed in Section 8.2.1. From the perception and localization outputs, the vehicle needs to assess the driving situation and infer the subsequent risk (Section 8.2.2). Then, different decision-making strategies are used (Section 8.2.3) to plan, considering a safe driver-vehicle interaction (Section 8.2.4), the most adapted vehicle motion (Section 8.2.5), which is processed and executed, at the end of the decision pipeline, by longitudinal and lateral control algorithms (Section 8.2.6).

8.2.1 Control Architectures

Every control system is articulated based on a functional block architecture, which describes the relationships and dependencies between each of the elements necessary for the control action to be performed correctly. The complex tasks of control of a vehicle must be structured in the form of logical steps that are built on each other and whose complexity can be simplified by a decomposition into functional blocks. In the case of autonomous vehicles, it is also possible and necessary to define all the elements that must be taken into account when carrying out the task of autonomous driving, as well as their relationships and the exchange of information to be shared among them. In this way, we can define the control architecture of an autonomous vehicle as the organization of the different systems of an autonomous vehicle, perception, computation, and

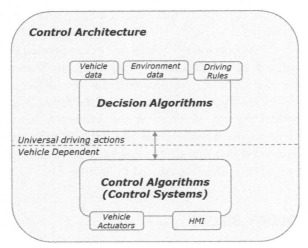

Figure 8.3 General schema of the control architecture.

actuation, to achieve the objectives for which that system has been designed.

Fig. 8.3 shows the general structure of the control architecture of an autonomous vehicle, where the two fundamental elements from the point of view of the management of the driving system are presented: the High-Level control or decision algorithms and the Low-Level control or control algorithms. The first one aims at guiding the autonomous vehicle based on the information supplied by the sensors, regardless of the type of vehicle being piloted. This guiding system generates high level commands such as "turn the steering wheel x degrees," "stop the vehicle" or "select a speed of x kmh." These orders are received by the low-level control, which is the one that is directly related to the actuators of the vehicle, and is able to handle them at its convenience in order to follow the instructions that come from the higher layer. This functional separation is present in most autonomous vehicle control systems, being an inheritance of conventional robotic control systems. The fundamental advantage is that with this structure, the high-level control system is independent of the vehicle in which it is installed, with its actuators being transposable and therefore can be moved from one vehicle to another without making any modifications, easing the interoperability.

The high level control can be subdivided into three elements as shown in Fig. 8.4. In the guidance system, the appropriate algorithms are executed so that the autonomous vehicle tracks the route that has been preset

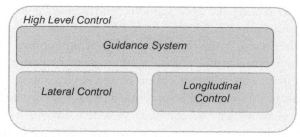

Figure 8.4 Detail of the High-Level Control.

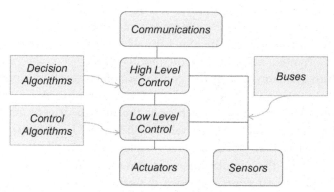

Figure 8.5 Detailed general schema of the control architecture.

to it, as a function of the digital cartography, the information of the sensors, the data of the individual vehicle, and the current driving regulations. This guiding system makes the appropriate decisions and sends them to the specific controllers that perform the lateral control and longitudinal control of the vehicle, who are in charge of managing the trajectory and the speed of the vehicle, whose outputs will be sent to the controller of the low-level to be executed by the actuators.

In this generic architecture for the control of an autonomous vehicle, four additional components must be added, as shown in Fig. 8.5. On the one hand, there is the functional block of actuation, where the actuators of the vehicle are: throttle, brake, steering, and gearbox, although many more systems can be added at convenience, such as lights, alarms, or safety systems. Within the actuation block, ad hoc automations are incorporated, if necessary, when the vehicle itself is not ready to be managed automatically. Once automated, all mechanical components of the vehicle must be able to receive the corresponding instructions generated by the low-level control system.

On the other hand, is the functional block of the sensors, which concerns the equipment that the control systems need to perform correctly the task of guidance and the perception of the environment. The task of guiding is basically through GPS receivers, inertial systems (speedometer, gyroscope, accelerometers, and compass), as well as advanced digital mapping. In recent times, this information for guidance has been enriched with information from the fusion of computer vision or Lidar in so-called visual odometry. The perception of the environment allows the autonomous vehicle to carry out the driving task, taking into account the possible obstacles that appear in its route, the traffic signals, pedestrians, and other vehicles that circulate along the way. For this, two types of sensors are mainly used: computer vision and 3D Lidar.

The third element, part of the control architecture of an autonomous vehicle, is the communications system. The communications allow surpassing the visual horizon to which both the human driver and the sensors installed in the autonomous vehicles are limited, allowing to reach the so-called electronic horizon. This enables the possibility of receiving information from the circulation environment, both infrastructure and other vehicles, so that control systems can take the appropriate decisions or take the necessary actions to prevent an accident, adapt to the circumstances of the road, or anticipate any kind of situation. In addition, communication systems enable autonomous cooperative driving, so that autonomous vehicles are able to collaborate with each other, allowing the emergence of high-level behaviors, such as platooning. In addition, communication systems allow autonomous vehicles to receive information and enable cooperative systems which, while initially designed for manually driven vehicles, must be compatible with autonomous technology.

Finally, the fourth additional element of the generic architecture of the autonomous vehicles is the data buses. These connect all the elements that make up the functional blocks of the architecture in order to allow the exchange of data, sensory information, and control orders in real time. These communication buses are basically two types: on the one hand, Controller Area Network buses (CAN), which allow the transmission of information in real time with messaging defined by priorities. Given their small bandwidth (maximum 1 Mbps), they are mainly used for the interconnection of physical components with low-level control. The other type are the Local Area Network (LAN) networks, networks of computer equipment that operate with a high bandwidth (maximum 1 Gbps), allow the interconnection of all the equipment and sensors with enough capacity.

Once this general autonomous vehicle control architecture is described, it is possible to present the different particular approaches of the different research groups that participate in the autonomous vehicle area.

8.2.2 Situation Awareness and Risk Assessment

Driving is a matter which needs the continuous evaluation of two main factors: the vehicle current state (position, velocity, acceleration, direction) and the environment conditions (near vehicles, obstacles, pedestrians, etc.). To the extent that these two factors are accurately assessed, appropriate decisions can be taken towards reliable autonomous driving. The closer we get to the fact that the vehicle itself is capable of doing this evaluation, the closer we will be to a vehicle-centric approach (Ibañez-Guzmán et al., 2012) to autonomous driving, where the main goal is the safe movement of the vehicle.

This section presents the current state of technology capable of providing onboard situation awareness (SA) and risk assessment (RA) capabilities.

To assess the driving situation, a highly automated vehicle needs the following main capacities (Urmson et al., 2009): global positioning, vehicle tracking, obstacle detection, and self-location in a road model.

The Global Positioning System (GPS) technology is established as the most useful one for answering the Where am I? question (Skog et al., 2009). To face problems such as GPS performance degradation or GPS signal occlusion, it is common to add inertial sensors like Inertial Navigation Systems (INS) or Dead Reckoning (DR) systems (Zhang and Xu, 2012; Tzoreff and Bobrovsky, 2012), with low-cost inertial sensors based on Micro Electromechanical Systems (MEMS) being the most suitable for autonomous driving application. However, due to different drawbacks like the presence of noise in the MEMS-based sensors or failures in the integrating software, INS and DR applications are susceptible to drift, thus causing a loss of accurate vehicle location (Bhatt et al., 2014). The solution that is usually adopted to combat this problem is the use of the above technology combined with sensing technologies such as computer vision, radar, or LIDAR (Jenkins, 2007; Conner, 2011). These, combined with artificial intelligence algorithms, have been developed to

Table 8.1 Main features of different sensing technologies

	Range for optimal operation	Spatial info	Sample rate	Speed measurement	Operation under bad weather	Operation at night
VISION	0–25 m	2D/3D	High	No	Bad	Weak
Radar	1–200 m	–	High	Yes	Excellent	Excellent
LIDAR 2D	1–20 m	2D	Medium	No	Weak	Excellent
LIDAR 3D	1–100 m	3D	Medium	No	Weak	Excellent

put into the market reliable sensing and control systems employed for SA and RA assessment. Table 8.1 shows the main properties of different sensing technologies commonly used in the vehicle-centric approach to autonomous driving.

More recently, some companies have deployed high-performance autonomous cars, with Google Car (Google Inc, 2015) being the most outstanding example. It is fair to say that its astonishingly good behavior in urban environments is not owed so much to its sensing capacities as to very accurate a priori information about the route (ultraprecision 3D maps where the positions of every element—curbs, lights, signals, etc.—are registered with centimeter precision). Nevertheless, it is necessary an accurate detection of relevant information to on-line decide on the safety of the current trajectory. Among aforementioned sensing technologies, it is an accepted fact that 3D LIDAR is the more outstanding one for vehicle-centric autonomous driving, although some manufacturers already provide devices combining different technologies (vision and radar (Delphi Inc., 2015), vision and LIDAR (Continental, 2012)).

Table 8.2 shows the main features of some currently available 3D LIDAR devices for autonomous driving purposes.

Safety is a key piece of the autonomous driving paradigm. Risk (the probability of a vehicle suffering some damage in the future) must be assessed in every particular vehicle situation.

For this purpose, recent works propose mathematical models (motion models) to predict how a situation will evolve in the future. For autonomous driving, those motion models which consider interaction (among vehicles, and pedestrians) are the most useful ones. The challenges include detecting interactions and identifying interactions; the commonly used variables are communications, joint activities, or social conventions and the common tools include rule-based systems. Different motion models are considered and Lefevre et al. (2014) have summarized their main characteristics, as shown in Table 8.3.

Table 8.2 Some commercial 3D LIDAR devices

	Range	Type of data Data rate	FOV (H)-(resolution) FOV(V)-(resolution)	Weight	Accuracy
All-purpose 3D LIDAR					
VLP-16	100 m	Dist./reflect. 0.3 M points/s	360°–(0.1°) ±15°–(0.4°)	0.8 kg	±3 cm
HDL-32	80–100 m	Dist./reflect. 0.7 M points/s	360°–(0.1°) +10°/−30°–(0.4°)	1.3 kg	±2 cm
HDL-64	120 m	Dist./reflect. 2.2 M points/s	360°–(0.018°) +2.0/−25°–(0.4°)	13.5 kg	<2 cm
RobotEye	160 m	Dist./reflect. 0.5 M points/s	360°–(0.01°) on-line adjust. ±35°–(0.01°) on-line adjust.	2.8 kg	±5 cm
Minolta SingleBeam	100 m	Dist./reflect. 0.4 M points/s	180°–(0.01°) 25°–(0.01°)	N.A.	N.A.
Specific purpose 3D LIDAR					
DENSO Pedestrian	200 m	Dist./reflect. 8k points/s	40°–(0.1°) 2°–(1°)	0.2 kg	±6 cm
DENSO Lane	120 m	Dist./reflect. 16k points/s	36°–(0.1°) −2/+2°–(1°)	0.2 kg	±6 cm
SRL1 Obstacle	10 m	Distance 2.3k points/s	27°–(1°) 11°–(1°)	<0.1 kg	±10 cm

Table 8.3 Two motion models

	Variables	Challenges	Tools
Physics-based	Kinematic and dynamic data	State estimation from noisy sensors Sensitivity to initial conditions	Kalman filters Monte Carlo sampling
Maneuver-based	Intentions Perception Surrounding objects and places	Complexity of intentional behavior	Clustering Planning as prediction Hidden Markov models Goal oriented models Learning

8.2.3 Decision Making

Human driving capacity requires not only the ability to properly maneuver the steering wheel, brake, and accelerator according to a set of traffic rules, but also must assess social risks, health, legal consequences, or the life-threatening results of driving actions (e.g., What should the vehicle do if a pedestrian does not stop at a red light?). The resolution of driving requires high levels of human knowledge, for this reason science uses complex artificial intelligence systems to emulate them. The decision-making system has the role of interpreting the abstract information supplied for the perception system of the vehicle and generates actions to carry out sustainable and safe driving.

To operate reliably in the real world, an autonomous vehicle must evaluate and make decisions about the consequences of its potential actions by anticipating the intentions of other traffic participants. The first decision-making systems in autonomous vehicles appeared in 2007 in DARPA Urban Challenge (Urmson et al., 2008). These systems allowed the vehicles to operate in the urban scenarios in which they were involved, U-turns, intersection, parking areas, and real traffic among others. These early decision systems used common elements such as planners where systems were implemented using finite state machines, decision trees, and heuristics. More recent approaches have addressed the decision-making problem for autonomous driving through the lens of trajectory optimization. However, these methods do not model the closed-loop interactions between vehicles, failing to reason about their potential outcomes (Galceran et al., 2015). Nowadays, there are no real systems that outperform a human driver. Advances in decision making are aimed at increasing the intelligence of the systems involved in decision making. Cognitive systems (Czubenko et al., 2015), agents systems (Bo and Cheng, 2010), fuzzy systems (Abdullah et al., 2008), neural networks (Belker et al., 2002), evolutionary algorithms (Chakraborty et al., 2015), or rule-based methods (Cunningham et al., 2015) compose the Intelligent-Decision-Making Systems (IDMS) (Czubenko et al., 2015). Fig. 8.6 shows the location of IDMS from a point of view of the functional architecture for autonomous driving (Behere and Törngren, 2015).

A general intelligent system in the autonomous driving functional architecture will contain processing units of sensory processing, a world model, behavior generation, and value judgement with information flow as shown in Fig. 8.7 (Meystel and Albus, 2002).

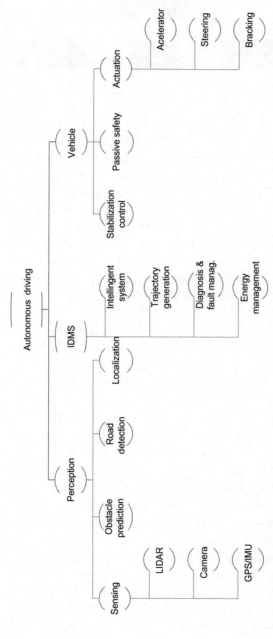

Figure 8.6 Autonomous driving functional architecture.

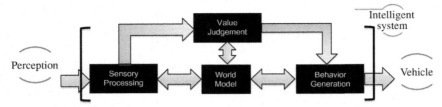

Figure 8.7 General ANSI intelligent system description.

Autonomous driving in complex scenarios where multiple vehicles are involved (e.g., urban areas) with inherent risk to the integrity of people, require real-time solutions. In this case decision making demands reliability, safety, and a fault tolerance system. In this direction it is necessary to mention, a real-time motion planning algorithm, based on the rapidly-exploring random tree (RRT) approach proposed by (Kuwata et al., 2009). The proposed algorithm was at the core of the planning and control software for Team MIT's entry for the 2007 DARPA Urban Challenge, where the vehicle demonstrated the ability to complete a 60 mile simulated military supply mission, while safely interacting with other autonomous and human driven vehicles.

A Multicriteria Decision Making (MCDM; Furda and Vlacic, 2010) and Petri nets (Furda and Vlacic, 2011) are proposed for solving the problem of the real-time autonomous driving. MCDM offers a variety of benefits such as:

- The hierarchy of objectives allows a systematic and complete specification of goals to be achieved by the vehicle.
- The utility functions can be defined heuristically to reflect the choices of a human driver, or, alternatively, learning algorithms can be applied.
- MCDM allows the integration and evaluation of a very large number of driving alternatives.
- Decision flexibility can be achieved by defining the set of attribute weights depending on the road conditions.
- Additional objectives, attributes, and alternatives can be added without the need for major changes.
- The driving maneuvers are modeled as deterministic finite automata.
- The decision making unit is modeled as a Petri net.

The author concluded that the method was highly based on heuristics, but the application of MCDM in this new research area offers a variety of benefits with respect to the problem specification, decision flexibility, and scalability.

8.2.3.1 Simulation and Software tools for IDMS

The first tasks of testing and implementation of IDMS are strongly relational with software simulation and implementation. There are numerous works about framework and middleware capable of the testing and development of IDMS (Veres et al., 2011; Behere, 2013). Table 8.4 shows a set of software environments and tools capable of participating in the process of modeling, simulation, and implementation of a decision-making system. The table shows the name of the environment, a short description, its purpose and features, and a website where the software packages can be found.

8.2.4 Driver—Vehicle Interaction

As shown in Fig. 8.8, a car's cockpit has constantly increased its complexity and evolved, according to the implemented systems, available technologies, and driver needs present at each specific moment in automotive history.

At the end of the 1990s and the beginning of 21st century the increase of in-vehicle implemented functions and the associated human interaction complexity, obliged OEMs to pay more attention to the proper design of the driving place and the driver—vehicle interaction. At that time, new transversal HMI departments were introduced in most R&D automotive manufacturing organizations to manage this new challenge. The role of this department was to take the leadership to develop, from a global vehicle perspective, the complete driving place in terms of the driver—vehicle interaction—of course, with the support of all other technical areas involved.

In recent years, the massive introduction of new driver assistance systems and infotainment and telematics applications has given even more relevance to the HMI design activities.

Nowadays, the introduction of automated driving functions in the next-generation vehicles has HMI and driver—vehicle interaction as one of the key challenges for its success in the market. In this sense, there is a need to rethink ant to redesign the driving place and the way the "driver" or "vehicle supervisor" is going to interact with the vehicle and the automated and autonomous functions.

Some of the key questions related to the HMI design for automated vehicles are the following:

• How can two different driving modes coexist at the same time?

Table 8.4 Analysis, simulation, and controller development tools for developing of IDMS

Software environment/ tools	Short description, purpose, or features	Website
Charon	Charon HS modeling language, supports hierarchy and concurrency, has simulator and interfaces to Java.	www.cis.upenn.edu/mobies/charon
Modelica/ Dymola	OO HS modeling language for multi-domain physics, has simulator, has object libraries.	http://www.modelica.org/ http://www.3ds.com/products/catia/portfolio/dymola
HyTech	Modeling and verification of hybrid automata, has symbolic model checker.	http://embedded.eecs.berkeley.edu/research/hytech/
HyVisual	Visual modeling (Ptolemy II) and simulation of HS, supports hierarchies.	http://Ptolemy.berkeley.edu/hyvisual/
Scicos/ Syndex	Modeling and simulation of HS, has toolbox, real-time code generation, provides formal verification tools.	http://www.scicos.org/
Shift	Has its own programming language for modeling of dynamic networks of hybrid automata, has extension to real-time control and C-code generator.	http://path.berkeley.edu/shift/
Simulink/ Stateflow	Has analysis, simulation, has libraries and domain specific block-sets, can compile C-code for embedded applications via the use of embedded MATLAB™, and Real Time Workshop™.	http://www.mathworks.com/
OROCOS	Portable C++ libraries for advanced machine and robot control. Including kinematics chains, EKF, Particle Filters, RT software components, state machine, etc.	http://www.orocos.org/
ROS	It is a set of software libraries and tools that help you build robot applications. From drivers to state-of-the-art algorithms, and with powerful developer tools. It is NOT RT framework.	http://www.ros.org/

(Continued)

Table 8.4 (Continued)

Software environment/ tools	Short description, purpose, or features	Website
YARP	It is a communication middleware, or "plumbing," for robotic systems. It supports many forms of communication (tcp, udp, multicast, local, MPI, mjpg-over-http, XML/RPC, tcpros, . . .).	http://www.yarp.it/
CLARATy	It consists of a Functional Layer that provides abstractions for various subsystems and a Decision Layer that can do high level reasoning about global resources and mission constraints.	http://ieeexplore.ieee.org/ lpdocs/epic03/wrapper.htm? arnumber=1249234
BALT & CAST	It is a middleware for cognitive Robotics development.	http://ieeexplore.ieee.org/xpls/ abs_all.jsp?arnumber= 4415228; http://ieeexplore. ieee. org/lpdocs/epic03/ wrapper.htm?arnumber= 4415228

Figure 8.8 Evolution of car's driving place.

- How would the driver select the automated mode?
- Which functions should be controlled during automated mode and how to manage them?
- How should the vehicle give back the control to the driver?
- How should be the transitions between different automation levels be achieved?
- Which kind of information should the vehicle provide to the driver in automated mode to make him/her feel safe and comfortable?
- Which present and future technologies can be used and are more appropriate to make all this happen?

Figure 8.9 Example of an augmented reality evaluation study for an automated driving prototype (carried out at CTAG's driving simulator).

To approach in a consistent manner all these questions, there are several issues which are very relevant and that must be taken into account:

- To follow a systematic methodology that allows to identify relevant functions and HMI parameters for automated driving, to design innovative HMI solutions, and to evaluate the developed concepts from a human factors perspective.
- To implement a multidisciplinary approach that takes into account, from the very first moment, the opinion of engineers, specialists, and technicians coming from different disciplines.
- To prepare and to adapt new methods and new development and evaluation tools (driving simulators, mock-ups, vehicle prototypes, ...) that allows to perform clinics with users to measure in detail ergonomics and technical aspects.
- To test the driver—vehicle interaction proposed solutions in very early development stages to get the voice of the user at the beginning of the development process.
- To use, in an intelligent manner, the future possibilities of new HMI technologies such as augmented reality (see Fig. 8.9), state-of-the art displays, new control elements, gesture recognition, interior lighting, reconfiguration possibilities, etc.
- To improve, in a very safely manner, the in-vehicle onboard user experience.

8.2.5 Motion Planning

Motion planning first for mobile robots and then for autonomous vehicles has been extensively studied over the last few decades. The resulting strategies were designed to meet, under different hypotheses, a variety of kinematic, dynamic, and environmental constraints. In this section, path and speed planning are presented under a specific approach, but different strategies could be alternatively considered (c.f. Paden et al., 2016 or Katrakazas et al., 2015 for more details)

8.2.5.1 Path Planning

General techniques to obtain optimal paths can be grouped into two categories: indirect and direct. Indirect techniques discretize the state/control variables, and convert the path planning problem into one of parameter optimization which is solved via nonlinear programming (Dolgov et al., 2010) or by stochastic techniques (Haddad et al., 2007). The latter use Pontryagin's maximum principle and reexpress the optimality conditions as a boundary value problem, whose approximate solutions have been investigated under a large set of possibilities and constraints. The subsequent description is based on the latter family, where a local planner is encapsulated into a generic procedure to cope with complex scenario topologies and obstacle avoidance.

The path planning system is composed of several subsystems that operate separately as standalone processes depending on one another. These subsystems are: *Costmap Generation, Global Planner,* and *Local Planner.* The former is in charge of the computation of the costmap that will be used by the other two methods in order to compute the trajectories, considering the safety existing for the different possibilities (attending to the obstacles in the environment, as well as to the estimations of the expected changes in the near future); the second is used for the computation of a trajectory that allows the vehicle to travel between the current position and the goal in the unstructured map. The third provides the system with the mechanisms needed to follow it, computing the commands required by the low-level controller to move the prototype.

8.2.5.1.1 Costmap Generation

The costmap maintains information about occupied/free areas in the map in the form of an occupancy grid. It uses sensor data and information from the static map to store and update information about obstacles in the world, which are marked in the map (or cleared, if they are no longer there). Costmap computation is supported on a layered costmap, which will be used for the integration of the different information sources into a single-monolithic costmap. At each layer, information about occupied/

free areas in the surroundings of the vehicle is maintained in the form of an occupancy grid, using the different observation sources as input. Using this information, both dynamic and static obstacles are marked in the map. For example, let us suppose each cell in the map can have 255 different cost values. Then, at each layer, costmap is represented as follows:

- A value of 255 means that there is no information available about a specific cell in the map.
- 254 means that a sensor has marked this specific cell as occupied. This is considered as a lethal cell, so the vehicle should never enter there.
- The rest of cells are considered as free, but with different cost levels depending on an inflation method relative to the size of the vehicle and its distance to the obstacle.

Cost values decrease with the distance to the nearest occupied cell using the following expression:

$$C(i,j) = \exp\left(-1.0 \cdot \alpha \cdot \left(\left\|c_{ij} - \vec{o}\right\| - \rho_{\text{inscribed}}\right)\right) \cdot 253 \qquad (8.1)$$

In this expression, α is a scaling factor that allows increasing or decreasing the decay rate of the cost of the obstacle. $\left\|c_{ij} - \vec{o}\right\|$ is the distance between the cell $c_{ij} \in C$ (where C is the set of cells in the costmap) and the obstacle. Finally, $\rho_{\text{inscribed}}$ is the inscribed radius, which is the inner circle of the limits of the car.

Despite all of them being free cells, normally different distance thresholds are defined in order to set different danger levels in the map. For example, it is possible to define four thresholds:

- ζ_{lethal}: There is an obstacle in this cell, so the vehicle is in collision. It would be represented by the cost level 254.
- $\zeta_{\text{inscribed}}$: Cell distance to the nearest obstacle is below $\rho_{\text{inscribed}}$. If the center of the vehicle is in this cell, it is also in collision, so areas below this distance threshold should be avoided. Cost level would be 253.
- $\zeta_{\text{circumscribed}}$: If the vehicle center is on this cell, it is very likely that the car is in collision with an obstacle, depending on its orientation. A cell with a distance to an obstacle below this threshold should be avoided, but there are still chances of being in one of them without colliding an obstacle.
- The rest of cells are assumed to be safe (except from those with unknown cost, for which it is not known if they are occupied or not, being considered as lethal).

In the presented approach, just those paths passing through cells with a cost below $\zeta_{\text{circumscribed}}$ are considered. This cost is obtained using the

Eq. 8.1 and other cost factors that will be explained later. Paths passing through the cells over this threshold will be truncated at the last safe point.

For the computation of the costmap and the costs associated to each cell, ROS plugin costmap 2d (http://wiki.ros.org/costmap_2d) could be used, which implements some of the functionalities described in this section.

Layered Costmap

Nowadays, in the layered costmaps, four different layers are usually being considered:

- A first layer represents the obstacles in a static map previously captured. This map represents the static obstacles in the whole area in which the vehicle will move. This layer is the only one used by the nonprimitive-based global planner, since nonstatic obstacles are not being considered for nonprimitive-based trajectory generation (meaning by nonstatic those obstacles that are not already included into the map). These are supposed to be avoided at local planning level.

- A second layer, also based on a static map, is included. For optimization reasons, in this and following layers, the costmap is not computed for the whole map at each iteration. Instead, just the cells in an area centered into the current car position are updated. The goal is not to update the whole map, since these layers are just used for local planning or local maneuvering. Static obstacles are also included for local planning, since the vehicle is not desired to pass along restricted areas while avoiding obstacles. This allows the vehicle to know which areas are forbidden, also at local planning level.

- A third layer is used to represent the dynamic obstacles detected by the different sensors. Using this input, ground is detected and removed, extracting just the vertical obstacles to which the vehicle could collide. Parameters in this layer are chosen so the obstacle inflation is stronger than the one computed for the second layer. This gives more priority to the obstacles being detected in real time over those in the static map.

- The last layer provides an estimation of the future motion of the dynamic obstacles. To do so, input point clouds are segmented using a voxel grid, in order to reduce dimensionality. The world surrounding the vehicle is divided into a discrete number of voxels of equal size. For each voxel, an occupancy probability is assigned, based on the number of points from the input point cloud in its neighborhood.

Using this probability, valid voxels (with a higher occupancy probability) are distinguished from the noisy ones (with a smaller probability).

All these layers are combined into a single costmap. Note that it is interesting to include the motion of the obstacles in the costmap because then the vehicle tries to avoid the obstacle by the side in which it is not crossing its trajectory

8.2.5.1.2 Global Planner

Usually there are two global planners in use in autonomous vehicles: the primitive-based planner and the nonprimitive global planner. These planners are intended to obtain a feasible path going from the vehicle's current position to a determined goal.

Although both methods are included in this section, their aims in the system are completely different. The nonprimitive-based global planner is used for regular navigation, while the primitive-based global planner, is used for recovering the vehicle in situations in which there is an obstacle in the way for a long time, or the vehicle is performing some complex maneuver, like parking.

Primitive-Based Global Planner

The primitive-based global planner constructs a path from the vehicle's position to a desired goal. The path is generated by combining "motion primitives," which are short, kinematically feasible motions. These motion primitives are generated using a model of the vehicle in order to comply with the curvature restrictions of the vehicle.

The computation of these primitives is performed as follows: a set of predefined orientations is considered. For each orientation, the model is evolved until it reaches one of the predefined orientations, at different speeds. This process is done both forwards and backwards. After this process, a set of small trajectories that fulfill vehicle restrictions is obtained, which will be used as the building blocks for the planner.

Having these, an ARA algorithm is used for the search of a feasible path. At each node expansion, a new x, y, and θ position is explored, until the best path is found or the exploration time finishes (if so, the best path found until then is used). During this search, the cost of backward primitives is set higher than the cost of forward ones to prevent the vehicle from using backward paths as much as possible, without decreasing the performance. Also the original search algorithm can be improved by adding a new cost that penalizes the concatenation of forward and backward primitives. This is done with the intention of planning more natural paths.

Nonprimitive-Based Global Planner

The nonprimitive based global planner computes the minimum cost path from the vehicle's position to the goal using, e.g., the Dijkstra's algorithm. Given the speed of the search algorithm to obtain the global plan, this planner is being used as a rough estimate of the route that the vehicle is going to follow. The static obstacles of the costmap are then overinflated in order to make the planner construct smooth paths, feasible for being followed by an Ackermann vehicle.

If the generated routes are not constructed bearing in mind the non-holonomic restrictions of the vehicle, it is frequent that the initial angle between the vehicle's orientation and the orientation of the global plan is larger than the maximum angle required by the local planner to generate feasible paths. That is the reason why the nonprimitive-based planner is used in combination with a local planner state machine that takes into account this circumstance and reorients the vehicle properly before using the Frenet-based local planner.

8.2.5.1.3 Local Planner

Once the global path is defined, a method is required that is able to compute the steering and speed commands needed to control the vehicle, in order to follow that path. This method should be also able to avoid the obstacles present in the road. This must be done in a safe and efficient way.

The basic idea of the local path generation is to define a set of feasible paths and choose the best option in terms of their cost. The winner path defines the steering and speed commands that the vehicle will use. Having options among local paths is useful to overcome the presence of unforeseen obstacles in the road.

Usually the current Euclidean coordinate system is transformed into a new system based on the Frenet space. This space is computed as follows: the global path is considered as the base frame of a curvilinear coordinate system. The feasible local paths are defined in terms of this base frame in the following way:

- The nearest point (where the distance is computed perpendicular to the global path) to the main trajectory will be the origin of the curvilinear coordinate system.
- The horizontal axis will be represented by the distance over the global path, in its direction.
- The vertical axis is represented by the vector perpendicular to the origin point, which is pointing to the left with respect to the path direction.

In this schema, trajectories can be computed easily in the curvilinear space (that is, maneuvering information is generated). These are then transformed to the original Euclidean space, in which the obstacles information is added by assigning costs to each path.

Based on this idea, the method can be divided in five stages:

1. Generation of the costmap. Using the information generated by the sensors or by the methods described in previous sections, the system constructs a costmap in which costs are related to the distance to obstacles.
2. Base frame construction. Based on the global path constructed in the previous section, the base frame of the curvilinear coordinate system is generated.
3. Candidate paths generation. Candidate paths are generated into the curvilinear space. Then, they are transformed to the Euclidean space.
4. Selection of the winner path. Costs for all the paths are assigned, and the one with the lowest value is selected.
5. Computation of the vehicle commands. Vehicle speed and steering angles are computed based on the characteristics of the winner path.

Base Frame Construction

In this stage, the base frame of the curvilinear coordinate system is defined, so the algorithm is able to compute the trajectories in this space as if the global plan was a rectilinear trajectory. At this point, the potential presence of obstacles or the restrictions associated to the vehicle's motion model are not considered, limiting this stage to the generation of trajectories.

The origin of coordinates of the base frame is the nearest point in the global plan to the vehicle's position.

The base frame's arc length is obtained as the distance of each point along the global plan (represented as a green line) to the origin of coordinates. This distance is represented in the x-axis of the curvilinear system. y-axis, q, represents the perpendicular lateral distance respect to the path. The left side is represented by positive values and the right by negative values.

For the computation of the transformation between the Euclidean and the curvilinear coordinate system, the path curvature κ is needed:

$$\kappa = \frac{S}{Q} \cdot \left(\kappa_b \cdot \frac{(1 - q \cdot \kappa_b) \cdot (\partial^2 q / \partial s^2) + \kappa_b \cdot (\partial q / \partial s)^2}{Q^2} \right) \qquad (8.2)$$

where

$$\begin{cases} S = \text{sign}(1 - q \cdot \kappa_b) \\ Q = \sqrt{\left(\dfrac{\partial q}{\partial s}\right)^2 + (1 - q \cdot \kappa_b)^2} \end{cases} \qquad (8.3)$$

A generated path will be rejected if $q > \frac{1}{\kappa_b}$. In this case, the path curvature and sense is opposed to that of the base frame. The path violates the nonholonomic condition of the movement of the vehicle, so the vehicle enters in a recovery state.

Only paths with a lateral offset q equal or smaller to the curvature radius of the base frame $\frac{1}{\kappa_b}$ are accepted.

If $q = \frac{1}{\kappa_b}$, that means that the path passes through the center of curvature of the base frame. Also, the maximum curvature a path can have in order to be feasible by the vehicle is limited by the maximum steering angle. If this restriction is violated, the corresponding path is rejected. Curvature is directly related to the movement of the vehicle, which can be described through several models.

Candidate Paths

As seen, path generation is performed in the curvilinear space, without considering the obstacles in the environment. These will be taken into account later, once the tentative trajectories are transformed to the Euclidean space.

Maneuvering paths generation. The curvature of the generated paths is defined by the lateral offset q with respect to the base frame. First and second order derivatives of q are needed for the computation of κ (see Eqs. 8.2 and 8.3), so a function dependent on the lateral offset is needed to compute a smooth lateral change.

Candidate paths generation. Once the paths in the curvilinear coordinate system are computed, they are transformed to the Euclidean space. In this new space, their associated costs will be evaluated. Now the paths are in Euclidean coordinates, the maximum distance they can reach individually (if obstacles are considered) can be calculated. To do that, the cells C_{ij} of the costmap associated to the points of the trajectory are checked. If this cost is over the value associated to the threshold circumscribed, the path is truncated at this point.

When a path collides with an obstacle, it is not completely removed. The reason is that there are certain situations in which the maximum length cannot be reached with any path. However, it is still desirable to approach slowly towards the maximum reachable point, with the hope that the obstacles that are blocking the way will disappear in the next iterations. In crowded areas with many pedestrians this is a typical situation: the way is blocked, but when pedestrians see a vehicle that is approaching, they move away. However, if the vehicle reaches a point in which it can not move for a long time, the recovery behavior is triggered. The problem with this strategy is that one of the colliding paths could win even if there is a path able to go through a clear area. In order to avoid that, a weighted cost function based schema is implemented. This schema, which permits a smart selection of the winner path, is explained in the next section.

Winner Path

The winner path is selected through the use of a linear combination $J[i]$ of weighted cost functions, related to the following parameters: occlusion, length, distance to the global path, curvature and consistency of the path. $J[i]$ is evaluated as follows:

$$J[i] = \omega_0 C_0[i] + \omega_l C_l[i] + \omega_d C_d[i] + \omega_\kappa C_\kappa[i] + \omega_c C_c[i] \qquad (8.4)$$

Here, i is the path index, and C_0, C_l, C_d, C_κ, and C_c are the costs of occlusion, length, distance to the global path, curvature, and consistency, respectively. Their relatives ω_i, $i \in \{o, l, d, \kappa, c\}$ are the associate weights that allow to adjust the influence of each of the costs to the final cost value. All these costs are normalized to 1.0, and

$$\sum_{i \in \{o,l,d,\kappa,c\}} \omega_i = 1.0 \qquad (8.5)$$

so it is easy to determine the proportional influence of each weight

Occlusion. The occlusion cost is related to the safety of the path. This cost estimates the goodness of a path, with the bests paths being those passing far enough from the obstacles. To do that, the method iterates along the path, simulating the footprint of the car at each position. The occlusion cost corresponding to the trajectory point i will be the maximum cost of each of the cells $c_{ij} \epsilon C$ under the footprint of the vehicle at that position. Based on this, the occlusion cost will be:

$$C_0 = \frac{\max\{c_i\}}{255}, \quad i = 1, ..., L \tag{8.6}$$

In this expression, L is the length of the current path being evaluated. $\max\{c_i\}$ is the maximum value of all the costs, associated to a point in the path. If the maximum value of each cost is 255, so C_0 is divided by this value, in order to normalize it to 1.

Length. This cost represents the length of the current path. By iterating along the points in the path, the distance between them is accumulated, so the real distance traveled in Euclidean coordinates is known. The longer a path is, the better, as it is assumed that it will traverse an obstacle-free zone. Thus, long paths should produce low cost values. This is done through the expression:

$$C_l = 1 - \frac{\sum_{i=1}^{L} \|p_i - p_{i-1}\|}{q_{f_{max}} + s_f} \tag{8.7}$$

Here, p_i is a certain point inside the evaluated path. $q_{f_{max}}$ is the maximum value that a q_f can have for a certain path. Lengths are normalized to a value that a path will never reach. This cost is subtracted from 1.0, in order to make it comparable to the rest of costs (as said, lower values are preferred respect to the higher ones).

Distance to the global path. In the presented implementation, information about the average lateral o_{set} with respect to the global path has been also considered. The use of this cost will benefit the choice of those paths that are allowed to come back to the global path after an occasional obstacle is avoided. It is computed as follows:

$$C_d = \frac{\sum_{i=1}^{L} \|p_i - \mathrm{nearest}(p_i, g)\|}{L \cdot q_{f_{max}}} \tag{8.8}$$

where nearest(p,g) is the nearest point in the global path g to the point p. This cost is normalized with respect to the maximum expected offset $q_{f_{max}}$.

Curvature. This cost gives priority to the smoother paths. Let $p(x_i, y_i)$, $i = 1, ..., L$, be a point in the path. Then,

$$C_\kappa = \max \left\{ \frac{\dot{x}_i \cdot \ddot{y}_i - \ddot{x}_i \cdot \dot{y}_i}{\left(\dot{x}_i + \dot{y}_i\right)^{3/2}} \right\}, i = 1, ..., L \tag{8.9}$$

Consistency. This cost avoids the continuous changes in the winner paths between iterations. Once the vehicle starts a maneuver, the idea is to keep this behavior in the following iterations. This is done through the following expression:

$$C_c = \frac{1}{s_2 - s_1} \int_{s_1}^{s_2} l_i ds \tag{8.10}$$

The lateral cost $l_i(s)$ is the distance between the current and the previous winner path at the same longitudinal position s.

Selection of the winner path. Once all costs are computed, the expression described in Eq. (8.4) is applied. In those paths for which it is impossible to advance due to the presence of a nearby obstacle or because the car is incorrectly aligned to the global path (meaning that no valid paths can be generated in this situation), the cost will be negative (invalid path). From all paths, that path with the smallest cost (winner path W) is selected. If for any reason there are no valid paths, a recovery maneuver is initiated.

8.2.5.2 Speed Planning

The speed reference is commonly assumed to be continuously differentiable, and is often designed by optimizing an appropriate performance index (minimum time is the commonest criterion, but minimum acceleration and/or jerk has also been used). For most of them, as topological semantic maps are not used, iterative or optimization processes are required to satisfy a certain number of driving comfort constraints—maximum speed, longitudinal and lateral acceleration, and jerk. The strategy described below tackles this problem, approximating the considered path by well-known primitives, from which a speed profile can be derived.

Indeed, any path in an unstructured environment can be decomposed, with the help of a path planning algorithm, into a succession of turns—composed of clothoids and arcs of circles—and straight lines. The clothoid is chosen because an arc of a clothoid has variable curvature, in every point proportional to the arc length, and it provides the smoothest link between a straight line and a circular curve. It is used in roads and railroads design: the centrifugal force actually varies in proportion to the time, at a constant rate, from zero value (along the straight line) to the maximum value (along the curve) and back again.

This decomposition is extremely useful in finding closed-form optimal speed profiles because both straight line segments and circle arcs can be associated with constant speeds. More precisely, when a turn is initiated the maximum velocity will be constrained by the comfort lateral acceleration threshold, and when a straight segment is being tracked, the maximum longitudinal speed, acceleration, and jerk will be the limits imposed on the reference speed.

The speed profile can be defined as follows:
- Constant speed curves at a minimum value V_{min} when the curvature profile is a circular arc or its preceding clothoid.
- A smooth transition from the minimum value V_{min} to a maximum allowed speed V_{max} and back again to V_{min} that fulfills the acceleration and jerk constraints.
- A set of one or two smooth transition curves (of type 2 above) that go from zero to the maximum speed, and vice versa.

In order to obtain closed-form expressions for the second type of curve, the speed trajectory is divided into a number of intervals. Let us suppose seven intervals $[t_{i-1}, t_i]$, $i = 1 \ldots 7$ and represented in terms of the arc length s_r as follows:

$$\dddot{s}_r(t) = \begin{cases} \dddot{s}_{r_{max}}, & t \in [t_0, t_1] \ \text{or} \ t \in [t_6, t_7] \\ 0, & t \in [t_1, t_2] \ \text{or} \ t \in [t_3, t_4] \ \text{or} \ t \in [t_5, t_6] \\ -\dddot{s}_{r_{max}}, & t \in [t_2, t_3] \ \text{or} \ t \in [t_4, t_5] \end{cases}$$

$$\ddot{s}_r(t) = \ddot{s}_r(t_{i-1}) + \dddot{s}(t)_r \cdot (t - t_{i-1})$$

$$\dot{s}_r(t) = \dot{s}_r(t_{i-1}) + \ddot{s}_r(t_{i-1}) \cdot (t - t_{i-1}) + \frac{1}{2} \dddot{s}_r(t_{i-1})(t - t_{i-1})^2$$

$$s_r(t) = s_r(t_{i-1}) + \dot{s}_r(t_{i-1}) \cdot (t - t_{i-1}) + \frac{1}{2!} \ddot{s}_r(t_{i-1})(t - t_{i-1})^2 + \frac{1}{3!} \dddot{s}_r(t_{i-1})(t - t_{i-1})^3$$

$$(8.11)$$

The arc length will go from the initial point of the closing clothoid in a turn $(s_r(t_0))$ to the final point of a straight line segment $(s_r(t_7))$, the initial and final speeds $(\dot{s}_r(t_0), \dot{s}_r(t_7))$ will be set by the minimum speed V_{min}, and the initial and final accelerations $(\ddot{s}_r(t_0), \ddot{s}_r(t_7))$ and jerks $(\dddot{s}_r(t_0), \dddot{s}_r(t_7))$ will be both equal to zero. Concerning the comfort constraints, the maximum speed will be V^*_{max} and the maximum speed and acceleration will be determined by design parameters γ_{max} and J_{max}.

Note that the value of V^*_{max} corresponds to the V_{max} previously defined if there is enough distance to reach the target. If the available arc length is less than some critical value, the maximum speed will be set equal to the initial speed V_0 resulting in the generation of a constant speed profile. Otherwise, a maximum speed between V_0 and V_{max} will be computed. The closed form polynomial expression of equations (8.11) permits the maximum speed to be computed as follows:

$$V^*_{max} = \begin{cases} V_{max}; & \text{if condition1 is satisfied} \\ V_0; & \text{if condition2 is satisfied} \end{cases} \tag{8.12}$$

condition1

$$\Delta_s \geq (V_{max} + V_0)\sqrt{\frac{V_{max} - V_0}{J_{max}}} + (V_{max} + V_{min})\sqrt{\frac{V_{max} - V_{min}}{J_{max}}} \tag{8.13}$$

condition2

$$\Delta_s < \frac{V_0}{2} \cdot \left(\frac{V_0}{\gamma_{max}} + \frac{\gamma_{max}}{J_{max}}\right)$$
$$- \frac{1}{2J_{max}}\left(\gamma^2_{max} - \sqrt{\gamma^4_{max} + 8J^2_{max}\gamma_{max}\Delta_s + 4J^2_{max}V^2_0 - 4\gamma^2_{max}J_{max}V_0}\right) \tag{8.14}$$

where $\Delta_s = s(t_7) - s(t_0)$

An alternative algorithm can be implemented to reduce the overall time needed to cover the path by slightly compromising the passenger comfort. Instead of reducing the speed to V_{min} in each turn, only the nondegenerate turns are taken into account for this purpose.

8.2.6 Vehicle Control

Mathematical models are of great importance in the analysis and control of automotive vehicle dynamics. Several mathematical models are available in the literature with different levels of complexity and accuracy according to the physical phenomena captured. Usually the motion of the vehicle is considered in the yaw plane, mainly describing the longitudinal and lateral vehicle motion. In the description of the vehicle motion, different longitudinal and lateral dynamic couplings must be considered:

- Dynamic and kinematic couplings are due to the motion in the yaw plane caused by wheel steering.
- The interaction between tire and road is at the origin of another important coupling.
- The longitudinal and lateral accelerations cause a load transfer between the front and rear axles as well as the right and left wheels.

The complexity degree is used to obtain a trade-off between complexity and accuracy. A complexity model can provide a good accuracy level but remains too complex for controller synthesis. For this reason, usually a nonlinear bicycle model is used for lateral control and a one-wheel vehicle model for longitudinal control design.

A nonlinear bicycle model considers the longitudinal (x), lateral (y), and yaw motion (θ). For this model, it is assumed that the mass of the vehicle is entirely in the rigid base of the vehicle, and it considers the pitch load transfers while neglecting the lateral load transfer caused by roll motion.

In the Fig. 8.10, α is the steer angle, and a and b represent the distance between wheels and the gravity center of the vehicle. The indexes f and r indicate front and rear.

The dynamic equations are:

$$m(\ddot{x} - \dot{y}\dot{\theta}) = \sum_{i=f,r} F_{xi} + F_r$$
$$m(\ddot{y} + \dot{x}\dot{\theta}) = \sum_{i=f,r} F_{yi} \qquad (8.15)$$
$$I_z\,\ddot{\theta} = F_{yf} \cdot a - F_{yr} \cdot b$$

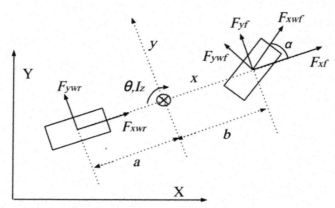

Figure 8.10 Nonlinear bicycle model.

where m is the vehicle mass, I_z is the yaw moment of inertia, F_r is the sum of resisting forces, and F_{xi}, F_{yi} are longitudinal and lateral tire forces along the x-axis and y-axis, respectively. These forces can be relationed with the longitudinal tire force F_{xwi}, lateral force F_{ywi}, and the wheel steer angle α as:

$$\begin{cases} F_{xf} = F_{xwf}\cos(\alpha) - F_{ywf}\sin(\alpha) \\ F_{yf} = F_{xwf}\sin(\alpha) - F_{ywf}\cos(\alpha) \end{cases} \tag{8.16}$$

When a driving torque T_d and a braking torque T_b are applied the rotational motion can be derived as:

$$I_z \dot{\omega}_{wi} = T_{di} - F_{xi}R - T_{bi}(i = f, r) \tag{8.17}$$

where R is the radius of the wheel and ω_{wi} is the yaw angular velocity.

The trajectory of vehicle center of gravity in an absolute inertial coordinate system is given by:

$$\begin{cases} \dot{X} = \dot{x}\cos(\theta) - \dot{y}\sin(\theta) \\ \dot{Y} = \dot{x}\sin(\theta) + \dot{y}\cos(\theta) \end{cases} \tag{8.18}$$

8.2.6.1 Longitudinal Motion Control

For controller synthesis, usually the longitudinal model is based on a one-wheel vehicle model. So, the sum of the longitudinal forces acting on the vehicle center of gravity is:

$$m\dot{v} = F_p - F_r \tag{8.19}$$

where $v = \dot{x}$ is the vehicle speed, F_p is the propelling force, and F_r is the sum of resisting forces. The propelling force is the controlled input resulting from brake and driving actions.

The equation describing the wheel dynamics is:

$$I_z \dot{\omega} = T_d - F_x R - T_b \tag{8.20}$$

For longitudinal controller synthesis, a nonslip rolling is assumed. Then

$$v = R\omega; \quad F_p = F_x \tag{8.21}$$

So, the longitudinal dynamics is:

$$\left(m + \frac{I_z}{R^2}\right)\dot{v} = \frac{T_d - T_b}{R} - F_r \tag{8.22}$$

A Lyapunov-based approach is frequently used to synthetize the longitudinal control. Consider the speed tracking error given by:

$$e = v_{ref} - v \qquad (8.23)$$

where v and v_{ref} are the actual and reference speeds. The derivative of the error is:

$$\dot{e} = \dot{v}_{ref} - \dot{v} = \dot{v}_{ref} - \frac{1}{M_t}(T_d - (T_b + RF_r)) \qquad (8.24)$$

where $M_t = (mR^2 + I_w)/R$, using the expression of $\dot{v}$ given by the nonlinear longitudinal model. Note that T_b can be considered zero, since that when throttle is active the brake is inactive.

As is known, in the Lyapunov methodology to ensure the convergence of the tracking error towards zero it is neccesary to propose a Lyapunov candidate function, which verifies two conditions: it must be definite positive and its derivative with respect to time must be negative.

Usually the following function is proposed:

$$V = \frac{1}{2}e^2 \qquad (8.25)$$

Its time-derivative will be:

$$\dot{V} = e\dot{e} \qquad (8.26)$$

To ensure the convergence to zero, the following condition is imposed, where $c > 0$:

$$\dot{V} = -cV \qquad (8.27)$$

Substituting the value of $\dot{e}$, the following expression can be obtained for $\dot{V}$

$$\dot{V} = e\left(\dot{v}_{ref} - \frac{1}{M_t}(T_d - RF_r)\right) \qquad (8.28)$$

Then the control law is:

$$\hat{T}_d = M_t(ce + \dot{v}_{ref}) + RF_r \qquad (8.29)$$

with the parameter $c > 0$.

It is important to highlight that the stability condition assumes that the model matches with the real system. This is a very strong assumption. The uncertainties in the real parameters of the system must be considered in the controller synthesis. A robustification term must be added to the control law to ensure the robust convergence of the tracking error.

8.2.6.2 Lateral Motion Control

The lateral control problem is complex due to the longitudinal and lateral coupled dynamics as well as the tire behavior. These phenomena are well captured in a simplified way by the nonlinear bicycle model.

An algorithm chosen to perform the steering control tasks of the vehicle is fuzzy logic. Another algorithm frequently chosen to perform the steering control tasks of the vehicle is the predictive controller. When the reference trajectory is a priori known, a predictive algorithm has important advantages compared with other algorithms and is simpler to implement as PID controllers.

The precepts contained in the control strategies included under the term predictive control are:

- This kind of algorithm uses an explicit plant model that is able to predict the system output until a given time (prediction horizon).
- The future control signals obtained by the controller are calculated minimizing an objective function to a certain number of steps (control horizon).
- Sliding horizon concept. The prediction is carried out and the objective function is minimized in order to obtain input commands to the plant. The first control command obtained in the minimization is applied, discarding the rest, and slides the horizon into the future, repeating this steps in every sampling period.

The different predictive control algorithms differ in the models used to describe the system and in the cost function to be minimized. Fig. 8.11 shows the general structure of Model Predictive Controller. Many successful implementations of predictive controllers have been proposed in the literature. In particular, and for simplicity reasons we show the Dynamic Matrix Control (DMC) algorithm. The mathematical model used in this method to represent the system is the step response of the piecewise linearized system. The cost function used is intended to minimize future errors and control efforts. The name of the algorithm comes from the fact that the dynamic of the system is represented in a single matrix formed by the step response elements.

Mathematical expressions for the prediction and the cost function are:

$$\hat{y} = Gu + f \tag{8.30}$$

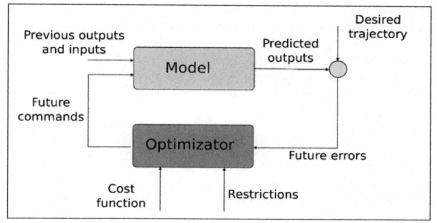

Figure 8.11 MPC structure.

with u, proportional to the lateral torque.

$$J = \sum_{j=1}^{p} \left[\hat{y}(t+j|t) - w(t+j)\right]^2 + \sum_{j=1}^{m} \lambda\left[\Delta u(t+j-1)\right]^2 \qquad (8.31)$$

where $\hat{y}$ is a vector with dimension equal to the prediction horizon containing the predicted outputs until the prediction horizon p, w is the future output desired value, u is a vector with dimension equal to the control horizon m containing the future control actions, G is the dynamic matrix control of the locally linearized system, and f is the vector of free response, with dimension equal to the prediction horizon. The free response is the prediction of how the system will behave if the command keeps constant and equal to the last command calculated. The λ parameter allows to carry out the weighing of the path tracking errors and the control efforts separately, in the way that we could design a controller to try to adjust to the desired trajectory, regardless of usage command, or on the other hand the controller could be more permissive with the path tracking errors and has a more soft use of command, saving energy in the control.

The main methodology of a predictive controller could be summarized as: the model of the process is used to predict the future outputs using the information of the past input signals, past control commands, as well as the future control actions calculated by an optimizer. To calculate the future control signals, the optimizer uses the cost function mentioned previously. Keeping in mind this explanation, the model process is fundamental to the correct functioning of the system.

Note that if in the optimization process it doesn't include the restrictions of the physical model, it is possible to obtain the minimization of the next cost function analytically:

$$J = ee^T + \lambda uu^T \tag{8.32}$$

where e is the vector of predicted errors until the prediction horizon and u is the vector of future signal control increments until the control horizon. The mathematical expression to calculate the future commands is obtained taking the derivative of J and equating to zero:

$$u = \left(G^T G + \lambda I\right)^{-1} G^T (w - f) \tag{8.33}$$

The optimizer will be able to calculate the steer angle in the way to minimize the differences between the free response and the desired trajectory. In other words, the optimizer will calculate the steer angle in order to produce the best path tracking.

The software reads the sensors and sets the values of the internal states of the system in each iteration. These states are the position, the orientation, and the velocity of the vehicle. With these values the step response of the system is calculated. The parameters of the step response form the dynamic matrix G. The prediction of vehicle behavior is calculated using the read values of the sensors at the beginning of the iteration. The prediction of vehicle movement is compared with the desired trajectory from the point closest to the prototype. The future errors vector is the result of the previous comparison. The future commands are obtained using the equation, but only the first term of future commands vector is applied, keeping in mind the concept of sliding horizon. Finally, the variables of the algorithm are updated.

8.3 COOPERATIVE AUTOMATED DRIVING

To tackle the current traffic congestion problems it is essential to improve road capacity and safety while reducing travel time. With a cooperative approach, individual vehicles relate to the environment communicating with other individual vehicles or road infrastructures. Indeed, using wireless communication, potential risk situations can be detected earlier to help avoiding crashes and more extensive information about other vehicles' motions can help to improve traffic throughput.

The extension of the commercially available Adaptive Cruise Control (ACC) system toward the Cooperative ACC (CACC) system has a high potential to be the first cooperative system to be deployed in the market. By introducing V2V communications, the vehicle gets information not only from its preceding vehicle—as occurs in ACC—but also from the vehicles in front of the preceding one.

This section pays particular attention, on the one hand, to platooning, based on the CACC concepts, and on the other hand, to the early-stage developments on Urban Road Transport.

8.3.1 Platooning

Platooning is a particular example of connected (and cooperative) adaptive cruise control (CACC), where a single driver may be in control of an entire "road train," potentially also including their lateral (lane changing) behavior. It combines the use of exteroceptive sensors (mainly 76 Ghz radars), the automation of pedals (or even steering wheel) with the use of a V2V secure wireless communication (using mostly 5.8 GHz DSRC) between the involved vehicles to synchronize braking and acceleration, obtaining a reaction time that is unattainable by humans.

Platoons, deeply investigated for freight transportation systems using heavy-duty trucks, are usually managed following a dynamic assignment for the interested vehicles before the trip. Then, a platoon formation stage leads to the nominal platoon operation mode. A machine state usually handles different events, such as an emergency break, the interference introduced by an intermediate vehicle, the special management of a motorway entrance or exit. The latest advance in truck platooning proposes the use of fault-tolerant systems (Companion, 2016) to perform an eventual recalculation of the assignment when significant deviations from the original plan are detected (Fig. 8.12).

Truck platooning is likely to be one of the earliest applications of road vehicle automation to be commercially viable. It is highly likely that it would mostly materialize on highways, where traffic is less turbulent than on city streets, with a deployment where automation would gradually increase, going from driver assistance up to highly (or even fully) automated vehicles. To make this happen, regulations governing how long

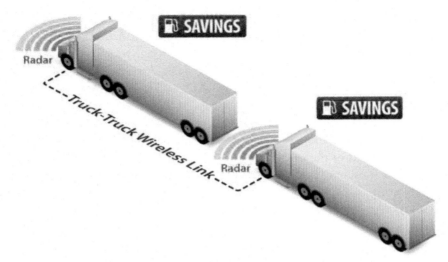

Figure 8.12 Platoon scheme (from Peloton Tecnology).

truckers can drive (or supervise) before taking breaks may have to be modified to consider situations where drivers are in the sleeper berth while an autonomous truck is in operation.

Although the different studies show certain variability in their conclusions, it is widely accepted that the deployment of platooning might have a positive impact on the following aspects:

Capacity: The accordion effect that generates traffic jams could be significantly reduced using constant spacing, increasing thus the capacity of roads by closer spacing of vehicles, narrower lanes, reduction in the wave effect of braking, faster average speeds, and fewer accidents (Swaroop and Hedrick, 1996; Rajamani, 2011). In Fig. 8.13A, a research work (Fernandes, 2012) shows how the maximal flow can be increased up to five times with respect to the peak of the flow/density curve in the classical traffic model. However, to be effective with respect to traffic flow, platooning should be performed with vehicles evolving on dedicated tracks and operating on a nonstop basis from origin to destination (Anderson, 2009). As such, by eliminating the stop-and-go problem of common car and transit systems, platooning could contribute to a faster and more comfortable mobility with higher energy efficiency.

Conversely, other research has suggested that cooperation may negatively impact capacity in merge or lane-drop situations, creating bottlenecks. In this connection, some regulatory bodies are likely to require

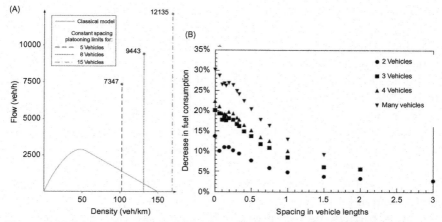

Figure 8.13 (A) Road capacity with different number of vehicle trains (Fernandes, 2012), (B) Decrease in fuel consumption (Zabat et al., 1995).

dedicated truck lanes. The main reason behind this idea is that even a three-truck platoon will function as a visual and physical barrier for cars needing to get on or off the road, and cars attempting to dart between trucks in a platoon represent a new safety hazard.

Fuel consumption: For heavy-duty trucks, the potential fuel savings obtained by platooning are particularly large, ranging from 4%–10% with conservative gaps (Al Alam et al., 2010; Janssen et al., 2015) up to over 20% when spacing is 1/10 vehicle length (see Fig. 8.13B). Thus, 2 trucks driving 100,000 miles annually can save €6,000 on fuel by platooning, compared to driving on cruise control.

Employment: In the long term, fully automated trucks may provide a solution to the growing driver shortage. The American Trucking Associations estimates that the industry will need more than 96,000 new drivers annually for the next 10 years to keep pace with current consumer spending rates (Costello and Suarez, 2015).

Vehicle platooning is an active research area and many contributions have been reported over the last decades. Early theoretical results on the control of platoons were presented in Levine and Athans (1966) and Melzer and Kuo (1971), focusing on a centralized optimal control scheme. Safety is typically addressed by the concept of string stability (Swaroop and Hedrick, 1996; Ploeg et al., 2014), which is related to the suppression of disturbances in vehicle position, velocity, or acceleration, as they propagate through the platoon. More recently, research on implementation aspects has been emerging, herein analyzing the aspects of heterogeneous vehicle strings (Shaw and Hedrick, 2007), intervehicular communication constraints (Al Alam et al., 2010), and implementation issues (Naus et al., 2010).

Although there are still some technological barriers to overcome (e.g., V2V safety and security or the stable platoon control under any circumstances), the main risks for the soft deployment of this type of cooperative systems come from the legal, business, deployment/timing, and user acceptance aspects. The interoperability between service providers, the absence of commitment and corresponding deficient market take up from stakeholders, or the potential boycott by driver-representation lobbies are some of the most significant risks in the exhaustive list of barriers and risks towards platooning (Janssen et al., 2015).

8.3.2 Urban Road Transport

One of the most difficult and challenging scenarios in implementing automated driving is the urban environment, since there are many complex and changing situations with different moving actors and infrastructure elements (vehicles, pedestrians, bikes, intersections and crossing areas, traffic lights, etc.) that must be taken into account at any moment. Considering this complexity, cooperative communication technologies have a very high potential to support and to enhance automated driving strategies, in its different automation levels, in a holistic approach that could cover vehicles, vulnerable road users, traffic infrastructure, shared digital data, and mobility management centers.

Through the exchange of cooperative information among all actors involved, it could be possible, for example, to extend the sensing capabilities of automated and autonomous vehicles beyond the perception of their own physical sensors. Moreover, urban traffic authorities can also enhance the information gathered at their mobility management centers with the one coming from automated vehicles that would be acting as moving sensors, increasing therefore its capacity to implement new traffic management strategies. In this connection, some interesting applications of cooperative urban road transport are listed in the next section.

Nevertheless, due to its specific high complexity, urban automation deserves still a big R&D joint effort among all involved stakeholders. Specific research in the domains of environmental sensing, Internet of Things, cloud computing, Big Data, or artificial intelligence will significantly contribute to the progress of automation in urban environments.

8.4 VERIFICATION AND VALIDATION

While Automated Driving is becoming nowadays a key topic for the future of the automotive industry, the technology behind it has been evolving since some years ago and it is reaching maturity in some of the Advanced Driver Assistance Systems (ADAS) that can be found right now in serial vehicles (e.g., ACC, AEB, or LKA).

In parallel to this evolution, the processes and procedures for testing ADAS functions have been also developed and established during the previous decades, according to the functions' requirements and the normative established. Therefore, today, it is possible to find standard procedures for testing such functions, e.g., Euro NCAP procedures (Fig. 8.14).

In the case of the Automated Driving functions, the work is still to be done. Several research projects in various stages of development can be found in this field, including extended on-road testing with vehicle fleets, but these demonstrations are only the beginning of the Verification and Validation steps, and the challenges are still not solved.

Among others, some of the main questions appearing that have to be clarified are the following:

- How Automated Driving functions should be tested (methods and tools) to achieve the levels of safety and confidence required?
- How much testing will be needed at each development stage?

Test approaches supporting massive and specific tests in the different technologies and at the different levels and points of the lifecycle are needed, and should cover concepts and algorithms, software units and physical components, integrated systems, and in-vehicle functions.

The V-Model development cycle (see Fig. 8.15) has been used in the development of vehicle functions already for some time. More recently, it has been adopted as the reference model that can be used for the ISO

Figure 8.14 AEB testing following EuroNCAP procedures.

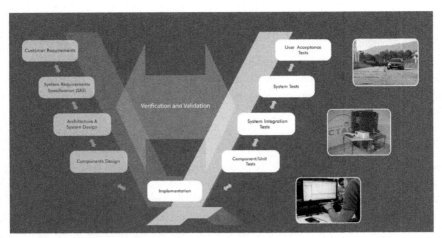

Figure 8.15 Traditional V-Model development cycle.

26262, for functional safety critical systems. But, although ISO 26262 and the V-Model provide a generic methodological framework for assuring automotive safety and also a reference for development of automated driving functions, automated driving presents unique challenges for the application of both approaches, that are now under discussions in the different working groups and fora dealing with both topics.

With regards to the performance of tests at early phases, the use of modeling and simulations presents a clear advantage for achieving acceptable levels of safety and assurance for an autonomous vehicle. Virtual testing allows performing many test cases with variation of parameters in order to assure a correct performance of the algorithms, bringing the system into a good maturity level.

However, there is still a need for extensive testing of the overall system in the real environment, on first the proving ground and as a second step in real world campaigns. In this sense, the development of new specific proving grounds devoted to testing this kind of vehicle will have a major relevance in the coming years.

In-vehicle quantitative validation also requires a set of tools that allows to control a wide range of different parameters (related to the adjustment of the system to be tested, but also of the other targets in the scenario). On the other side, systems for repetitiveness of the maneuvers instead of using human drivers, in addition to the minimization of risks, allows being more efficient in the testing process. With this aim, several systems

Figure 8.16 Example of new tools for automated driving testing.

have been proposed in the recent years to bring the vehicle under test in exactly the same test conditions while guaranteeing safety. In this case, it is possible to find in the market reference systems with enough precision, driving robots, or new movement platforms and dummies to support the new needs appearing in the market (Fig. 8.16).

The amount of testing needed for covering the needs of the Automated Driving functions is the other key topic, and subject to much debate. On the one hand, the discussion is focused on the number of kilometers that must be driven in order to establish a safe operation of an Automated Driving function, and, on the other hand, to arrive at a reasonable and efficient testing cost to meet the time to market in an affordable manner.

As a main conclusion, the verification and validation processes for the Automated Driving functions should meet, in an efficient way, the strong safety and acceptance objectives required to deploy the systems in millions of vehicles around the world and, in parallel, should be efficient enough in terms of costs and time to arrive to the market with the right product. Both facts suppose the modification of current frameworks and development processes, the increase on the usage of virtual testing and modeling, and finally, to be able to perform real environment tests in the most appropriate and repetitive conditions to evolve and fine-tune the correct system performances.

8.5 MAIN INITIATIVES AND APPLICATIONS

8.5.1 Prototypes

8.5.1.1 Relevant Prototypes at International Level

Besides some pioneer works at the University of Ohio in the 1970s, at the Carnegie Mellon University in the 1980s, in specific Programmes in

Europe and Japan (i.e., Prometheus and Advanced Safety Vehicles projects, respectively), the turning point of automated vehicles are the DARPA Grand and Urban Challenges (2005 and 2007).

Although more than 20 teams designed and built automated vehicles to meet the challenges posed by the organizers, the most successful prototypes were the following:

- *Stanley, Stanford University* (Thrun et al., 2006): it won the 2005 Grand Challenge. The vehicle is a Volkswagen Touareg, where the native drive-by-wire control system was adapted to be run directly from an onboard computer without the use of actuators or servo motors. It used five roof mounted Lidars to build a 3-D map of the environment, supplementing the position-sensing GPS system. As in many other prototypes, an internal guidance system utilizing gyroscopes and accelerometers monitored the orientation of the vehicle and also served to supplement GPS and other exteroceptive sensor data. Additional guidance data was provided by a video camera used to observe driving conditions out to 80 m (beyond the range of the LIDAR) and to ensure room enough for acceleration. Includes a Planning and Control layer with a local path planner, obstacle detection, health monitor module in the case of critical system or GPS failures, emergency stop remote control. The sequel of Stanley, Junior, obtained second place in the Urban Grand Challenge.

- *SandStorm, Carnegie Mellon University* (Buehler et al., 2007): the 2005 version of Sandstorm, mounted in a Hummer, used six fixed Lidars, a steerable LIDAR, and short- and long-range radar. It implements a global route planning, path planning, satellite images, and other topography data to generate the global path. The A* heuristic function together with cubic B-splines were used to smooth a reference path. Velocity and steering were controlled using classic PID controllers derived from Simulink models.

- *Boss, Carnegie Mellon University* (Urmson et al., 2008): winner of the Urban Grand Challenge. It is a Chevrolet Tahoe that uses perception, planning, and behavioral software to reason about traffic and take appropriate actions while proceeding safely to a destination. It is equipped with more than a dozen lasers, cameras, and radars to view the world. It allows the tracking of other vehicles, detecting static obstacles, and localizing itself relative to a road model. Planning system combines mission, behavioral, and motion planning to drive in urban environments

More recently, other relevant prototype examples from Universities have achieved very significant milestones:

- *The CMU autonomous vehicle research platform* (Wei et al., 2013): based on a Cadillac SRX. Since the car is not equipped with drive-by-wire controls for operation, several mechanical and electrical modifications were necessary to enable computer control of the required inputs that operate the vehicle. The vehicle is capable of a wide range of autonomous and intelligent behaviors, including smooth and comfortable trajectory generation and following; lane keeping and lane changing; intersection handling with or without V2I and V2V; and pedestrian, bicyclist, and work-zone detection. Safety and reliability features include a fault-tolerant computing system and smooth and intuitive autonomous-manual switching
- *Shelley, Stanford's self-driving Audi TTS* (Funke et al., 2012): it managed to autonomously ascend Pikes Peak in 2010. More recently, they took the vehicle to Thunderhill Raceway Park, and let it go on track without anyone inside, hitting over 120 miles per hour. The goal of this prototype was to push autonomous driving to the vehicle's handling limits. To that end, a high speed, consistent control signal is used in combination with numerous safety features capable of monitoring and stopping the vehicle. The high level controller uses a highly accurate differential GPS and known friction values to drive a precomputed path at the friction limits of the vehicle
- Vislab (Broggi et al., 2012), from the University of Parma, with which the International Autonomous Challenge was accomplished. The platforms are small and electric vehicles produced by Piaggio. The automated driving technology did not affect its performance since the sensors, the processing systems, and the actuation devices are all powered by solar energy, thus they do not drain anything from the original batteries. The vehicle managed to run almost 16.000 km on a 100-day trip, combining automated and manual mode in very challenging driving zones.
- Spirit of Berlin and Made in Germany, prototypes from the Freie Universität Berlin (Berlin, 2007). They have a modular sensor setup with most of its sensors mounted on top of the car on a flexible rack. Obstacle processing is done by a combination of rotating and fixed Lidars with stereo camera systems. In addition, the car localizes itself with an integrated GPS/INS unit and RTK correction signals.

In addition to these representative prototypes, many others have appeared in recent years from OEMS (Daimler, BMW, Audi, GM, Ford,

Volvo, ...), Tier 1 providers (Bosch, Delphi, Valeo, Continental ...), and new incomers providing either the embedded intelligence or the whole automated vehicle (Waymo, Tesla, Peloton Technologies, EasyMile, Navya, Otto, Cruise, Zoox, Baidu, Aimotive). In addition, big players working on new solutions for the mobility as a Service paradigm are intensively working on highly/fully automated driving solutions (Lyft, Uber, Nutonomy, Didi Chuxing).

8.5.1.2 Relevant Prototypes in Spain

There are some R&D centers and universities in Spain working since several years intensively in the domain of automated driving, such as CSIC, UPCT, CTAG, and INSIA. Some of them have prepared real functional automated driving prototypes to test the progress of their different developments.

In 2012 the AUTOPIA Program, from CSIC, showcased their technology in communications and control performing a widely publicized demonstration, from El Escorial up to Arganda, in Madrid (see Fig. 8.17A). One automated prototype vehicle (Platero, Citroën C3) ran driverless for 100 km following a leading manual car (Clavileño, Citroën C3 Pluriel), with sensing and communicating devices, which dynamically generated a high precision map to be tracked by the automated following car. The journey covered a wide range of driving scenarios, including urban zones, secondary roads and highways, in standard traffic conditions. To that end, V2X communications were combined with onboard sensors to achieve a centimetric localization and a safe and smooth motion planning (Godoy et al., 2015).

Probably the most relevant activity performed so far in Spain was carried out by PSA, in collaboration with CTAG, and took place in

(A)   (B)

Figure 8.17 (A) Platero in the El Escorial demonstration, (B) PSA autonomous trial Vigo-Madrid.

November 2015. A level 2 and level 3 PSA C4 Picasso prototype (see Fig. 8.17B), equipped with different sensors and enriched digital maps, covered in automated mode the distance of 599 km from Vigo to Madrid, showing the feasibility to deploy automated driving vehicles in the coming years. DGT was also deeply involved in this trial, facilitating all required authorizations to perform this test in open roads.

Other relevant prototypes have been built in Spain, such as the Renault Twizy automated by the Technical University of Cartagena (Navarro et al., 2016). It has a large set of exteroceptive sensors connected to three different computing platforms running on VxWorks. It includes a global route planning based on maps, a machine learning (SVM) for detecting pedestrian and vehicles, local path planning, and an obstacle avoiding system based on Bezier curves trajectories.

CTAG has also developed some automated driving prototype vehicles in the framework of European, national, or bilateral cooperative R&D projects. To mention two interesting examples (see Fig. 8.18), it is possible to mention a prototype developed in the framework of the project $CO^2PERAUTOS^2$ (INNTERCONECTA Spanish R&D Program), implementing some cooperative automated driving functions such as cooperative highway chauffeur and cooperative urban chauffeur, including also some cooperative sensing use cases (Sánchez et al., 2015). A second relevant example, in this case in cooperation with PSA, is the MobilLab automated driving prototype, devoted to explore the challenges of HMI and driver-interaction for automated and autonomous vehicles.

INSIA (University Institute for Automobile Research of Technical University of Madrid) is also involved in several projects regarding intelligent vehicles, specifically in automated and connected experiences, including cooperative systems, like AUTOCITS (Regulation Study in the Adoption of the autonomous driving in the European Urban Nodes

Figure 8.18 Examples of CTAG prototypes: Cooperative Automated Driving prototype, Mobil-Lab HMI prototype.

Figure 8.19 INSIA's autonomous vehicles prototypes.

funded by European Commission), focused on the deployment of pilots of Cooperative Systems and Autonomous Vehicles in the cities of Madrid, Paris, and Lisbon. The first projects on autonomous driving began with SAMPLER and ADAS-ROAD projects in which automatic evasive maneuvers were performed when risky situations were detected. Then, the Cooperative and Autonomous Vehicles (CAV) project was focused on the integration of autonomous vehicles with C-ITS in critical environments such as complex crossings, roundabouts, and tunnels. AUTOMOST (Automated Driving for dual transport systems) project is aimed to the development of autonomous city buses. INSIA also have strong links with the Spanish automotive sector and with other important actors such as the Spanish Ministry of Defence with the project REMOTE-DRIVE (drive-by-wire for tactic vehicles in surveillance missions) where a military tactic vehicle has been automated to act in emergency and defence missions. This center includes a testbed private circuit, two fully automated vehicles (see Fig. 8.19), two instrumented vehicles, one instrumented motorcycle, electronics and instrumentation lab, V2X proprietary communication technology, and a granted patent of a device to automatically control the steering of a vehicle from a computer.

8.5.2 Projects

Automated and connected driving has become one of the technological mega-trends, recognized by several reports of consulting firms (Manyika et al., 2013). In addition to that, strategic roadmaps from different international organizations merely confirm the importance of these technologies:

- The Amsterdam Declaration (European Union, 2016), signed on April 2016 by all 28 EU member states during the informal meeting of the Transport Council, lays down agreements on the steps necessary for the development of self-driving technology in the EU. In this document the EU member states and the transport industry pledge to draw up rules and regulations that will allow autonomous vehicles to be used on the roads.

- La Nouvelle France Industrielle (NFI, 2013) is a French strategic document whose aim is to focus economic and industrial stakeholders around common goals, to align government means more effectively to these goals, and to harness local ecosystems to build a new, competitive French industrial. It has borne fruit with the presentation of 34 industrial renewal initiatives, among which driverless vehicles, pushing the French automotive sector to be a pioneer in vehicle automation, notably by removing regulatory barriers to growth
- The German Association of the Automotive Industry (VDA) proves that the adaptation to legal provisions and parameters is required, since the corresponding regulations always assume that a driver is actively steering and controlling the vehicle at all times. However, this is not the case in higher automation levels. Their position paper (der Automobilindustrie eV, 2015) on automated paper argues why The Vienna Convention of 1968 must be amended accordingly in order to create a basis for compliance with the national road traffic regulations of the respective signatories.
- NHTSA (National Highway Traffic Safety Administration) released a policy document for Automated Vehicles in 2016 (U.S. Department of Transportation, 2016), where it recognizes three realities that necessitate some sort of guidance: (1) the rise of new technology is inevitable; (2) more significant safety improvements will be achieved by establishing an approach that translates knowledge and aspirations into early guidance; (3) as this area evolves, the "unknowns" of today will become "knowns" tomorrow. The overall intention is therefore to establish a foundation and a framework upon which future Agency action will occur.

Supported by this trend and guidelines, the European Commission and public authorities of the EU Members States have already funded an important number of research and innovation projects (see Fig. 8.20) that seek to set the basis for a sustainable and competitive development of automated driving technologies in Europe.

The first attempts in autonomous driving permitted to see that taking the driver out of the loop in the evolution process from automated to autonomous driving will not happen easily and probably will not happen

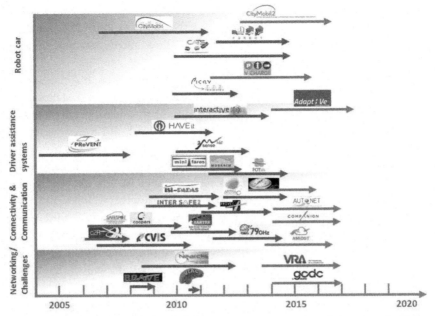

Figure 8.20 Overview of EU-funded Project on automated and connected driving (Dokic et al., 2015).

at all. As a result of this, some initiatives guided his steps through highly automated vehicles. Among them, the project HAVEit, Highly Automated Vehicles for Intelligent Transport (2008−11), aimed at a higher level of automation to be used on existing public roads in mixed traffic.

Following this path, the project interactIVe (2010−13) involved 29 entities from 10 countries working together towards the increase of an accident-free traffic in Europe and developed advanced assistance systems for safer and more efficient driving. The driver is continuously supported by these systems, that warn the driver in potentially dangerous situations. The systems do not only react to driving situations, but are also able to actively intervene in order to protect occupants and vulnerable road users. The continuation of interactIVe was AdaptIVe, an European project launched in 2014 that is about to conclude. It is a lighthouse project on automated driving, involving most of the major OEMs and Tier ones. Its main goal is to demonstrate automated driving in complex traffic environments, taking into account the full range of automation levels. In addition, it is working on (1) providing guidelines for the implementation of cooperative controls involving both drivers and automation; (2) defining and validating specific evaluation methodologies; (3) assessing the impact of automated driving on European road transport; and (4) evaluating the legal framework with regards to existing implementation barriers.

Another example for the successful integration of driverless intelligent vehicles (level 5) in urban environments is the project of CityMobil2. As a successor of CityMobil, the project is implementing intelligent transportation systems (ITS) for automated transport in protected environments. The applied vehicles are based on the "CyberCars" concept defined and promoted by INRIA in France, but extended to new marketable platforms (Easymile, Navya).

As already mentioned, cooperative driving addresses automotive and road traffic systems that make use of information and communication technologies (ICT), in conjunction with automated or nonautomated driving vehicles. These technologies are used to exchange specific information between vehicles (vehicle-to-vehicle communication, or V2V) and between vehicles and road infrastructure (V2I).

In the last years, several European projects and initiatives have marked important milestones in the deployment of these technologies. SAFESPOT (2006−10) developed dynamic cooperative networks where the vehicles and the road infrastructure communicated to share information gathered onboard and at the roadside, to enhance the driver's perception of the vehicle surroundings. Intersafe 2 (2008−11) aimed to develop and demonstrate a cooperative Intersection safety system able to significantly reduce injury and fatal accidents at intersections. The Grand Cooperative Driving Challenge (GCDC, 2011), and its sequel i-Game (2013−16), aims at accelerating the development and implementation of cooperative driving technologies, by means of a competition between international teams.

Several experimental case studies have shown the feasibility of platooning. One of the earliest demonstrations was given in the US in 1997 by the PATH project at UC Berkeley using a platoon of eight cars, followed then in the 2000s by experiments with trucks that had only automated longitudinal control (Fig 8.21A). Regarding truck automation, the first studies were "Chauffeur" within the EU project T-TAP from the mid-1990s to the beginning of 2000 (Fritz et al., 2004), where driving experiments were conducted with three heavy trucks along the Brenner Pass through the Alps between Austria and Italy.

Figure 8.21 From top left to bottom right: (A) PATH Program (PATH, 2017); (B) Konvoi Project (Kunze et al., 2009); (C) SARTRE Project (Sartre, 2016); (D) European Truck Platooning Challenge (Eckhardt, 2015); (E, F) COMPANION Project (Companion, 2016).

From 2005 to 2009 the Aachen University developed a platoon of four heavy trucks (Fig. 8.21B) in their project KONVOI (Kunze et al., 2009) with the objective of increasing transportation capacity while reducing fuel consumption. In 2008 Japan started a 5-year project "Energy ITS" aiming at reducing energy consumption by truck platooning (Tsugawa, 2014).

The most recent European projects SARTRE (Sartre, 2016) and COMPANION (Companion, 2016) have showed significant advances in platooning. The former (Fig. 8.21C) developed strategies and technologies to allow heterogeneous vehicle platoons (car and trucks) to operate on normal public highways, showing its potential with a five vehicle demonstration. The latter (Fig. 8.21E,F) provided solutions for handling the lack of holistic solutions including the creation, coordination, and operation of platoons. As a result, a real-time coordination system can dynamically create, maintain, and dissolve platoons, according to a decision-making mechanism, taking into account historical and real-time information about the state of the infrastructure. More recently, in 2016 about a dozen trucks from major European manufacturers completed a week of largely autonomous driving across Europe in the European Truck Platooning Challenge (Eckhardt, 2015), the first major exercise involving multibrand platooning on the continent (see Fig. 8.21D). In parallel in the USA, Peloton Technologies is the first company to provide automated vehicle technology for commercial truck platooning.

Another interesting and completely different project on cooperative and automated vehicle technology is AutoNet2030. It has worked towards a decentralized decision-making strategy which is enabled by mutual information sharing among nearby vehicles. It considers the gradual introduction of fully automated driving systems, which makes the best use of the widespread existence of cooperative systems in the near-term and makes the deployment of fully automated driving systems beneficial for all drivers already from its initial stages.

8.5.3 Special Applications

The potential of autonomous driving goes beyond standard road applications. There are several areas where the use of this technology can be applied. Researchers, scientists, universities, and R&D departments of the best automotive companies have explored this idea and other possible special applications based on this type of vehicles have been implemented along years. In this way, there are specific applications for off-road environments, nonurbanized scenes, where the useful information for the interpretation of the surroundings is much more limited: military missions, rescue, supervision and surveillance, land exploration, agricultural applications, among others. Furthermore, an additional field of broad development is the specific applications for public passenger transportation, taxi services, car-sharing, or for freight transport and, in a general way, any application that either cannot be performed by a human operator due to imminent danger exposure or provides a new solution for a specific service.

The military environment is one of the main promoters of this type of applied technology. The first military applications were carried out in 1930 with the so-called "Nagayama tank," a tank that received the orders of movement via radio. Subsequently, and more famously, the Soviet Union developed one during the late 1930s and early 1940s, the so-called "Teletank." These military platforms used in the well-known "Winter War," at the beginning of World War II, consisted of tanks controlled by radio without any operator, carrying out missions to approach the enemy while they were guided from another tank at a safe distance. Another military platform that became popular, not for the technology used, but for laying the groundwork for post-World War developments in tele-operation technologies, was "The Goliath," a small-tracked tank, used by the German army during World War II, in battles such as the Normandy landings. "The Goliath" was controlled remotely by wires up to 650 meters and its mission was to demolish buildings and infrastructures of the enemy through the explosive charge that was carried.

More recently, the American organization DARPA, founded the "DARPA Grand Challenge," an autonomous vehicles' competition where they had to travel long distances in a totally autonomous mode. In first instance, it was intended to promote driving and vehicles with a high level of automation, with new developments in technology, but the final goal had a strong military character. In 2004, the first edition of this competition was celebrated, in which vehicles had to complete a route of 240 km along an off-road environment. No participant completed the route. In 2005, the second edition was held, where five vehicles successfully completed the test. This edition continued to consist of an off-road environment, where the participating vehicles had to go through different complex situations like narrow tunnels, roads where it is difficult to delimit the boundaries, or perform complicated turns. The third edition of the DARPA Grand Challenge took place in 2007, better known as "Urban Grand Challenge"; in this case, the environment was urban and 96 km long, respecting traffic rules and the other vehicles.

Since then, different armies around the world have been developing their own Unmanned Ground Vehicles (UGV's). For example, UGVs have been developed such as the "Guardium-LS UGV," a military platform used by Israel equipped with a large amount of ballistic material, capable of being tele-operated from another mobile platform and able to detect and avoid obstacles that appear in their way. Another interesting example is the Tank Automotive Research Development and Engineering Center (TARDEC) group of U.S army that develops UGVs for military applications whose purpose is to be able to be tele-operated from anywhere in the world. On the other hand, in this type of application, unlike when urban environment are considered, it is difficult to predict the state of the road through which the vehicle circulates, besides not being able to rely on elements that serve as reference, such as lane lines, curbs, buildings, crossings, etc. With the LiDAR technology, obstacles that are above the ground level could be easily recognized; however, in these off-road conditions it cannot be assumed that the condition of the terrain is always in good condition and the identification of negative obstacles is of great importance. That is why research projects are being developed especially dedicated to the identification of such obstacles, such as the case of Shang et al. (2015), where a different set up for the LiDAR sensors is presented in order to identify the negative obstacles in nonurbanized environments. They were the winners of the "Overcome Danger 2014," a ground vehicle challenge supported by the Chinese army, similar to the DARPA Grand Challenge of the USA.

Another interesting application of autonomous vehicles is their use for space missions. For example, the vehicle Mars Rover can be guided by a human operator remotely but the obstacles avoidance and best trajectory finding for reaching the destination are tasks that the vehicle performs autonomously.

Apart from these projects supported by governments or state administrations, some companies from the private sector are also paying special attention to developing new solutions for special off-road applications such as Jaguar Land Rover. Throughout 2016, they have been working on a self-driving off-road connected vehicle. On the one hand, the off-road vehicle can identify the terrain where it circulates based on recognition of the environment and it can offer semi-autonomous driving. On the other hand, it is intended that these vehicles can talk to each other and communicate, in such a way that, if several vehicles circulate in a convoy, the first one can communicate to the other vehicles located behind it the state of the terrain, the speed, and location, etc., through DSRC communication modules.

In the agricultural sector, the autonomous navigation of the industrial machinery is promoting a great increase of the productivity in tasks of plowing, mowing, harvesting, etc. All these tasks become more efficient when a fleet of tractors works at the same time throughout the day and the operator manages all the work in remote. There are several developments in this sector. One of the most outstanding is the manufacturer of agricultural machinery Case IH with its concept of autonomous vehicles, capable of following a preloaded route and calculating the most optimal paths to perform. Furthermore, since the vehicle has sensors for the recognition of the environment, it can identify obstacles and make the decision to stop and send a warning signal if necessary. In a similar line, other agriculture machinery manufacturers have developed vehicles that can follow a predefined trajectory.

In addition, autonomous navigation serves as a catalyst for other services that are carried out in an urban or interurban scene. One of the most successful special applications among the manufacturers and software companies is the taxi service. In fact, the first self-driving taxi was tested in Singapore in 2016 by nuTonomy. This software startup for autonomous vehicles has launched a taxi service that currently operates in a specific area of Singapore with specific destinations, using electric cars, providing, therefore, a solution to decongesting the cities of traffic and pollution. On the other hand, Uber is currently developing its autonomous taxi service, conducting tests in the city of Pittsburgh, through the sensor fusion of different LiDARs, Radars, stereovision, and computer vision for the recognition of the environment and making use of machine learning in their control algorithms.

Besides the specific aforementioned initiative, there are many potential interesting cooperative applications related to urban automation. To summarize some of them, the following examples could be found in the future:

- Automated parking and valet parking cooperative services.
- Cooperative services to support and to manage autonomous vehicles in dedicated lanes, such as last-mile autonomous vehicles.
- High precision positioning cooperative services.
- Cooperative services for urban chauffeur and urban autopilot, for example, to manage intersection scenarios in a safely and efficient manner.
- Cooperative services for robotaxis and autonomous car sharing fleets.
- Cooperative sensing services to extend perception capabilities of AVs.

Parking scenarios usually represent a kind of relatively stable - environment where vehicle automation is already happening. With the support of cooperative communication technologies, the next step will be in the direction of autonomous valet parking applications. In this case, cars and garages will cooperate to park autonomously the vehicle, without the presence of the driver, which will interact with the car through his/her smart device. All vehicles manufactures and main Tier 1 suppliers in this domain are in a development phase to bring this functionality into the

Figure 8.22 Remote valet parking assistance demonstration (BMW).

Figure 8.23 Last-mile driverless shuttles testing trial that will run from 23rd January 2017 until 7th April 2017 in a dedicated lane in Paris (Easymile).

market, with the involvement also of parking management companies (Fig. 8.22).

Other important area, very much related to connectivity and cooperative services, which have an interesting potential to make urban mobility more sustainable, is linked to the implementation of electric autonomous last-mile vehicles. There are many R&D projects and predeployment activities running at this moment to support the future introduction of these kind of solutions (Fig. 8.23).

Talking about urban scenarios that automated vehicles will have to cover, intersections are very complex situations where cooperative services can really provide support. As an example, cooperative traffic lights can exchange Green Light Optimal Speed Advisor (GLOSA) information with automated vehicles to let them adapt their speed to pass the intersection in the safest and most efficient manner (Fig. 8.24).

Moreover, V2V, V2I, V2VRU, or V2Cloud cooperative sensing strategies can enhance the perception capabilities of automated vehicles, allowing them to manage better these complex urban scenarios (Fig. 8.25).

Figure 8.24 Example of automated GLOSA cooperative service (CTAG).

Figure 8.25 C2C pedestrian detection demonstration at Bordeaux ITS World Congress (PSA-CTAG).

8.6 SOCIOREGULATORY ASPECTS

8.6.1 Legal Pathways

There are important questions to be answered regarding autonomous vehicles, like security, ethics, use of data, and coexistence of autonomous technology with conventional vehicles (manually controlled vehicles).

Firstly, the system must be safe for both the driver and the other users of public roads. In addition, autonomous vehicles must comply with the traffic laws of the region in which they operate, in the same way as all other vehicles in circulation.

Secondly, it is necessary that the regulation on autonomous vehicles is as homogeneous as possible. So that there should not be differences between states that prevent the same model of vehicle from operating in all of them (which would be an important obstacle to the deployment of this technology).

Then, appropriate education and training are essential to ensure the safe deployment of automated vehicles. Therefore, manufacturers and other entities should develop, document, and maintain some form of training for employees, distributors, and users. The differences between the use and operation of autonomous vehicles and conventional vehicles should be addressed. In addition, these programs should be designed to provide users with the level of understanding necessary to use them properly and safely.

A critical aspect in this type of technology is the changing between manual control and autonomous vehicle control. Adequate mechanisms and procedures must be provided to ensure the change is made safely, comfortably, and efficiently.

In addition, the data generated by the use of connected and/or autonomous vehicles may be useful, for example in case of accidents, to analyze the causes of them. It is necessary to clarify the conditions and availability for the use and exchange of data generated by connected and automated vehicles, as well as the responsibility of each of the parties involved.

Moreover, manufacturers and other entities are expected to develop software upgrades for automated vehicles or new vehicle versions incorporating different and/or upgraded hardware. If these software or hardware upgrades substantially change the operation of the vehicle, an evaluation or additional certification process may be necessary. The purpose of these updates may be to improve performance, security, or other aspects of the system. In addition, in case of changes in the software, the download of these updates or patches could be done through "over-the-air updates" or other methods that should be regulated.

Finally, it is necessary to contemplate the possibility of when two situations of risk happen simultaneously, requiring a consideration of "dilemma situations." It may be necessary to define the procedure in these kinds of situations.

8.6.1.1 General Framework: Vienna and Amsterdam

There are two traditional international framework agreements on Road Traffic: the Geneva Convention and the Vienna Convention. Firstly, the Geneva Convention on Road Traffic was signed in 1949 by 95 states to promote the development and safety of international road traffic by establishing certain uniform rules. Secondly, the Vienna Convention on Road Traffic from 1968 is an international treaty ratified by 74 countries. It was designed to facilitate international road traffic and to increase road safety through the adoption of uniform traffic rules. At European level, the need to modify the Vienna Convention to promote the use of autonomous vehicles in road traffic has intensified, mainly due to the need of a "driver" controlling the vehicle at all time. Article 1, paragraph (v): "Driver means any person who drives a motor vehicle or other vehicle (including a cycle), or who guides cattle, singly or in herds, or flocks, or draught, pack or saddle animals on a road." Article 8, paragraph 5: "Every driver shall at all times be able to control his vehicle or to guide his animals."

Then, on April 2016, European Union transport ministers, as well as a number of car manufacturers, signed the European Declaration of Amsterdam. The main objective of this initiative is the cooperation between governments and industry to develop a legal framework and to boost research and development on connected and automated driving.

In addition, there are currently informal discussions on the analysis and development of regulations for the autonomous vehicles at the United Nations Economic Commission for Europe's "Working Party on Road Traffic (W.P.1)."

8.6.1.2 Legal Framework and Regulation About Autonomous Vehicles

This section summarizes the state of the regulation of autonomous vehicle driving in early 2016. Initiatives in USA and Europe are mainly cited, but others are also discussed.

The US Federal Government released in September 2016 an autonomous vehicle policy designed to help the safe development of driverless technology, while also allowing enough flexibility so development of the technology can continue.

In addition, in the United States the number of states working on legislation related to autonomous vehicles has gradually increased. The enacted autonomous vehicles legislations in the USA are listed below.

Nevada was the first state to authorize the operation of autonomous vehicles in 2011. AB 511 authorizes operation of autonomous vehicles it also defines "autonomous vehicle" and directs state Department of Motor Vehicles to adopt rules for license endorsement and for operation (including insurance, safety standards, and testing). SB 140 permits use of cell phones or other handheld wireless communications devices for persons in a legally operating autonomous vehicle (these persons are deemed not to be operating the vehicle for the purposes of this law). SB 313 requires certain conditions that human operators and autonomous vehicles must meet in order to being registered, or tested or operated on a highway within the state.

California, in 2012 (SB 1298), permits the operation and testing of autonomous vehicles pending the adoption of safety standards and performance requirements that would be adopted under this bill. In 2016 (AB 1592) California authorized the Contra Costa Transportation Authority to conduct a pilot project (only at specified locations and speeds) "for the testing of autonomous vehicles that do not have a driver seated in the driver's seat and are not equipped with a steering wheel, a brake pedal, or an accelerator."

Florida declared in 2012 (HB 1207 and HB 599) "desires to encourage the current and future development, testing, and operation of autonomous vehicles on the public roads of the state" and found that it "presently does not prohibit or specifically regulate the operation of autonomous vehicles." In 2016 (HB 7027) legislation expands the allowed operation of autonomous vehicles on public roads and eliminates the requirement that the vehicle operation is being done for testing purposes and the presence of a driver in the vehicle.

Florida House Bill 7061 (2016) defined driver-assistive truck platooning technology and required a study on the use and safe operation of this kind of technology and allows for a pilot project upon conclusion of the study.

Through House Bill 1143 (2016), Louisiana defined "autonomous technology" for purposes of highway regulatory provisions and related matters.

In the Michigan Senate Bills 169 and 663 (2013), issues like "automated technology," "automated vehicle" and "automated mode" were defined. Automated vehicles were allowed to be tested by certain parties under certain conditions. By 2016 Michigan Senate Bills 995, 996, 997, and 998, modified aspects such provide immunity from liability that arises out of any modification made by another person without the autonomous technology manufacturer's consent.

Through House Bill 1065 (2015), North Dakota provided for a legislative management study of automated motor vehicles. The study might include research into the degree that automated motor vehicles could reduce traffic fatalities, crashes and congestion.

In 2015 Tennessee prohibited (SB 598) local governments from banning the use of motor vehicles equipped with autonomous technology if the motor vehicle otherwise complies with all safety regulations. SB 2333 (2016) allows an operator to use an electronic display (integrated with the vehicle) for communication, information, and other uses enabled by the display only if the autonomous technology isn't disengaged. SB 1561 (2016) establishes certification program through department of safety for manufacturers of autonomous vehicles before such vehicles may be tested, operated, or sold; creates a per mile tax structure for autonomous vehicles.

The Utah House Bill 373 (2015), modified the Motor Vehicles Act by authorizing the Department of Transportation to deploy connected vehicle technology tests.

The HB 280 (2016) requires a study related to autonomous vehicles, including evaluation of the different standards, best practices, regulatory strategies, and schemes implemented by other states.

Washington, D.C. through DC B 19–0931 (2013) defined "autonomous vehicle" as a "vehicle capable of navigating District roadways and interpreting traffic control devices without a driver actively operating any of the vehicle's systems." It authorizes autonomous vehicles to operate on public roadways if a driver can assume control of the autonomous vehicle at any time.

At European level, there are different countries that have carried out initiatives for the development of regulation related to autonomous driving. The most significant, are listed below.

The Ministry of Transport and Communications of Finland is preparing an amendment to the Road Traffic Act that would allow for driverless robotic cars to drive within a restricted area on public roads. The act in question would constitute experimental legislation that would be in force for five years starting at the beginning of 2015.

France (*L'Etat*) announced in July 2014 that the necessary regulations to guarantee road safety in the first experiments of autonomous vehicles on public roads should be developed. On August 3, 2016, the *Conseil des ministres* of France announced an Ordinance (Ordonnance: *experimentation de vehicules a delegation de conduite sur les voies publiques*), which allows the deployment of (partially or completely) autonomous vehicles tests, but only if safety is ensured.

Germany does not have specific (ad hoc) legislation on autonomous vehicles due to the strict interpretation of the Vienna Convention followed in Germany. In addition, Germany's Minister of Transport has announced that a section of the A9 autobahn that connects Berlin and Munich is to be set up for testing autonomous vehicles and connected vehicles (V2V and V2I).

In the Netherlands, the Ministry of Infrastructure and the Environment amended the Dutch regulation to allow large-scale road tests. Companies that wish to test autonomous vehicles must submit an application for admission to the RDW (Dutch Vehicle Authority) and demonstrate that the tests will be conducted in a safe manner.

In Spain, by means of instruction 15/V-113 of November 2015, the regulation for the authorization of tests with autonomous vehicles on open roads to traffic in general was published. In addition, in January 2016, the regulation on assisted parking of motor vehicles (INSTRUCTION 16 TV/89) was made public.

Sweden started in 2014 a project (Drive Me) which has given Volvo permission to test 100 autonomous vehicles in the city of Gothenburg by 2017−2018. It will be the world's first large-scale autonomous driving project. This initiative, which is the result of the collaboration between Volvo, the Swedish Department of Transport, the Swedish Transport Agency, the Lindholmen's Science Park, and the city of Göteborg, is supported by the Swedish Government.

UK Department for Transport released in February 2015 a regulatory review. Testing automated vehicles is allowed on any road in the UK without needing to seek permission from a network operator, report any data to a central authority, or put up a surety bond. In July 2015 a Code of Practice for testing autonomous vehicles was published. It explains to testers how to comply with the UK laws: "testers must obey all relevant road traffic laws; test vehicles must be roadworthy; a suitably trained driver or operator must be ready, able, and willing to take control if necessary; and appropriate insurance must be in place."

Finally, some of the initiatives developed in other countries are indicated. In Australia, there is an initiative called Australian Driverless Vehicle Initiative (ADVI) that includes different companies, government bodies, and research centers. The main objective of this initiative is to "build momentum by rapidly exploring the impacts and requirements of this new technology in a truly Australian context and making recommendations on ways to safely and successfully bring self-driving vehicles to Australian roads."

There are several initiatives developed in Singapore, related to the autonomous car. In 2014, the Committee for Autonomous Road Transport for Singapore (CARTS) was launched by the Singaporean Ministry of Transport. One of the tasks of this team is to investigate and create a framework for autonomous vehicles to work safely and efficiently on public roads. Besides this, the Land Transport Authority (LTA) jointly Agency for Science, Technology and Research (A*STAR) announced Singapore Autonomous Vehicle Initiative (SAVI), which one of their focus areas is to prepare technical and statutory requirements for future deployment of autonomous vehicles in Singapore. In addition, in August 2016, Singapore's government gave NuTonomy permission to test self-driving Taxis in a business park called "one-north." The tests began in the third quarter of 2016.

8.6.2 Ethical Aspects

Automated vehicles will be able to make precrash decisions, overcoming thus many of the limitations experienced by humans. However, there will be fatal car crashes that are unavoidable.

In these situations, a computer can quickly compute the best way to crash taking into consideration the likelihood of the outcome. One major disadvantage of automated vehicles is that, unlike a human driver who can decide how to crash in real time, an automated vehicle's decision of how to crash is a priori designed by a programmer ahead of time (Goodall, 2014). And there are many challenging driving situations where a dilemma may appear, requiring actions that are legally and ethically acceptable to humans.

To illustrate this complexity, consider for instance a sort of modified trolley problem, called the tunnel problem (Open Roboethics Initiative, 2016). A self-driving car just before entering a tunnel encounters a child that attempts to run across the road, but trips in the center of the lane, effectively blocking the entrance to the tunnel. The car has but two options: hit and kill the child, or swerve into the wall on either side of the tunnel, thus killing you. How should the car react? Or even more subtle. An autonomous car is facing an imminent crash, but it could select one of two targets to swerve into: either a motorcyclist who is wearing a helmet, or a motorcyclist who is not. What is the right way to program the car?.

Both outcomes will certainly result in harm as there is no obvious "correct" answer to these kind of dilemmas. If crash–optimization is con-sidered as the most relevant criterion, the outcome may result in unfair actions, as the most responsible potential victim would be penalized, somehow awarding careless road actors and stakeholders.

An alternative and apparently elegant solution would be not to make a deliberate choice. However, such a random decision mimics human driving, which is completely against one of the key reasons to deploy autonomous cars: to avoid the human factor, responsible for 95% of accidents. Even worse, while human drivers may be forgiven for making a poor split-second reaction, robot cars will not enjoy that freedom, as such an action might be the difference between premeditated murder and involuntary manslaughter.

Some others argue that instead of assuming designers are the right people to decide in all circumstances how a driverless car should react, alternative methodologies may be explored allowing drivers to decide on the preferred unfortunate outcome.

In any case, to optimize crashes, designers/programmers would need to face an ethics problem, as they will need to design optimization algorithms that calculate the expected costs of various possible options, selecting the one with the lowest cost. It is therefore legitimate to ask the question of whether control algorithms of automated vehicles can be designed a priori to embody not only the laws but also the ethical principles of the society in which they operate (OCR Software blog, 2016). The basis for such complex decision system could be inspired from Isaac Asimov's three laws of robotics: (1) property damage takes always precedence of personal injury; (2) there must be no classification of people, for example, on size, age, and the like; (3) if something happens, the manufacturer is liable (Gerdes and Thornton, 2016).

Another very relevant ethical aspect derived from the pervasive connectivity of the new generation of vehicles is the challenge to preserve data security, and more in particular, privacy. Indeed, all the data that circulates within the transportation system will be subject to an intense reflection in order to regulate the data to be collected, owned, and shared; who will keep it, why, and for how long.

REFERENCES

Abdullah, R., Hussain, A., Warwick, K., Zayed, A., 2008. Autonomous intelligent cruise control using a novel multiple-controller framework incorporating fuzzy-logic-based switching and tuning. Neurocomputing. 71, 2727—2741.

Al Alam, A., Gattami, A., Johansson, K.H., 2010. An experimental study on the fuel reduction potential of heavy duty vehicle platooning. s.l., s.n., pp. 306—311.

Anderson, J.E., 2009. An intelligent transportation network system: Rationale, attributes, status, economics, benefits, and courses of study for engineers and planners. PRT International.

Behere, S., Törngren, M., 2015. A Functional Architecture for Autonomous Driving. ACM, New York, NY, USA, pp. 3—10.

Behere, S., 2013. Architecting Autonomous Automotive Systems: With an Emphasis on Cooperative Driving, s.l.: s.n.

Belker, T., Beetz, M., Cremers, A.B., 2002. Learning action models for the improved execution of navigation plans. Rob. Auton. Syst. 38, 137–148.

Berlin, T., 2007. Spirit of Berlin: an autonomous car for the DARPA urban challenge hardware and software architecture. Retrieved Jan, 5, p. 2010.

Bhatt, D., Aggarwal, P., Devabhaktuni, V., Bhattacharya, P., 2014. A novel hybrid fusion algorithm to bridge the period of GPS outages using low-cost INS. Expert Syst. Appl. 41, 2166–2173.

Bo, C., Cheng, H.H., 2010. A review of the applications of agent technology in traffic and transportation systems. IEEE Trans. Intell. Transp. Syst. 11, 485–497.

Broggi, A., et al., 2012. Autonomous vehicles control in the VisLab intercontinental autonomous challenge. Annu. Rev. Control 36, 161–171.

Buehler, M., Iagnemma, K., Singh, S., 2007. The 2005 DARPA grand challenge: the great robot race. s.l. Springer Science & Business Media.

Chakraborty, D., Vaz, W., Nandi, A.K., 2015. Optimal driving during electric vehicle acceleration using evolutionary algorithms. Appl. Soft Comput. 34, 217–235.

Companion, 2016. COMPANION - COoperative dynamic forMation of Platoons for sAfe and eNergy-optImized gOods transportatioN. [Online] Available at: <http://www.companion-project.eu/>.

Conner, M., 2011. Automobile sensors may usher in self-driving cars. EDN Mag.21–24.

Continental, A.G., 2012. SRL-CAM400 CMOS camera and infrared LIDAR in a compact unit. s.l.:s.n.

Costello, B., Suarez, R., 2015. Truck driver shortage analysis 2015. Am. Truck. Assoc. 206, 2015, Retrieved from http://www.trucking.org/ATA\%20Docs/News\%20 and\%20Information/Reports\%20Trends\%20and\%20Statistics/10.

Cunningham, A.G., Galceran, E., Eustice, R.M., Olson, E., 2015. MPDM: Multipolicy decision-making in dynamic, uncertain environments for autonomous driving. s.l. IEEE1670–1677.

Czubenko, M., Kowalczuk, Z., Ordys, A., 2015. Autonomous driver based on an intelligent system of decision-making. Cognit. Comput. 7, 569–581.

Delphi Inc., 2015. Integrated Radar and Camera System. [Online].

der Automobilindustrie eV, V, 2015. Automation: From Driver Assistance Systems to Automated Driving. VDA Magazine-Automation.

Dokic, J., Müller, B., Meyer, G., 2015. European roadmap smart systems for automated driving. s.l.:European Technology Platform on Smart Systems Integration.

Dolgov, D., Thrun, S., Montemerlo, M., Diebel, J., 2010. Path planning for autonomous vehicles in unknown semi-structured environments. Int. J. Robot. Res. 29, 485–501.

Eckhardt, J., 2015. European Truck Platooning Challenge 2016. The Hague, Delta3.

ERTRAC, 2015. ERTRAC Roadmap for Automated Driving. s.l., s.n.

European Union, 2016. Declaration of Amsterdam - Cooperation in the Field of Connected and Automated Driving. s.l.:s.n.

Fernandes, P., 2012. Platooning of IVC-Enabled Autonomous Vehicles: Information and Positioning Management Algorithms, for High Traffic Capacity and Urban Mobility Improvement, s.l.: s.n.

Fritz, H., Bonnet, C., Schiemenz, H., Seeberger, D., 2004. Electronic Tow-Bar Based Platoon Control of Heavy Duty Trucks Using Vehicle-Vehicle Communication: Practical Results of the CHAUFFEUR2 Project. s.l., s.n.

Funke, J. et al., 2012. Up to the Limits: Autonomous Audi TTS. s.l., s.n., pp. 541–547.

Furda, A., Vlacic, L., 2010. Multiple criteria-based real-time decision making by autonomous city vehicles. IFAC Proc. 43, 97–102.

Furda, A., Vlacic, L., 2011. Enabling safe autonomous driving in real-world city traffic using multiple criteria decision making. IEEE Intell. Transp. Syst. Mag. 3, 4–17.

Galceran, E., Cunningham, A.G., Eustice, R.M., Olson, E., 2015. Multipolicy decision-making for autonomous driving via changepoint-based behavior prediction. Robot. Sci. Syst.

Gerdes, J.C., Thornton, S.M., 2016. Implementable ethics for autonomous vehicles. Autonomous Driving. s.l. Springer, pp. 87–102.

Godoy, J., et al., 2015. A driverless vehicle demonstration on motorways and in urban environments. Transport 30, 253–263.

Goodall, N., 2014. Ethical decision making during automated vehicle crashes. Transp. Res. Rec. J. Transp. Res. Board 58–65.

Google Inc, 2015. The Google Self-Driving Car Project. [Online] Available at: <http://www.google.com/selfdrivingcar/>.

Haddad, M., Chettibi, T., Hanchi, S., Lehtihet, H.E., 2007. A random-profile approach for trajectory planning of wheeled mobile robots. Eur. J. Mech. A/Solids 26, 519–540.

Holguín, C., 2016. 4 misconceptions on "autonomous" cars that can lead cities to disaster, and how to remedy. [Online].

Ibañez-Guzmán, J., Laugier, C., Yoder, J.-D., Thrun, S., 2012. Autonomous driving: context and state-of-the-art. In: Eskandarian, A. (Ed.), Handbook of Intelligent Vehicles. Springer, London, pp. 1271–1310.

Janssen, R., Zwijnenberg, H., Blankers, I., de Kruijff, J., 2015. Truck Platooning: Driving the Future of Transportation.

Jenkins, A., 2007. Remote Sensing Technology for Automotive Safety. s.l.:s.n.

Katrakazas, C., Quddus, M., Chen, W.-H., Deka, L., 2015. Real-time motion planning methods for autonomous on-road driving: State-of-the-art and future research directions. Transp. Res. Part C: Emerg. Technol. 60, 416–442.

Kunze, R., Ramakers, R., Henning, K., Jeschke, S., 2009. Organization and Operation of Electronically Coupled Truck Platoons on German Motorways. s.l., s.n., pp. 135–146.

Kuwata, Y., et al., 2009. Real-Time Motion Planning With Applications to Autonomous Urban Driving. IEEE Trans. Control Syst. Technol. 17, 1105–1118.

Lefèvre, S., Vasquez, D., Laugier, C., 2014. A survey on motion prediction and risk assessment for intelligent vehicles. ROBOMECH J. 1, 1.

Levine, W., Athans, M., 1966. On the optimal error regulation of a string of moving vehicles. IEEE Trans. Autom. Control 11, 355–361.

Manyika, J. et al., 2013. Disruptive technologies: Advances that will transform life, business, and the global economy. s.l.: McKinsey Global Institute San Francisco, CA.

Melzer, S.M., Kuo, B.C., 1971. Optimal regulation of systems described by a countably infinite number of objects. Automatica 7, 359–366.

Meystel, A., Albus, J.S., 2002. Intelligent Systems: Architecture, Design, and Control s.l.: Wiley.

Naus, G.J.L., et al., 2010. String-stable CACC design and experimental validation: A frequency-domain approach. IEEE Trans. Veh. Technol. 59, 4268–4279.

Navarro, P.J., Fernández, C., Borraz, R., Alonso, D., 2016. A machine learning approach to pedestrian detection for autonomous vehicles using high-definition 3D range data. Sensors 17, 18.

NFI, 2013. The new face of industry in France. s.l.:s.n.

OCR Software blog, 2016. OCR Software blog. [Online] Available at: <https://ocr.space/blog/2016/09/liability-for-self-driving-car-accidents.html>.

Open Roboethics Initiative, 2016. Open Roboethics Initiative. [Online] Available at: <http://www.openroboethics.org> (Accessed 12.16).

Paden, B., et al., 2016. A survey of motion planning and control techniques for self-driving urban vehicles. IEEE Trans. Intell. Veh. 1, 33–55.

PATH, 2017. <http://www.path.berkeley.edu>. s.l., s.n.

Ploeg, J., Shukla, D.P., van de Wouw, N., Nijmeijer, H., 2014. Controller synthesis for string stability of vehicle platoons. IEEE Trans. Intell. Transp. Syst. 15, 854—865.

Rajamani, R., 2011. Vehicle Dynamics and Control. s.l.: Springer Science & Business Media.

SAE, 2016. Taxonomy and Definitions for Terms Related to On-Road Motor Vehicle Automated Driving Systems, s.l.: SAE International.

Sánchez, F., et al., 2015. CO2PERAUTOS2—Challenges for Cooperative Automated Driving. s.l., s.n.

Sartre, 2016. The SARTRE project. [Online] Available at: <http://sartre-project.eu/>.

Shang, E., et al., 2015. LiDAR based negative obstacle detection for field autonomous land vehicles. J. Field Robot.

Shaw, E., Hedrick, J.K., 2007. String Stability Analysis for Heterogeneous Vehicle Strings. s.l., s.n., pp. 3118—3125.

Skog, I., Handel, P., 2009. In-car positioning and navigation technologies—a survey. IEEE Trans. Intell. Transp. Syst. 10, 4—21.

Swaroop, D., Hedrick, J.K., 1996. String stability of interconnected systems. IEEE Trans. Autom. Control 41, 349—357.

Thrun, S., et al., 2006. Stanley: the robot that won the DARPA grand challenge, J. Field Robot., 23. pp. 661—692.

Tsugawa, S., 2014. Results and Issues of An Automated Truck Platoon Within the Energy ITS Project. s.l., s.n., pp. 642—647.

Tzoreff, E., Bobrovsky, B.-Z., 2012. A novel approach for modeling land vehicle kinematics to improve GPS performance under urban environment conditions. IEEE Trans. Intell. Transp. Syst. 13, 344—353.

U. S. Department of Transportation, 2016. Federal Automated Vehicles Policy — September 2016. s.l.:s.n.

Urmson, C., et al., 2008. Autonomous driving in urban environments: Boss and the urban challenge. J. Field Robot. 25, 425—466.

Urmson, C., et al., 2009. Autonomous Driving in Urban Environments: Boss and the Urban Challenge. In: Buehler, M., Iagnemma, K., Singh, S. (Eds.), The DARPA Urban Challenge: Autonomous Vehicles in City Traffic. Springer, Berlin Heidelberg, pp. 1—59.

Urmson, C., Anhalt, J., Bagnell, D., Baker, C., 2008. Autonomous driving in urban environments: boss and the urban challenge. J. Field Robot.

Veres, S.M., Molnar, L., Lincoln, N.K., Morice, C.P., 2011. Autonomous vehicle control systems -- a review of decision making. Proc. Inst. Mech. Eng. Part I. J. Syst. Control Eng. 225, 155—195.

Wei, J. et al., 2013. Towards A Viable Autonomous Driving Research Platform. s.l., s.n., pp. 763—770.

Zabat, M., Stabile, N., Frascaroli, S., Browand, F., 1995. The Aerodynamic Performance of Platoons: Final Report. s.l. University of California, Berkeley.

Zhang, T., Xu, X., 2012. A new method of seamless land navigation for GPS/INS integrated system. Measurement 45, 691—701.

PART III

Additional Aspects

CHAPTER 9

Human Factors

Contents

Intelligent Vehicles
DOI: http://dx.doi.org/10.1016/B978-0-12-812800-8.00009-6

SUBCHAPTER 9.1

Human Driver Behaviors

Luis M. Bergasa[1], Enrique Cabello[2], Roberto Arroyo[1], Eduardo Romera[1] and Ángel Serrano[2]
[1]Universidad de Alcalá, Alcalá de Henares, Spain
[2]Universidad Rey Juan Carlos, Móstoles, Spain

9.1.1 INTRODUCTION

Intelligent vehicle technologies require a good understanding of human driver behaviors to guarantee safety, adjust to drivers' needs, and meet their preferences. Therefore, the study of driver states estimation, the recognition of driving style, and the driving intention inference are essential for the development of these systems. Three significant questions related to intelligent vehicle technologies are influencing the humans' relation with the automobile: (1) how to design technologies that can help drivers avoid a crash; (2) how to warn the driver of an imminent crash; and (3) how to take actions, which substitute for those the driver himself or herself would initiate. In order to answer these questions, from the human factors community, has been formulated three basic laws (Fisher et al., 2016).

The first and most fundamental law is: "intuition is not a reliable guide for Driver Assistant System (DAS) design." In the last decades, there has been an important decrease in fatalities due to vehicle crashes. Technologies such as antilock braking systems (ABS) have contributed to this decrease. But, curiously, vehicles with ABS were early on associated with an increase in crashes (Broughton and Baughan, 2002). It was hypothesized that drivers computed the cost and benefits associated with a given technology and kept constant the level of risk. This is known as risk homeostasis theory (Hedlund, 2000). It appears to be that early in the adoption of ABS drivers did not know how to operate a car equipped with such systems (Broughton and Baughan, 2002). What the ABS experience demonstrated was that when the human is in the loop with any technology one cannot a priori predict how the human will behave. Human behavior must be taken into account in the DAS design.

The second law of human factors that complement the first one is: "human behavior is largely predictable but the prediction depends on careful and well controlled research to be useful in the warning systems' design." Developing and evaluating what is predicted to get the optimal design of a Driver—Vehicle Interface (DVI) is crucial. This experimentation increases what the automation community refers to as the correct use of technology and decreases what this community refers to as the misuse (overreliance) and disuse (underutilization) of technology (Parasuraman and Riley, 1997). For instance, the early forward collision warning systems were often unreliable (Zador et al., 2000) and some lane departure warning systems still suffer from this problem (HLDI, 2012). What was needed was research to define the performance envelope within which the different types of warning systems would work for the drivers. For example, it has been found that if false alarms with imminent crash warning systems are limited to less than 0.5 per 100 miles, then they are not considered a nuisance (Kiefer et al., 1999).

The third law of human factors says: "failing to take into account human factors considerations in autonomous vehicles' design can result in systems that are suboptimal." This is what the general automation community often refers to as abuse of technology (Kiefer et al., 1999). Nowadays, we have advances in technologies which can be used to take over the functions normally reserved for the driver. In some systems (Levels 1 and 2) the driver is still supposed to be fully engaged in supervising the actions of the vehicle under all circumstances. In others cases (Levels 3 and 4), the driver is freed from supervision, either in limited situations or during the entire trip. Problems arise at these levels, because the driver is in-the-loop and not always aware of what is happening (Levels 1 and 2) or is out of the loop and needs quickly to be brought back into the loop for some unexpected reason (Levels 3 and 4).

In addition to road safety issues, driver behaviors are involved in fuel consumption reduction. Fuel efficiency is mainly, although not exclusively, influenced by vehicle characteristics, road type, traffic conditions, and driving pattern, but there is still limited knowledge of the direct relationship (Corti et al., 2013). Currently, advanced driver assistance systems (ADAS) are a promising field of research that contributes to safety and powertrain efficiency improvement (Karginova et al., 2012). These systems assist drivers, but are normally designed based on average driver characteristics, which are representative of a wide majority, but do not adapt to specific driver particularities (Guardiola et al., 2014). The next generation of ADAS will take into account driver style recognition to individualize the system performance potentially improving safety and fuel consumption (Bolovinou et al., 2014). On the other hand, autonomous vehicles are coming to the world with the hope of increasing safety, decreasing congestion, and reducing greenhouse gases in a general way. But, new open questions need to be answered: what will be the role of humans in such a rapidly approaching future? Would they sit as passive occupants, who fully trust their vehicles, or would there be a need for humans to "take over" control in some situations either triggered by the need perceived by the autonomous vehicle or desired by someone in the cabin?

Motivated by the explained challenges, this chapter makes a review of the literature related with the human driver behaviors, presenting the main sensors, signals, algorithms, and datasets used to detect driving styles as well as some applications into the current ADAS framework towards increasing levels of automation until autonomous driving is reached.

9.1.2 DRIVING STYLE: DEFINITIONS

Intelligent vehicles of the future may soon have the capability to detect driver states and driving style, warn the driver of potential problems, and take over control in situations that require it. But before studying how the human factor influences the development of these technologies, it is important to clarify some terms which lack an agreed definition in the state of the art and consequently require a concise description to avoid reader confusion. Hereafter we define the most used terms in this context (Wang and Lukic, 2011; Taubman-Ben-Ari et al., 2004; Bergasa et al., 2006), which are depicted in Fig. 9.1.1.

- Environmental factors are all the external variables related to the environment (traffic, road type, time of the day, weather, road conditions, etc.), which influence the driving style.
- Vehicle factors are all the variables related to the vehicle (brand, model, engine, transmission, dynamic, wheels, age, load, etc.), which condition a driving style.
- Human factors are all the variables related to the driver (conscious decision making, character, sociodemographic background, and driver condition) that influence his/her driving.
- Social-demographic background are the physical factors of the drivers (age, gender, disability level, etc.) as well as other variables, such as driving experience, familiarity with the vehicle, education level, country development level, etc., which impact their driving style.
- Driver condition is the level of attention of the driver in his/her driving, which can be influenced by external aspects as (drugs, alcohol, fatigue, distraction, stress, panic, etc.).
- Driving events are the maneuvers occurring during the driving task, which may be used to identify driving style (e.g., sudden acceleration, deceleration or turning, lane changes, overtaking, etc.).

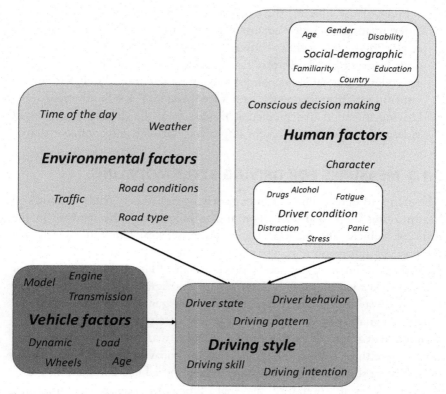

Figure 9.1.1 Terms related with human driving.

- Driving pattern is defined as position, speed, or acceleration profile and even includes driving events obtained from their analysis (e.g., number of accelerations, time at constant speed, number of lane changes, etc.). It is strongly related to road type, weather conditions, and driving style but does not include them specifically.
- Driver behavior is used to describe driving patterns but only focusing on driver's decisions and ignoring external influences.
- Driver state expresses the driver behavior classified among a predefined number of states.
- Driving skill is usually defined as the driver's ability to maintain vehicle control and it is normally used to analyze professional versus average drivers or young versus old drivers.
- Driving style definition has different interpretations caused by driver's subjective factors associated to it (mood, drossiness, etc.). It can be understood as the manner the driver operates the vehicle in the

context of the human, the vehicle, and the environment factors, as a differentiating indicator against driving pattern. In Taubman-Ben-Ari et al. (2004), the authors showed the high influence of driver conditions in driving style. In this case the same driver could exhibit different styles under different conditions (e.g., a day to day style for going/returning to/from the work, and a weekend style for family trips).

- Driving intention uses prediction models from the driving style ones in order to anticipate behaviors for improving warnings and controls systems.

9.1.3 MEASURES FOR DRIVING STYLE MODELING

The ability to detect different human driving behaviors greatly depends on the measures used to model them and improves with the increase in the level of vehicle automation, because the sensors required to advance the level of automation are also the ones that provide the measures to the algorithms often used to detect driving states. Then, an initial step for driving style identification is determining the signals to measure in order to provide a robust detection. There is no general agreement on the required signals, this is why a wide variety of alternatives are used. This disagreement responds to the sensors availability and the different applications where these systems are applied (driver behavior analysis, driver assistant, driving score, fuel consumption reduction, vehicle automation, etc.).

In the scientific literature, five main types of measures are commonly used: driver biological, driver physical, vehicle dynamics, sociodemographic, and hybrid. Table 9.1.1 summarizes some representative signals typically used for each measure type.

9.1.3.1 Driver Biological Measures

Biological signals include electroencephalogram (EEG), electrocardiogram (ECG), electro-oculography (EOG), surface electromyogram (sEMG), galvanic skin response (GSR), and respiration (Berka et al., 2007; Oron-Gilad et al., 2008; De Rosario et al., 2010). These signals are usually collected through electrodes in contact with the skin of the human body and are used to detect the mental state of the driver. EEG is widely accepted as a good indicator of the transition between wakefulness and sleep and is often referred to as the gold standard (Dong et al., 2011). The main drawback of such signals is that conventional measuring instruments are intrusive. In addition, most of the biological patterns vary between individuals. However, some biological variables, such as ECG, GSR, and respiration, can be measured with nonintrusive devices

Table 9.1.1 Approaches for driving style modeling

Measure type	Selected signals	References
Driver biological measures	Electroencephalogram (EEG), electrocardiogram (ECG), electro-oculography (EOG), surface electromyogram (sEMG), galvanic skin response (GSR), respiration.	Berka et al. (2007), Oron-Gilad et al. (2008), De Rosario et al. (2010), Baek et al. (2009), Bal and Klonowski (2007)
Driver physical measures	Dynamics of driver body pose (head, hand, eyes, foot), gaze zone classification using head and eyes cues, facial gestures, object interaction analysis and secondary task classification with hand cues or a fusion of them (head, hand, eyes, etc.), cabin occupant activity.	Tawari and Trivedi (2014a), Ohn-Bar and Trivedi (2014a), Bergasa et al. (2006), Tran et al. (2012), Jimenez et al. (2012), Bergasa et al. (2008), Ohn-Bar and Trivedi (2013), Ohn-Bar et al. (2014b), Yebes et al. (2010), Supancic et al. (2015)
Vehicle dynamics measures	Vehicle speed, acceleration, braking, pedals operation, jerk, steering wheel position, GPS, car-following, lateral position, lane changing and cornering, overtaking, driving in roundabouts, allowed speed, number of lanes, location of roundabouts, density of the traffic.	Huang et al. (2010), Mudgal et al. (2014), Romera et al. (2016a), Johnson and Trivedi (2011), Bergasa et al. (2014), Wang et al. (2014), Neubauer, Wood (2013), Kurz et al. (2002), Filev et al. (2009), Miyajima et al. (2010), Huang et al. (2012)
Socio-demographic measures	Gender, age, driver experience, familiarity with the vehicle and the environment, education level, country development level, personality.	Taubman-Ben-Ari et al. (2004), Ericsson (2000)
Hybrid measures	Fusion between driver physical measures (head, hand, foot, etc.) and surrounding cues (vehicle dynamic, pedestrian detection, salient object, etc.).	Ohn-Bar and Trivedi (2016), Ohn-Bar et al. (2015), Daza et al. (2014), Tawari et al. (2014b), Doshi et al. (2010a) Tawari et al. (2014c)

in the vehicle, as e.g., sensors based on seat pressure (De Rosario et al., 2010), flying instrumented with electrodes (Baek et al., 2009) or nano-sensors (Bal and Klonowski, 2007). These devices are based upon the same technology as conventional devices, but the body contact is achieved indirectly by taking advantage of the driver's own position while driving. Their main limitation is that the quality of contacts is worse than those made clinically, which means using more sophisticated methods of signal analysis to remove noise and extract the relevant parameters. This causes a performance decrease that makes the proposal impossible in most cases. These signals are normally used in psychological studies and as ground truth for other nonintrusive methods (Daza et al., 2014).

9.1.3.2 Driver Physical Measures

This approach is based on monitoring the driver and occupants behaviors inside the cabin, employing mainly cameras and image processing techniques, to obtain some measures related to: dynamics of driver body pose (head (Tawari and Trivedi, 2014a), hand (Ohn-Bar and Trivedi, 2014), eyes (Bergasa et al., 2006) and foot (Tran et al., 2012)); gaze zone classification using head and eyes cues (Jimenez et al., 2012); facial gestures (Bergasa et al., 2008); object interaction analysis and secondary task classification with hand cues (Ohn-Bar and Trivedi, 2013); head, hand, and eye cue fusion for secondary task activity analysis (Ohn-Bar et al., 2014b); cabin occupant activity (Yebes et al., 2010) etc. This approach is effective due to these measures being employed for in-cabin analysis of secondary tasks and model driver behavior in a nonintrusive way. Certain types of secondary tasks, such as gaze zone estimation and head gesture analysis, are more usually studied than others, such as driver—object interaction. Although passenger-related secondary tasks were found critical in the literature, there are very few studies on such tasks. Driver and passenger hand gesture and user identification have been widely studied, but works relating to analysis of interaction activity lack references. Most of the explained measures are obtained from cameras in an easy, inexpensive, and nonintrusive way. However, real applications must address the typical problems of vision systems working in outdoor environments (lighting conditions, sudden movements, etc.) (Daza et al., 2014). In this sense, depth sensors may also be used to improve behavior recognition (Supancic et al., 2015).

9.1.3.3 Vehicle Dynamics Measures

Vehicle dynamic signals are used to detect driving style through sensors which need to be installed in the vehicle platform. A commonly used source is the vehicle network, which contains data that can be directly obtained through the Controller Area Network (CAN) bus from the vehicle's own sensors. Signals available include vehicle speed, acceleration, braking, force on the pedals, steering wheel position, etc. (Huang et al., 2010). Other internal sensors are the Inertial Measurement Unit (IMU) or Global Position System (GPS) (Wang and Lukic, 2011), the latter being particularly beneficial as it provides information about the vehicle position and therefore it is an indirect measure of speed and acceleration. More advanced approaches include supplementary sensors, such as cameras, radar or LIDAR, already available in vehicles featured with ACC (Adaptive Cruise Control) or LDW (Lane Departure Warning) that provide signals related to car-following, lateral position or lane changes. On-line Internet access opens new possibilities, such as the use of Geographical Information System (GIS) to detect the location of roundabouts (Mudgal et al., 2014) or to know the allowed speed or the number of lanes of each position (Romera et al., 2016a). The advantage of this approach is that the signals have physical meaning, and their acquisition and posterior computation are relatively easy and inexpensive. This is the reason why commercial systems currently available are mainly based on these measures. The problem is that car manufacturers often protect access to most of their advanced signals and therefore it is necessary to know the access protocols of each manufacturer, which differ among them.

When signals are not available from the vehicle, smartphones can be used instead to obtain them. Due to the increasing popularity of these devices, smartphones have become a simple and cheap alternative to internal vehicle sensors and they don't need vehicle prior installation. In Johnson and Trivedi (2011), a fusion of accelerometers, gyroscope, GPS, and rear camera are used to detect driving style. Similarly, in Bergasa et al. (2014), a smartphone app called DriveSafe is presented, which uses all the available sensors on the smartphone (inertial sensors, GPS, camera, and internet access) to log and recognize driving maneuvers and infer behaviors from them.

Speed, acceleration, and deceleration are the most commonly used signals in the literature to detect driving styles. A more holistic approach defined 10 conventionally used driving behavior signals: average speed, average speed excluding stop, average acceleration, average deceleration, mean length of a driving period, average number of acceleration/ deceleration changes within one driving period, proportion of standstill time, proportion of acceleration time, proportion of deceleration time, and proportion of time at constant speed. However, authors conclude that by using only four of the previously listed variables satisfactory results can be obtained (Wang et al., 2014). The use of other variables such as jerk, braking pressure, and usage of the accelerator and brake pedal is also popular and considered to be better indicators by many researchers. These signals are based on longitudinal and lateral dynamics, and can capture the majority of the vehicle motions (Neubauer and Wood, 2013). However, these signals were only suitable when the driving situation (road type) and the environmental conditions (weather, traffic, etc.) were also known (Kurz et al., 2002). Alternative approaches focus on driving maneuvers instead of longitudinal and lateral dynamics. Some of them are braking (Filev et al., 2009), pedals operation (Miyajima et al., 2010), lane changing and cornering (Huang et al., 2012), car-following (Miyajima et al., 2010), overtaking (Filev et al., 2009), and driving at roundabouts (Mudgal et al., 2014). These maneuvers give information about the surrounding of the vehicle.

The commented approaches respond to discrepancies for sensors installation and signals capturing in terms of electronics installed and control access in vehicles currently produced. Given the tendency of further vehicle automation, it can be presumed that future vehicles will be provided with an integrated sensor system and a standard protocol able to provide all required information.

9.1.3.4 Sociodemographic Measures

As already mentioned in the definition section, driving behavior is a complex process influenced by human and external factors. Regarding the first factor there are different variables that influence it, with sociodemographic factors being some of them. Research on the relationship between some sociodemographic characteristics (gender, age, experience, familiarity, etc.) and driver driving style was presented in Taubman-Ben-Ari et al. (2004) and Ericsson (2000). These publications revealed that parameters such as familiarity with the vehicle and the environment, driving experience, gender (women have a more dissociative and anxious driving style, whilst being more careful, compared with men), age (tendency for more careful and patient driving for increasing ages), or even level of education (tendency for more careful driving for increasing level of education), or drivers' personality can influence driving style. Self-esteem was associated with patient driving and related to careful driving. Extraversion was inversely associated to adopting dissociative and anxious driving styles. Drivers with sensation seeking personalities were characterized by risky behavior, while drivers with a need for control developed angry as well as careful driving styles. The sociodemographic approach opens an interesting research perspective, although the implementation in real vehicles require large amounts of data that for now are not available. On-line Internet access, big data, and deep learning techniques can help to make it happen in the near future.

9.1.3.5 Hybrid Measures

Intelligent vehicles are able to integrate information coming from multiple domains for better scene understanding and improved driver forecasting. As drivers interact with their surrounding continuously, driver activities are often related to surrounding cues. Maneuver prediction often requires integrating surround and cabin measures for an improved model of the driver state and consequently better early event detection (Ohn-Bar and Trivedi, 2016). Therefore, combining driver measures with vehicle dynamic measures could intuitively increase the driving style recognition confidence. Road scene analysis and observations of the driver's pose would make it possible to estimate what the driver knows, what the driver needs to know, and when the driver should know it. In Ohn-Bar et al. (2015), both driver physical measures (head, hand, and foot) and surround cues related to the vehicle dynamic (e.g., distance to other vehicles) were integrated to identify intent of the ego-vehicle driver to overtake. Driver attention estimation is another common research topic in integrative frameworks where driver and surrounding measures are integrated. Relating these last ones, Daza et al. (2014) focused on vehicle dynamics measures, Tawari et al. (2014b) on pedestrian detection, and Doshi et al. (2010a) on salient objects. On the other hand, in Tawari et al. (2014c), situational need evaluation and driver alertness levels are used as cues for an assistive braking system.

9.1.4 DRIVING STYLE CLASSIFICATION

The signals selection, explained in the before section, are connected to the classification criteria. Signals, classification criteria, and algorithm definition are linked. Given the large amount of available signals and the different potential applications, there are many classification alternatives. Table 9.1.2 depicts a survey of the different approaches for driving style classification found in the revised literature and divided into two main groups (discrete classes and continuous scoring).

Table 9.1.2 Approaches for driving style classification

Papers	Classification basis	Classes labeling
Johnson and Trivedi 2011, Karginova et al. (2012); Vaitkus et al. (2014)	2 classes	Aggressive, calm
Xu et al. (2015), Dörr et al. (2014), Romera et al. (2016a)	3 classes	Aggressive, calm, moderate
Fazeen et al. (2012)	4 classes	Aggressive, calm, moderate, no speed
Stoichkov (2013)	5—7 classes	Range of styles from nonaggressive to aggressive
Vaitkus et al. (2014), Romera et al. (2016a)	Continuous scoring	Continuous scoring from calm to aggressive
Manzoni et al. (2010), Corti et al. (2013)	Continuous scoring	Overconsumption percentage
Neubauer and Wood (2013)	Continuous scoring	Efficiency and fuel consumption

9.1.4.1 Discrete Classes

The state-of-the-art revision reveals a general tendency to group driving styles into classes based on the selected driving measures. The classes must be defined prior to the classification algorithm and they need to take into account the trade-off between classification finesse and complexity to be understandable for the final users, as well as to guarantee the algorithm robustness (Bolovinou et al., 2014). Class labeling is conditioned for the applications where they are used. Whilst ADAS usually labels driving style according to the safety level, fuel-oriented classification generally uses terminology related to efficiency. Although multiple driving style classes can be identified, most approaches employ a rather simple classification basis using only two classes: aggressive and nonaggressive (normal driving) (Johnson and Trivedi, 2011; Karginova et al., 2012; Vaitkus et al., 2014). Aggressive drivers often show careless handling behavior (fast acceleration and deceleration changes, changes in throttle position operations with large magnitudes and frequent low diversity) which potentially contributes to compromised safety. Other authors preferred the division into three classes: Xu et al. (2015) includes aggressive, mild, and moderate (inbetween); Dörr et al. (2014) defines sporty (similar to aggressive), comfortable (similar to mild), and normal (inbetween); and Romera et al. (2016a) sets aggressive, normal, and drowsy (characterized by swinging around the sides of the lane instead of keeping a straight way and by involuntary lane changes). Some other authors opted for a more precise classification using a higher amount of classes. Fazeen et al. (2012) used four clusters: calm, normal, aggressive, and no speed driving (without information). Stoichkov (2013) proposed a five to seven classification method, covering a range of styles from nonaggressive to aggressive driving. A larger number of classes is not usual due to the complication in the algorithm development, in the learning process, and the interpretation of the classes.

9.1.4.2 Continuous Scoring

Another classification approach consists of using a continuous scoring from calm to aggressive driving avoiding a priori clustering, which eventually could be used in a threshold-based approach to transform this continuous score into classes (Murphey et al., 2009). In Romera et al. (2016a) a continuous scoring is calculated for its three classes (aggressive, drowsy, and normal) and the highest value defines the winning class. An alternative strategy for classification is to account for the relative fuel consumption instead of driver level of aggressiveness. In this case continuous scores are mostly used as alternatives to direct classification into enclosed classes. In Manzoni et al. (2010), the authors used an estimation of the fuel consumed along the trip and compared it to a benchmarked value to evaluate an overconsumption percentage, which is indicative of extra cost and can be used to estimate and assess the drivers' driving style. A fuel consumption simulation was also used in Neubauer and Wood (2013) to obtain the vehicle efficiency and use it as indicative of driving style. Similarly, Corti et al. (2013) assessed driver driving style with an energy-oriented cost function. In this approach, the estimated power overconsumption was used for classification instead of the absolute value of the fuel consumed.

9.1.5 ALGORITHMS FOR DRIVING STYLE MODELING

In this section we analyze the state-of-the-art algorithms typically used for driving style recognition, taking into account the selected signals and the classification strategy. Table 9.1.3 shows a summary of the revised techniques including their main characteristics and references.

Table 9.1.3 Methods proposed to classify driving behaviors

Method	Technique	Characteristics	References
Unsupervised methods (data should be label to define the classes)	Principal components analysis	Converts a set of observations of possibly correlated variables into a set of values of linearly uncorrelated variables	Constantinescu et al. (2010)
	Hierarchical cluster analysis	Seeks to build a hierarchy of clusters. Strategies can be: Agglomerative (bottom up) or Divisive (top down)	Constantinescu et al. (2010)
	Gaussian mixture method	Probabilistic model for representing the presence of subpopulations within an overall population	Miyajima et al. (2010)
	Dynamic time warping	Method that calculates an optimal match between two given sequences (e.g., time series) with certain restrictions	Johnson and Trivedi (2011)
	Markov- or histogram-based models	Markov is a stochastic model used to randomly model changing systems. It is assumed the Markov property: future states depend only on the current state not on the events that occurred before it	Guardiola et al. (2014)
Supervised methods	Rule based	Method that is domain-specific, expert system that uses rules to make deductions or choices	Engelbrecht et al. (2015), Romera et al. (2016a), Corti et al. (2013), Manzoni et al. (2010)
	Fuzzy logic	A form of many-valued logic in which the truth values of variables may be any real number between 0 and 1	Dörr et al. (2014), Gilman et al. (2015), Syed et al. (2007), Castignani et al. (2015), Kim et al. (2012), Arroyo et al. (2016), Syed et al. (2010)

(Continued)

Table 9.1.3 (Continued)

Method	Technique	Characteristics	References
	k-nearest neighbor (KNN)	Object is assigned to the class most common among its k nearest neighbors	Van et al. (2013)
	Neural network (NN)	Computational approach based on a large collection of neural units loosely modeling the way a biological brain solves problems	Siordia et al. (2010)
	Decision tree	A decision support tool that uses a tree–like graph or model of decisions and their possible consequences, including chance event outcomes, resource costs, and utility	Karginova et al. (2012)
	Random forest	An ensemble learning method that operate by constructing a multitude of decision trees at training time and outputting the class that is the mode of the classes of the individual trees	Karginova et al. (2012)
	Support vector machine (SVM)	Supervised learning models with associated learning algorithms that analyze data used for classification	Bolovinou et al. (2014)
	Bayesian based methods	Method of statistical inference in which Bayes' theorem is used to update the probability for a hypothesis as more evidence or information becomes available	Mudgal et al. (2014)
	Convolutional neural networks (CNNs)	It is a type of feed-forward artificial neural network but with a huge number of neurons	Romera et al. (2016b), Jain et al. (2015)

Like any other classification problem, driving style selection could be afforded using machine learning and pattern recognition techniques from a statistical point of view. The key point in these methods is the distinction between supervised or unsupervised methods. In the first one, labeled training data are required in the training phase to learn the classes. Then, in the test phase, new data are classified accordingly with the learned classes. To get a correct classification with unseen instances the training has to be able to generalize data (sample a representative subset of data). In an unsupervised method, a function to describe hidden structure from unlabeled data should be inferred. Since the examples given to the learner are unlabeled, there is no error or reward signal to evaluate a potential solution.

When considering supervised or unsupervised learning, there are some prior issues that have to be considered. One well-known problem in supervised learning is the definition of the classes. Different experts could assign the same data to different classes, different research could define a class in a slightly different way, and therefore direct comparison could be difficult. Another issue is that if input variables contain redundant or correlated information, some learning algorithms could perform poorly due to numerical instabilities. The most important problem for supervised learning is overfitting. This problem arises when the model fits very closely the set of training data. It appears in complex models, where there are too many parameters relative to the number of observations, and presents poor predictive performance, as it overreacts to minor fluctuations in the training data. The main drawback for unsupervised learning is the lack of information about the classes or clusters obtained. Since a high information label could not be assigned to a class, there is not a semantic information that can represent a class. Even more, unsupervised algorithms could assign an input data to a class in a controversial way for an expert. Therefore, as a rule of thumb, it is recommended to compare multiple learning algorithms and experimentally determine which one works best on the problem at hand. Tuning the performance of a learning algorithm can be very time-consuming, so selection of the algorithm with highest results from the beginning will be an important issue. It is often better to spend more time collecting additional training data and more informative variables than spending extra time tuning the learning algorithms. The emerging deep learning approach, based on big data, is revolutionizing the classical machine learning techniques explained so far, getting a breakthrough in the performance of complex classification problems. Romera et al. (2016b) presented a Convolutional Neural Network (CNN) to perform full pixel-wise semantic image segmentation in real time as an approach to unify most of the detection tasks required in the perception for autonomous vehicles based on cameras. The deep temporal reasoning approach has also shown similarly impressive performance (Jain et al., 2015), and is useful for a variety of learning tasks related to intelligent vehicle applications.

Results obtained for machine learning algorithms (supervised or unsupervised) depends on the correct variable selection and representation. Since dynamic events like driving require time series processing, which is highly computational demanding, and usually signals are very noisy (Liu, 2011), statistical models (Taniguchi et al., 2012) and robust representation of low-level data is considered crucial. In the excellent survey paper by Bolovinou et al. (2014), the authors reviewed different approaches typically used in ADAS, such as Bayesian inference, Gaussian Mixture Method (GMM), Support Vector Machine (SVM), K-mean clustering, Bayesian Network, and Markov-based models.

A comparison between supervised (SVM) and unsupervised (K-means) methods can be obtained in Van et al. (2013). Input vectors were defined by three events: braking, acceleration, and turning. Two sets of data were defined: a full raw data set and a shorter version of the data, formed by the model's parameters related to the three events. Supervised methods achieved better results than unsupervised. The report also pointed out that the model's data performed as well as the full data set. The unsupervised method also suggested that the most representative parameter was the braking event and its combination with turning achieved better classification results than acceleration. The importance of data selection was highlighted. However, unsupervised learning algorithms have been extensively considered for driving style classification. Let us mention that unsupervised methods like K-means needs at least the number of the desired classes. In the algorithm the user has the option to select manually some representative examples of the classes (e.g., calm or aggressive) and then an unsupervised algorithm will determine all similar instances.

9.1.5.1 Unsupervised Learning Techniques

A classical unsupervised dimensional reduction method PCA (Principal Components Analysis) has been used to obtain clusters for driving style in

Constantinescu et al. (2010). The authors use another unsupervised method, Hierarchical Cluster Analysis (HCA), to define clusters in reduced dimension data. After a final visual identification and manual labeling cluster step, the authors can identify five categories of driving style from calm to very aggressive.

The Gaussian Mixture Method has been used in Miyajima et al. (2010) to identify driving style; input data were pedals operation analysis and spectral features of pedal operation. Experimental results show that the driver model based on the spectral features of pedal operation signals efficiently models driver individual differences and achieves an identification error reduction of 55% over driver models that use raw pedal operation signals without spectral analysis.

Dynamic Time Warping (DTW) has been used to detect aggressive driving behavior using a smartphone (Johnson and Trivedi, 2011). The algorithm compares actual data with templates classified and labeled according with different driving behaviors. Even when classification is unsupervised, labeling of the initial representative class was done in a supervised way.

Other unsupervised techniques like Markov- or histogram-based algorithms have been proposed to model driving patterns (Guardiola et al., 2014). Proposed methods are based on the statistical analysis of previous driving patterns to predict future driving conditions. The robustness of the method depends on hard-to-assure conditions such as chains irreducibility or the verification of Markov Property. Results with simulated data were very satisfactory.

9.1.5.2 Supervised Learning Techniques

A supervised rule-based method based on an initial training set labeled by experts has been considered in (Siordia et al., 2014) to detect risky situations. The four basic elements of traffic safety were considered in the development of the system, i.e., the driver, the road, the vehicle, and the experts' driving knowledge.

A comparison among different supervised learning approaches: k-nearest neighbor (KNN), Neural Network (NN), decision tree and random forest were shown in (Karginova et al., 2012), obtaining good results for KNN, despite being the simplest algorithm. One simple but effective scheme to measure similarity between one driving sample and one reference pattern is distance measurement (e.g., Euclidean). Based on this approach, KNN algorithm assigns the new sample to the class most common among its k nearest neighbors. If k = 1, then the object is simply assigned to the class of that single nearest neighbor. Neural networks need some initial parametrization (selection of number of nodes in the hidden layer) and has the problem of overfitting. In the experiment, neural networks proved to be data dependent, since results were obtained from different processed data that in training. Decision trees showed worst results; so random forest were considered to improve its performance. Best results were obtained with 40 trees. Results of the experiment were obtained using an input vector of 16 signals.

Supervised methods are also compared in Siordia et al. (2010), in which trees generator (CART), rule learner (RIP), k-nearest neighbors (KNN), neural networks (NN), and support vector machine (SVM) are studied. Authors evaluate numerical results and the generalization power of each method in a specific application problem: detection of aggressive driving. The test set was based on simulator environments. The best results were obtained for CART algorithm, followed by KNN and NN.

Bayesian-based methods have been also considered in Mudgal et al. (2014) to classify driving style at roundabouts. Aggressive driving behavior has been easily detected and notified in real time and as a secondary task, the emission of hotspots can be predicted.

A rule-based technique is one of the most straightforward approaches to the problem of driving style classification. In this approach an initial set of clusters is defined from selected data. Each cluster is assigned to a priori known class which represents a driving style. Cluster definition implies the determination of the threshold values for the measured signals. In this initial cluster definition, the knowledge of some experts is required, since classes have to be meaningful and thresholds definition is a complex task. In working conditions, according to the values of the selected signals and the relation with the thresholds, the rule-based algorithm will assign a class for each input data set. This approach is simple, fast to implement, and does not need high processing power, so it can be embedded on vehicles hardware. However, if the number of variables is high, rules will be complex and can produce undesired responses if the system is not well defined. Also, the relationship between classes and measured variables (and threshold) could not be easily determined. Rule-based methods can be easily implemented even in smartphones and, since their computing load is not very high, can obtain real-time results even when the smartphone is also performing other tasks (e.g., recording events) (Engelbrecht et al., 2015). Cluster definition is influenced not only by dynamic vehicle and driver variables, but also by external environmental factors (weather, type of road, traffic, etc.).

In Romera et al. (2016a) a smartphone app called Drivesafe is used to detect aggressive and drowsy driving using thresholds defined for seven different maneuvers (accelerations, breakings, turnings, lane-weaving, lane-drifting, overspeeding, and car-following). Using auditory signals in real time, the system can warn the drivers who are unaware of the risks they are potentially creating for themselves and neighboring vehicles. On the other hand, as driving style influences the overall fuel consumption, its measure could be an initial estimation of the cluster associated to the driver (Lee and Son, 2011). This initial estimation could be added to other vehicle measurements. Corti et al. (2013) uses an initial estimation of the driving style, taking into account a cost function based on energy consumption. Three indicators were added to obtain the final cluster: power waste, power loss, and the ratio between the power request and actual power required at the wheels. Manzoni et al. (2010) shows that reference measures should be carefully selected, since they reported that they were associated to each particular driver, and should be dynamically changed if driving is performed over long periods of time or traffic congestion appears.

As pointed out before, class definition based on thresholds has undesired effects when the number of variables is high. Keeping some of the advantages of the rule-based techniques, fuzzy logic (Zadeh, 1965) can be a mechanism to overcome their disadvantages. Fuzzy logic is famous for its linguistic concept modeling ability and because the fuzzy rules expression is close to an expert natural language. A fuzzy system then manages the uncertain knowledge and infers high-level behaviors from the observed data. Driving classes are now seen as fuzzy sets whose elements have degrees of membership. In rule-based algorithms, the membership of elements in a set is defined in binary terms according to a bivalent condition, an element either belongs or does not belong to the set. As seen before, this can be done using thresholds to define membership to a certain class. By contrast, fuzzy set theory permits a soft definition of the membership of an element in a class; this membership now can be described with real numbers, e.g., with values in the real unit interval (0, 1) instead of the classical two possible values: member−not member.

Fuzzy algorithms have been used to detect driving style (Dörr et al., 2014) in a simulator environment, with good results. Results were obtained in real time using data from CAN bus as inputs, combined with other information sources like the road type and other indicators depending of the selected scenario (highway, urban, or rural). In a similar way, fuel consumption has been used as input variable. Gilman et al. (2015) combined fuel consumption with other factors to obtain the trip driving style classification according to a fuzzy algorithm. Syed et al. (2007) reported a 3.5% improvement in fuel consumption for the mildest driving using fuzzy logic applied to throttle and braking pedal operations. Their system suggests corrections to the driver to obtain a better fuel consumption in a hybrid vehicle simulator. Based on a smartphone, Castignani et al. (2015) presented a driver behavior classification between calm and aggressive based on fuzzy logic sets. Driver behavior is identified independently of the vehicle's characteristics. Fuzzy logic has been applied too in hybrid vehicles to optimize fuel consumption. Won (2003) based a strategy in driving event detection combined with road type and driving style. A similar approach was used in Kim et al. (2012) to adapt the parameters of the hybrid vehicle to the driving conditions.

Methods based in rules, even fuzzy ones, are limited in terms of the amount of variables and data that can be processed, and are based on the expertise of the designer who defined the fixed thresholds and rules. This approach lacks adaptability and some authors have addressed this problem. Arroyo et al. (2016) use an online calibration method to adjust the decision thresholds of the Membership Functions (MFs) to improve the identification of sudden driving events (acceleration, steering, braking) as a base to classify three driving styles (aggressive, normal, drowsy). Syed et al. (2010) introduced another approach, by using very simple machine learning algorithms to complete an adaptive fuzzy logic strategy. This machine learning mechanism was used to detect long- and short-term driver preferences using a weighted moving average with forgetting factors to reduce fuel consumption.

9.1.5.3 General Observations

We begin by pointing out that one main drawback to perform a fair comparison between results obtained with different methods is the lack of a common database. Results obtained with one database could be biased by data collected and, maybe, the specific characteristics of driver environment. Performance of the algorithms depends also on data collected,

CAN bus data are widely recorded, but other acquired signals could lead to achieving the best results, independently of the selected method.

A second observation is that most research papers only show the correct recognition rates to measure the goodness of the algorithm. Correct recognition performance is a vague indicator of algorithm actuation. For example, if our test set is formed by 95% samples which are labeled "normal driving" behavior and 5% samples with "aggressive driving," a system with a constant output of "normal driving" will achieve 95% correct recognition. Even when this system is impossible to use, performance is in the state-of-the-art rates. More useful indicators to measure the performance of the algorithms are precision and recall. Precision is the fraction of retrieved instances that are relevant, let's say "how useful the search results are," while recall is the fraction of relevant instances that are retrieved, in other words, is "how complete the results are." A single measure that combines both quantities is the traditional F-measure. The use of both indicators, together with F factor is highly recommended and encouraged. Other similar rates that better describe recognition performance are the false acceptance rate and false rejection rates, that can be combined to define Receiver operating characteristic (ROC) or detection error tradeoff (DET) curves. To represent recognition rate with a single quantity obtained from ROC curves, Equal Error Rate (EER) is widely used.

A third observation is that some systems increase the number of parameters when the input vector increases (e.g., adding more signals). In this case, the system may produce undesirable results. Trying to adjust one parameter may lead to an unpredictable result for other input data. Also, having more parameters to adjust requires more data in the training set, just to guarantee that enough samples are provided to the algorithm (and usually, input data are not so easy to obtain). If the training database is small, the algorithm could try to be overfitted to this training set with reduced generalization power.

A fourth observation is that the training set should be rich enough to cover a high spectrum of situations. Training with data obtained in very similar environments and testing with data obtained in other situations may lead to tricky results. Factors that have to be considered in experiment definition should include the number of drivers and experience of these drivers, drivers familiar or not with the scenario or the simulator; drivers with experience in the specific vehicle may obtain better results than novice ones.

A fifth observation is related to information extraction and feature definition. During the experimental phase a usual tradeoff should be achieved between the number of possible input signals and the cost of these signals. Since experimental data recording is expensive (drivers' recruitment, simulator hours, vehicle consumption), a common criteria is to record "everything." Once recorded, an initial information extraction procedure is suggested since maybe some variables could be correlated or could be combined without losing generality (but reducing processing time). In this initial procedure it is also advisable to estimate noisy signals that can be filtered or noise reduction procedure for signals that may require it.

A sixth observation is that the precise definition of classes is a complex task. Even simple classes like aggressive, calm, and normal driving have different definitions in research. Calm driving could mean optimal driving (in the sense of fuel consumption for example) or the driving condition that is more repeated or achieved in the training set. As a reference signal or indicator of driving behavior, fuel or energy consumption has been widely adopted. When this is not possible, experts in the field have been considered (e.g., in traffic safety) or the researcher directly evaluates the performance. Experts or external evaluation are preferred to avoid bias based on a priori knowledge of experiments.

A seventh observation notes that the acquisition of labeled data for a supervised driving learning process can be a costly and time-consuming task that requires the training of external evaluators. For naturalistic driving, where the driver voluntarily decides which tasks to perform at all times, the labeling process can become infeasible. However, data without known distraction states (unlabeled data) can be collected in an inexpensive way (e.g., from driver's naturalistic driving records). In order to make use of these low cost data, semisupervised learning is proposed for model construction, due to its capability of extracting useful information from both labeled and unlabeled training data. Liu et al. (2016) investigate the benefit of semisupervised machine learning where unlabeled training data can be utilized in addition to labeled training data.

The last observation is related to implementation issues. Using onboard embedded hardware in the vehicle or smartphones as a recording and processing platform is now the state-of-the-art technology. They have enough recording space, processing power, and enough sensor devices at a very reasonable cost. They can be placed in any car to obtain naturalistic driving data using huge numbers of drivers and vehicle types.

In the next years, new emergent techniques related to big data, deep learning, and data analytics will take advantage of these systems to improve intelligent vehicles' capabilities.

9.1.6 DATASETS FOR DRIVING STYLE MODELING

A key component in intelligent vehicles research relies on access to naturalistic, high quality, and large datasets, which can provide insights into better algorithms and system designs. We therefore summarize the evolution of datasets available to the scientific community for the study of driving behaviors in recent years.

The starting points were local datasets collected from ad hoc research, which usually had small amounts of data, were focused on one research field, and were owned by data collectors (Kawaguchi et al., 2001). These datasets were collected in the framework of national research projects, and usually needed additional equipment installation. Therefore, once the project was finished, data collection finished and it is hard to reproduce again. For naturalistic driving studies (NDS), we can mention the pioneer Argos project, funded by the Spanish Traffic Directorate in which a vehicle was equipped to acquire data from vehicle and driver (Recarte and Nunes, 2003). Nagoya University has been recording data from more than 800 drivers using instrumented vehicles (Kawaguchi et al., 2001; Takeda et al., 2011). Common sensors in these vehicles included video and audio records, physiological data, sensors to test the mental workload while driving, and vehicle information.

In parallel with the precedent NDSs, simulators have also been considered for driving research. Driving simulators can range from simple systems to highly complex and very immersive environments. One main concern with the latter kind of simulators is that if they are owned by automotive manufactures, there can be use restrictions or they are not be available for external use. Simulator use needs to be carefully considered since realistic conditions should be obtained. This means using high quality graphics, surround sound, and motion feedback among others. Realistic environments also affect the capability of a simulator to test different kind of vehicles without divergences that can occur when the vehicle dynamics do not correspond to the simulated vehicle. The advantages of using simulators are, on the one hand, that the test can be carefully designed and controlled, and on the other hand, that a huge amount of data can be collected and stored. Unfortunately, few public databases are available, since most of them are problem oriented (Siordia et al., 2012).

The next step was to collect a reasonable amount of data for which more than one partner is required and, as a consequence, data sharing starts as an obligation. At this point, one of the main references is the NDS that collected 43,000 h of driving data from 100 cars and 241 drivers in 2006 (Dingus et al., 2006). This data was used to gain a better knowledge of vehicle crashes and their causes, so effective crash prevention measures could be considered (Fitch et al., 2009). International cooperation efforts allow to share databases across Europe and USA (Takeda et al., 2011). These initial sets were the origin of the following huge data bases collections.

The Second Strategic Highway Research Program (SHRP2) conducted in the United States from 2010 until 2013 collected a huge amount of driving data from more than 3000 drivers (Dingus et al., 2015). Using these data, the evaluation of technologies from different providers or research institutes was possible and studies like Satzoda et al. (2014) arose in which critical events related to lanes and road boundaries were detected. Similar projects were conducted in Europe, taking advantage of the several European Framework Programs in which collaboration and data sharing are mandatory. The European Field Operational Test (euroFOT) data collection can be mentioned, during the years 2010 and 2011 data from more than 1,000 drivers were recorded. This database was used to evaluate ADAS, such as Adaptive Cruise Control (ACC) and Lane Departure Warning (LDW) systems (Benmimoun et al., 2013).

UDRIVE, can be considered the first large-scale European project in which naturalistic driving data was recorded. This project started in 2014 and will finish in 2017, passenger cars, trucks, powered two-wheelers, and more than 290 drivers were recorded and measured (Eenink et al., 2014). The European Union funded a network and research platform called FOT-Net Data (FOT-Net Data, 2017), that promotes and develops the methodology to share and reuse materials by providing information of available FOT/NDS data sets and is also responsible for the organization of cooperative meetings. In addition, there are other research projects collecting large-scale real-world or naturalistic driving data (Regan et al., 2013; Raksincharoensak, 2013; RIHEQL, 2017), part of these collected data set are publicly available.

In recent years, smartphones have been incorporated into NDS research projects. Smartphones are not intrusive for drivers (most of them have one), are equipped with many sensors (GPS, accelerometers, cameras, micro, etc.) and have many communication possibilities (Bluetooth, WiFi, 4G, etc.). Smartphones have also computing power and record storage at a reasonable cost. Although smartphone-based systems cannot record the same data quality as an instrumented vehicle, they have the invaluable advantage that they can be easily placed in different vehicles, increasing the range of possible users. In Miyajima et al. (2013) data are provided by a smartphone and from the CAN-bus via a Bluetooth connection to the smartphone. The dataset is based on 50 drivers with different vehicles. The first reported public database using smartphones is the UAH-DriveSet (Romera et al., 2016a), making openly available more than 500 min of naturalistic driving data (raw data and processed semantic information), together with video recordings provided by the app DriveSafe. The application is run by six different drivers and vehicles, performing three different behaviors (normal, drowsy, and aggressive) on two types of roads (motorway and secondary road).

Finally, there are some other publicly available datasets focused on image processing in the context of intelligent vehicles that can be used for driving behavior analysis too. Among them, we highlight VIVA (VIVA, 2017) for the detection and pose estimation of driver hands and face; RS-DM (Nuevo et al., 2011) for the detection and tracking of driver face. Brain4Cars (Jain et al., 2016); KITTI (Geiger et al., 2012) and Cityscapes (Cordts et al., 2015) for the vehicle dynamics estimation.

9.1.7 APPLICATIONS FOR INTELLIGENT VEHICLES

Driving style recognition has several applications concerning driving style feedback and correction, ADAS performance enhancement, driver in the automation loop, and fuel efficiency consumption. In this section we will focus on the safety applications classified according to the four level of automation defined by NHTSA (NHTSA, 2013) and following the ideas presented in Fisher et al. (2016). There are three human factor aspects to be analyzed at each level of automation: (1) driver—vehicle interfaces; (2) driver role at each level and associated problems; and (3) trust and use of the technology. Additionally, we will include a point focused consumption efficiency in the manual driving mode.

9.1.7.1 Level 1

9.1.7.1.1 Driver−Vehicle Interface (DVI)

There have been hundreds of refereed journal articles focusing on how to design a DVI for level 1 that are both effective and user friendly, including interfaces designed simply to deliver warning information and those designed to control the status of the vehicle. DVI can be carried out through either on-line or off-line feedback. This can be passive, active, or a combination of both. Driver−vehicle interfaces are usually installed directly in the vehicle, but also can be included in smartphones making used of their available sensors and processor capabilities. Passive feedback strategies (auditory, visual) are easy to implement and can develop satisfactory results, but there always exists the possibility that the driver will ignore the guidance. An example is presented in Stoichkov (2013) where visual feedback is displayed in a screen to promote style improvements for the risk of accident and fuel consumption reduction. Similar ideas but using both screen and microphone to generate feedback were revised in Johnson and Trivedi (2011). In Doshi et al. (2010a) authors studied various drivers' responses to feedback so as to adjust the information facilitated to the drivers. The results revealed a general tendency for aggressive drivers to have a more predictive behavior but less receptivity to suggestions and corrections, whilst calm drivers presented more complex behavior prediction and better feedback following. Active feedback strategies generate a response that change the normal vehicle operation, which cannot be as easily ignored, reaching a clear improvement in its working. In Syed et al. (2010), the authors used a strategy based on both visual and haptic feedback, increasing gradually the resistance over the throttle pedal when the vehicle operation diverges from optimal use. This strategy achieved a good improvement in fuel economy keeping the driver's expectations of vehicle performance. In Gilman et al. (2015) a "driving coach" is presented, highlighting the necessity of context awareness to facilitate suitable feedback including knowledge over the trip (map data, driving data, weather, and traffic conditions).

Future trends of driver feedback will probably lean toward a personalized assistance, where the system will be able to detect the driver's preferences and generate consistent feedback to both encourage style correction and avoid driver annoyance. Ideally, this feedback should combine haptic and passive approaches with the appropriate amount of information to easily interpret and follow based on the driver's cognitive capacity.

NHTSA is in the process of developing principles and guidelines for single and integrated interfaces that will populate the cars of the future (Monk, 2017).

9.1.7.1.2 ADAS Performance Enhancement

ADASs could benefit from knowledge about driver behavior so as to predict and anticipate drivers' reactions and adjust to each user. By adjusting ADASs to individual drivers' behavior may result in better and more efficient performance (Doshi et al., 2010b). Implementing adaptive ADASs based on driving style characterization is a complex task. So far it has not been sufficiently investigated and it has been identified as a promising field of intense research for the near future (Bolovinou et al., 2014). For example, a collision avoidance system that includes driver-in-the-loop, would be able to anticipate the driver's reactions and actuate accordingly. Nowadays we can find some incipient systems in the literature. Bolovinou et al. (2014) presented a driver style recognition based on individualized historical data using a threshold-based approach. A cooperative driving system with driver status recognition to identify the leader was studied in Rass et al. (2008). Arroyo et al. (2016) showed an on-line fuzzy strategy to identify driving events adapted to each user in order to score driving behaviors.

9.1.7.1.3 Trust and Use of the Technology

In this level, individual vehicle controls can be automated (e.g., ACC, LDW, etc.), but only one is active at any moment in time. The driver must keep his/her hands on the wheel or feet on the pedals, but one or the other (hands or feet) must be in the control loop. In this level, the goal with respect to the driver is that he/she can understand and trust the technology enough to engage with it under all circumstances. One interesting question is whether drivers can realistically function as supervisors of an automated system in which the driver is only controlling the longitudinal or the lateral position of the vehicle. Regarding the trust issues, it could have a real danger of level creep if the driver assumes that the level of automation has capabilities which exceed its envelope (overtrust or misuse) or the driver assumes that he/she needs to take control when such is not the case (mistrust or disuse).

9.1.7.2 Level 2
9.1.7.2.1 Driver–Vehicle Interface
The human factor issues in the design of effective warnings and interfaces used for vehicle control at level 2 are the same as those that will be needed for level 1. Questions about what mode to use for a warning (passive or active) and what method to communicate with the controls (manual, voice) will be quite similar at level 2 as at level 1. For instance, drivers will need to be warned when a given automatic system (e.g., ACC) is not able to control the vehicle and they will need to be able to set the particular features they want to keep active in an optimal way.

9.1.7.2.2 Driver as Supervisor
For level 2 systems the major problem is that they put the driver in the role of a supervisor of systems in which the driver is not potentially receiving tactile feedback (the automation is controlling the speed and position of the vehicle) (Parasuraman et al., 2000). As the supervisor needs both to notice changes to the system (remain vigilant) and keep track of the changing roadway environment (remain situationally aware). From simulated studies, it is known that humans are quite poor at vigilance and that their ability to detect a failure is inversely related to the likelihood that failure will occur. Not only does the ability to remain vigilant decrease as the frequency of alerts decreases (i.e., the request to transfer control from the vehicle to the driver), but so too does a human operator's situation awareness. Basically, in order to be situationally aware a driver needs to perceive an event, comprehend what the event means for the driver, and then predict what actions the driver should take. A driver is necessarily in the loop if he/she is operating both the steering wheel and foot pedals. But such is not the case if control over both of these is taken by the vehicle (Fisher et al., 2016). Thus, despite instructions to monitor the roadway, it is very probable that the driver will fail to monitor adequately, increasing the likelihood of a crash (Endsley, 1995a). So far in the literature, there isn't any study that has tried to maintain drivers' situation awareness at high levels of automation. However, researchers have suggested steps that can be taken to increase awareness and vigilance in automated systems. These include changes to the display (e.g., make salient alerts of automation failures), to the procedures (e.g., impose periodic times where the automation is disengaged), and to the training (e.g., impose unexpected automation failures during training).

Another interesting issue is that monitoring automation is supposed to reduce the load on the driver by integrating the many warning and control functions in an DVI. But the consequence of doing such is that monitoring can actually become more difficult. Besides, if the driver has more free time, then it is only natural to populate the DVI with content. But at level 2 the driver is being asked to serve as a monitor and therefore the DVI needs to have less content, not more (or at least be less distracting). There are some visual interfaces as an alternative of the visual in-vehicle displays as head-up (e.g., windshield HUD) or head-mounted (e.g., Google Glass). They do not require the driver to look down inside the vehicle and have been shown to provide advantages over head-down displays for certain classes of drivers (Kiefer et al., 1999). The aviation human factors community has explored extensively the best way to display information on a HUD. The most critical problem with these visual displays (head-up and head-mounted), is that they can serve as a source of distraction, leading to inattentional blindness and change blindness. Inattentional blindness may be defined as the event in which a person fails to recognize an unexpected stimulus that is in plain sight. Change blindness refers to our inability to notice changes in the environment that occur as we shift attention from one location to another (the change occurs during the shift of attention). There are some experiments that suggest new ways of mitigating inattentional blindness and change blindness (Hannon, 2009).

Finally, another obvious alternative to placing the display controls inside the vehicle when the eyes are off the forward roadway is to use voice-based interactive technologies to activate the controls. Guidelines will soon appear for audio DVIs (Monk, 2017). While voice-based interactive technologies generally have a benefit over interfaces inside the vehicle a certain cognitive workload and an extra cost must be assumed.

9.1.7.2.3 Trust and Use of the Technology

At this automation level, having an informative and controllable interface is as important as the driver's trust in the automation. Some studies have shown that individuals are less willing to take on risks over which they have little control rather than risks over which they have control. The human factors community has a great knowledge about how the driver's confidence in his/her own ability determines the driver's trust in the automation (Horrey et al., 2015). However, only a small number of studies have been carried out to increase drivers' trust in automation.

A misuse of the technology appears when the driver trusts the capabilities of the automation more than he/she should (overtrust). Human factors researchers have addressed broadly this problem and how this can be overcome in the design stage, perhaps requiring of drivers some training periods, but additional research needs to be carried out to detail how to implement these design guidelines. As well as a driver overtrusting in automation, a driver can place too little trust, and this can be dangerous too (Lee and See, 1997). Even when a driver is looking toward the forward roadway, he/she may not perceive potential hazards when cognitively distracted. The automated vehicle may act more intelligently than a driver who becomes cognitively distracted while attending to the forward roadway in response to unexpected vehicle actions. The solution to this problem is more complex than the solution to the problem of overtrust, since the driver now needs to be made aware of his/her own fallibility. The human factors community has found that error training works well in this context in a laboratory setting (Ivancic and Hesketh, 2000), but how it would be implemented broadly over automated vehicles is not clear. So, the question is how to build trust between the driver and the vehicle just as trust is built between two humans. Regarding this, it has been found that when the vehicle takes over control or performs unexpected maneuvers, drivers trust it more if it provides continuous feedback to the driver than if it does not (Verberne et al., 2012).

9.1.7.3 Level 3

At this level, it is assumed that the vehicle will control all driving functions for major sections of each trip. Many of the problems arising this level will be similar to those of level 2. But the problems of usability and of trust at level 3 will be quite different. The usability questions related to the design of the DVI will be much wider and the trust issues much more specific ones. In addition, there are questions of what will happen to drivers' skills over time as they exercise them less and less often as the automation level increases.

9.1.7.3.1 Driver–Vehicle Interface

In this level the human factors problems that are more likely to occur are during the transfer of control. There are two important issues to be addressed: (1) how much time the driver needs before transfer of control can be made safely from the vehicle to the driver; and (2) what information the vehicle could provide to the driver to facilitate this transfer of control. The interface must be designed with this in mind. In addition, there are two likely scenarios where control must be transferred: expected and unexpected.

The first issue is with regard to how much time the transfer can be anticipated to take in expected scenarios. There is concern in the human factors community about how to measure dynamically the elements of situation awareness (Endsley, 1995b). The automation should remain in the loop after the driver has taken over control long enough for the driver to regain full situation awareness. There are two basic questions to be solved: (1) how long does it take after a transfer of control signal is issued for the driver to control the vehicle? and (2) how long does it take after a transfer of control signal has been issued for the driver to anticipate hidden hazards? In general, it is found that it takes at least 8 s after the driver has been told that he/she can take control for the driver to recognize hidden hazards as well as he/she did when driving in a normal way (looking at the forward roadway). Then, control should only be transferred when at least 8 s separates the signal to transfer control from potential latent hazards (e.g., a very close ahead vehicle). Moreover, the automation should remain in the loop as a lifeguard to take over control just in case something might happen during at least 8 s.

The second issue is related to how the control must be suddenly transferred in unexpected scenarios. The driver needs to be aware of the driving context as quickly as possible. This would require a combination of auditory, visual, and haptics alerts. The auditory alerts will potentially warn the driver to a possible hazard ahead more quickly than a visual alert and the visual displays can depict the critical visual features of the roadway much more quickly than can a spoken description. In addition, a directional haptic alert related to the direction of the possible hazards could be more effective than visual and auditory ones. From the best of our knowledge no one has yet addressed this issue.

9.1.7.3.2 *Trust and Use of the Technology*

The main issues of trust at this level are quite different than those issues with level 2 systems because the driver is out of the loop for the most part of the trip. The driver does not need to be concerned about monitoring the road, only about being warned sufficiently ahead of time in order to smoothly transfer the control. Considering the transfer of control is carried out without problems, trust will probably be high when the benefits of automation are high and the costs are low (e.g., when the driver is stuck in traffic moving very slowly or in highways trips). There are some other problems which cross the boundary between human factors and philosophy that need to be solved. Sometime the driver will need to decide how the vehicle should respond to life threatening situations when the driver's life or someone else's life is at stake. Should the vehicle cede control immediately to the driver, taking into account that the driver will be out of the loop, or should it take control of the situation? In the second case, which criteria must the vehicle adopt to minimize crash fatalities? These are not easy questions to be addressed in the near future.

9.1.7.3.3 Driver Skill Over Time

It is reasonable to think that as drivers pay less and less attention to the forward roadway because they are out of the loop, they will lose critical hazard anticipation skills to determine where a potential hazard might materialize. The level of a driver's hazard anticipation skills is inversely proportional to the risk of a crash (Horswill and McKenna, 2004). However, there has not been evaluation yet of the loss of skills among drivers using automation. There are interesting studies in aviation that conclude that pilots' cognitive skills are deteriorated after long exposure to automation. It was found that there was a correlation between pilots' retention of the cognitive skills and how long they were focused on supervising the flight when the automatic pilot was engaged. This suggests that it is necessary to have periodic time outs from the automation in order for a driver to maintain hazard anticipation skills.

Another interesting issue related to cognitive skills is how to keep the pilot engaged with the automation and ready to take over manually. Drivers should keep their main hazard mitigations skills in addition to important hazard anticipation skills even with extensive automation. But two important issues must to be addressed: (1) a driver can't mitigate a hazard when he does not anticipate it; and (2) the drivers of the future will not have had years of training with vehicles in manual mode. Previous training is the only solution, as it happens in aviation, to the acquisition and maintenance of critical hazard mitigation skills that may never be acquired for the great majority of drivers if level 3 autonomous vehicles become a reality.

9.1.7.4 Level 4

The usability and trust issues explained in level 3 are the same as those in level 4 because the only difference is that the autonomous system is designed to operate during all the trip (under every condition) instead of in the major sections of the trip. However, technologies can fail and research is necessary about what to do when the vehicle suddenly fails. Probably the same techniques for sudden failure that were developed for level 3 could be used, with the difference that in this level there could be more failures because the potential for uncontrolled conditions are higher here. More research is required to overcome this problem, but for that, it is necessary to drive autonomously millions of km around the world under all conditions. Additionally, critical driving skills will also decay over time. Presumably, there will be a need for training at this level to solve this problem in the same way as at level 3.

9.1.7.5 Applications for Consumption Efficiency

The increasing level of vehicle electrification has motivated research on the influence of driving style on the energy demand of battery electrical vehicles (BEVs) and hybrid electric vehicles (HEVs). The dominant influence of driving style on fuel consumption has been shown in different publications. In Nie et al. (2013), the authors demonstrated that different driving styles can decrease the efficiency by more than 50% or increase it by more than 20% with regard to an average driver. The effect of driver driving style on different vehicles platforms (combustion, BEV, and HEV) was presented in Neubauer and Wood (2013), showing the important influence of driving style in all of them. Driving style recognition can provide useful information for BEV to estimate more accurately the battery remaining range and potentially extend it. The importance of driving style over the remaining BEV range was highlighted in Bingham et al. (2012) claiming a divergence of 10% between drivers. On the other hand, the selection of hybrid modes in HEV is strongly related with the driver torque request and consequently slight changes in driving style can cause unnecessary mode switching and can develop suboptimal performance. For example, high acceleration could trigger the internal combustion engine, whilst slightly lower torque demand would keep the electric drive. In addition, charging sustaining mode can be enhanced with prior knowledge of driving style in HEVs (Syed et al., 2010).

REFERENCES

Arroyo, C., Bergasa, L.M., Romera, E., 2016. Adaptive fuzzy classifier to detect driving events from the inertial sensors of a smartphone. In: 19th International Conference on Intelligent Transportation Systems (ITSC), pp. 1896—1901.

Baek, H.J., Lee, H.B., Kim, J.S., Choi, J.M., Kim, K.K., Park, K.S., 2009. Nonintrusive biological signal monitoring in a car to evaluate a driver's stress and health state,. Telemed. e-Health 15, 182—189.

Bal, I., Klonowski, W., 2007. SENSATION—new nanosensors and application of nonlinear dynamics for analysis of biosignals measured by these sensors., World Congress on Medical Physics and Biomedical Engineering 2006, 14. Springer, Berlin/Heidelberg, Germany, pp. 735—736.

Benmimoun, M., Pütz, A., Zlocki, A., Eckstein, L., 2013. euroFOT: Field operational test and impact assessment of advanced driver-assistance systems: Final results. In: Proceedings of FISITA World Automotive Congr. Lecture Notes in Electrical Engineering, vol. 197, pp. 537−547.

Bergasa, L.M., Nuevo, J., Sotelo, M.A., Barea, R., López, E., 2006. Real-time system for monitoring driver vigilance. IEEE Trans. Intell. Transp. Syst. 7 (1), 63−77.

Bergasa, L.M., Buenaposada, J.M., Nuevo, J., Jimenez, P., Baumela, L., 2008. Analysing driver's attention level using computer vision. In: 11th International IEEE Conference on Intelligent Transportation Systems (ITSC 2008), pp. 1149−1154.

Bergasa, L.M., Almería, D., Almazán, J., Yebes, J.J., Arroyo, R., 2014. DriveSafe: an app for alerting inattentive drivers and scoring driving behaviors,. Intelligent Vehicles Symposium (IV). Dearborn, Michigan, USA.

Berka, C., Levendowski, D.J., Lumicao, M.N., Yau, A., Davis, G., Zivkovic, V.T., et al., 2007. EEG correlates of task engagement and mental workload in vigilance, learning, and memory tasks. Aviat. Space Environ. Med. 78, 231−244.

Bingham, C., Walsh, C., Carroll, S., 2012. Impact of driving characteristics on electric vehicle energy consumption and range. IET Intell. Transp. Syst. 6 (1), 29−35.

Bolovinou, A., Bellotti, F., Amditis, A., Tarkiani, M., 2014. Driving style recognition for co-operative driving: a survey. In: ADAPTIVE 2014: The Sixth International Conference on Adaptive and Self-Adaptive Systems and Applications, pp. 73−78.

Broughton, J., Baughan, C., 2002. The effectiveness of antilock braking systems in reducing accidents in Great Britain. Acc. Anal. Prev. 34, 347−355.

Castignani, G., Derrmann, T., Frank, R., Engel, T., 2015. Driver behavior profiling using smartphones: a low-cost platform for driver monitoring. IEEE ITS Mag. 7 (1), 91−102.

Constantinescu, Z., Marinoiu, C., Vladoiu, M., 2010. Driving style analysis using data mining techniques. Int. J. Comput. Commun. Control 5 (5), 654−663.

Cordts, M., Omran, M., Ramos, S., Scharwächter, T., Enzweiler, M., Benenson, R., et al., 2015.The Cityscapes dataset. In: CVPR Workshop on The Future of Datasets in Vision.

Corti, A., Ongini, C., Tanelli, M., Savaresi, S.M., 2013. Quantitative driving style estimation for energy-oriented applications in road vehicles. In: IEEE International Conference on Systems, Man, and Cybernetics, pp. 3710−3715.

Daza, I.G., Bergasa, L.M., Bronte, S., Yebes, J.J., Almazán, J., Arroyo, R., 2014. Fusion of optimized indicators from advanced driver assistance systems (ADAS) for driver drowsiness detection. Sensors 14 (1), 1106−1131.

De Rosario, H., Solaz, J., Rodríguez, N., Bergasa, L.M., 2010. Controlled inducement and measurement of drowsiness in a driving simulator. Intell. Transp. Syst. IET 4, 280−288.

Dingus, T.A., Klauer,, S.G., Neale V.L., Petersen,, A., Lee S.E., Sudweeks, J., et al., April 2006. The 100-car naturalistic driving study, Phase II: Results of the 100-car field experiment. In: National Highway Traffic Safety Administration, Washington, DC, Technical Report DOT HS 810 593.

Dingus,, T.A., Hankey J.M., Antin, J.F., Lee, S.E., Eichelberger, L., Stulce, K.E., et al., Mar. 2015. Naturalistic driving study: technical coordination and quality control. Transportation Research Board, Washington, DC, SHRP2 Rep. S2-S06-RW-1.

Dong, Y., Hu, Z., Uchimura, K., Murayama, N., 2011. Driver inattention monitoring system for intelligent vehicles: a review. 596 IEEE Trans. Intell. Transp. Syst. 12 (2), 596−614.

Dörr, D., Grabengiesser, D., Gauterin, F., 2014. Online driving style recognition using fuzzy logic. In: IEEE 17th International Conference on Intelligent Transportation Systems (ITSC), pp. 1021−1026.

Doshi, A., Trivedi, M.M., 2010a. Attention estimation by simultaneous observation of viewer and view. In: IEEE Conference on Computer Vision and Pattern Recognition Workshops, pp. 21–27.

Doshi, A., Trivedi, M.M., 2010b. Examining the impact of driving style on the predictability and responsiveness of the driver: Real-world and simulator analysis. In: Intelligent Vehicles Symposium (IV), IEEE, pp. 232–237.

Eenink, R., Barnard, Y., Baumann, M., Augros, X., Utesch, F., April 2014. UDRIVE: the European naturalistic driving study. In: Transport Research Arena Conference.

Endsley, M., 1995a. Toward a theory of situation awareness in dynamic systems. Hum. Factors 37, 32–64.

Endsley, M., 1995b. Measurement of situation awareness in dynamic systems. Hum. Factors 37, 65–84.

Engelbrecht, J., Booysen, M.J., van Rooyen, G.J., Bruwer, F.J., 2015. Survey of smartphone-based sensing in vehicles for intelligent transportation system applications. IET Intell. Transp. Syst 9 (10), 924–935.

Ericsson, E., 2000. Variability in urban driving patterns. Transp. Res. Part D Transp. Environ. 5 (5), 337–354.

Fazeen, M., Gozick, B., Dantu, R., Bhukhiya, M., Gonzalez, M., 2012. Safe driving using mobile phones. IEEE Trans. Intell. Transp. Syst. 13 (3), 1462–1468.

Filev, D., Lu, J., Prakah-Asante, K., Tseng, F., 2009. Real-time driving behavior identification based on driver-in-the-loop vehicle dynamics and control. In: IEEE International Conference on Systems, Man and Cybernetics, San Antonio, USA, pp. 2020–2025.

Fisher, D.L., Lohrenz, M., Moore, D., Nadler, E.D., Pollard, J.K., 2016. Humans and intelligent vehicles: the hope, the help, and the harm. IEEE Trans. Intell. Veh. 1 (1), 56–67.

Fitch, G.M., Lee, S.E., Klauer, S.G., Hankey, J., Sudweeks, J., Dingus, T.A., June 2009. Analysis of lane-change crashes and near-crashes. In: National Highway Traffic Safety Administration, Washington, DC, Technical Report DOT HS 811 147.

FOT-Net Data. (Online). Available: <http://fot-net.eu> (accessed January 2017).

Geiger, A., Lenz, P., Urtasun, R., 2012. Are we ready for autonomous driving? the KITTI vision benchmark suite. In: IEEE Conference in Computer Vision and Pattern Recognition.

Gilman, E., Keskinarkaus, A., Tamminen, S., Pirttikanga, S.S., Röning, J., Riekki, J., 2015. Personalised assistance for fuel-efficient driving. Transp. Res. Part C 58, 681–705.

Guardiola, C., Blance-Rodriguez, D., Reig, A., 2014. Modelling driving behaviour and its impact on the energy management problem in hybrid electric vehicles. Int. J. Comput. Math. 91 (1), 147–156.

Hannon, D., 2009. Literature review of inattentional and change blindness in transportation. In: Federal Railroad Admin., Washington, DC, USA, Technical Report DTOS5908X99094.

Hedlund, J., 2000. Risky business: Safety regulations, risk compensation, and individual behavior. Injury Prev. 6, 82–89.

Highway Loss Data Institute, 2012. They're working. Insurance claims data show which new technologies are preventing crashes. (Online). Available: <http://www.iihs.org/iihs/sr/statusreport/article/47/5/1> (accessed January 2017).

Horrey, W., Lesch, M., Mitsopoulos-Rubens, E.L.J., 2015. Calibration of skill and judgment in driving: development of a conceptual framework and the implications for road safety. Acc. Anal. Prev. 76, 25–33.

Horswill, M., McKenna, F., 2004. Drivers' hazard perception ability: Situation awareness on the road. In: Banbury, S., Tremblay, S. (Eds.), A Cognitive Approach to Situation Awareness. Ashgate, Aldershot, U.K, pp. 155–175.

Huang, J., Lin, W., Chin, Y.-K., 2010. Adaptive vehicle control system with driving style recognition based on vehicle passing maneuvers. Washington, DC: U.S. Patent and Trademark Office Patent U.S. Patent No. 0,023,181.

Huang, J., Chin, Y.-K., Lin, W., September 2012. Adaptive vehicle control system with driving style recognition. Washington, DC: U.S. Patent and Trademark Office Patent U.S. Patent No. 8,260,515.

Ivancic, K., Hesketh, B., 2000. Learning from errors in a driving simulation: Effects on driving skill and self-confidence. Ergonomics 43, 1966–1984.

Jain, A., Koppula, H.S., Raghavan, B., Soh, S., Saxena, A., 2015. Car that knows before you do: anticipating maneuvers via learning temporal driving models. In: IEEE International Conference on Computer Vision.

Jain, A., Koppula, H.S., Soh, S., Raghavan, B., Singh, A., Saxena, A., 2016. Brain4cars: Car that knows before you do via sensory-fusion deep learning architecture. CoRR, vol. abs/1601. 00740.

Jimenez, P., Bergasa, L.M., Nuevo, J., Hernandez, N., Daza, I.G., 2012. Gaze fixation system for the evaluation of driver distractions induced by IVIS. IEEE Trans. Intell. Transp. Syst. 13 (3), 1167–1178.

Johnson, D., Trivedi, M.M., 2011. Driving style recognition using a smartphone as a sensor platform. In: 14th International IEEE Conference on Intelligent Transportation Systems (ITSC), pp. 1609–1615.

Karginova, N., Byttner, S., Svensson, M., 2012. Data-driven methods for classification of driving styles in buses. SAE Technical Paper, No. 2012-01-0744.

Kawaguchi, N., Matsubara, S., Takeda, K., Itakura, F., 2001. Multimedia data collection of in-car speech communication. In: Proceedings of Eurospeech, pp. 2027–2030.

Kiefer, R., LeBlanc, D., Palmer, M., Salinger, J., Deering, R., 1999. Development and validation of functional definitions and evaluation procedures for collision warning/ avoidance systems. National Highway Traffic Safety Administration, Washington, DC, USA, Technical Report DOT HT 808 964.

Kim, J., Sim, H., Oh, J., 2012. The flexible EV/HEV and SoC band control corresponding to driving mode, driver's driving style and environmental circumstances. In: SAE Technical Paper. Available at: <http://papers.sae.org/2012-01-1016/>.

Kurz, G., Müller, A., Röhring-Gericke, T., Schöb, R., Tröster, H., Yap, A., September 2002. Method and device for classifying the driving style of a driver in a motor vehicle. Washington, DC: U.S. Patent and Trademark Office Patent U.S. Patent No. 6,449,572.

Lee, J., See, K., 1997. Trust in automation: designing for appropriate reliance. Hum. Factors 39, 230–253.

Lee, T., Son, J., 2011. Relationships between driving style and fuel consumption in highway driving. SAE Technical Paper, No. 2011-28-0051.

Liu, A.M., 2011. Modeling differences in behavior within and between drivers. Human Modelling in Assisted Transportation (Models, Tools and Risk Methods)15–22.

Liu, T., Yang, Y., Huang, G.-B., Yeo, Y.K., Lin, Z., 2016. Driver distraction detection using semi-supervised machine learning. IEEE Trans. Intell. Transp. Syst. 17 (4), 1108–1120.

Manzoni, V., Corti, A., De Luca, P., Savaresi, S.M., 2010. Driving style estimation via inertial measurements. In: 13th International IEEE Conference on Intelligent Transportation Systems (ITSC), Madeire Island, Portugal, pp. 777–782.

Miyajima, C., Nishiwaki, Y., Ozawa, K., Wakita, T., Itou, K., Takeda, K., et al., 2010. Driver modeling based on driving behavior and its evaluation in driver identification. Proc. IEEE 95 (2), 427–437.

Miyajima, C., Ishikawa, H., Kaneko, M., Kitaoka, N., Takeda, K., September 2013. Analysis of driving behavior signals recorded from different types of vehicles using CAN and Smartphone. In: 2nd International Symposium on Future Active Safety Technology Toward Zero Traffic Accidents.

Monk, C., 2013. Driver-vehicle interface design principles. (Online). Available: <http:// slideplayer.com/slide/9041933/> (accessed January 2017).

Mudgal, A., Hallmark, S., Carriquiry, A., Gkritza, K., 2014. Driving behavior at a round-about: a hierarchical Bayesian regression analysis. Transp. Res. Part D20−26.

Murphey, Y.L., Milton, R., Kiliaris, L., 2009. Driver's style classification using jerk analysis. Comput. Intell. Veh. Veh. Syst.23−28.

Neubauer, J., Wood, E., 2013. Accounting for the variation of driver aggression in the simulation of conventional and advanced vehicles. In: SAE Technical Paper, No. 2013-01-1453.

NHTSA, 2013. National highway traffic safety administration preliminary statement of policy concerning automated vehicles. Washington, DC, USA.

Nie, K., Wu, L., Yu, J., 2013. Driving Behavior Improvement and Driver Recognition Based on Real-time Driving Information. Stanford University, Stanford.

Nuevo, J., Bergasa, L.M., Llorca, D.F., Ocaña, M., 2011. Face tracking with automatic model construction. Image Vision Comput. 29 (4), 209−218.

Ohn-Bar, E., Trivedi, M.M., 2013. The power is in your hands: 3D analysis of hand gestures in naturalistic video. In: IEEE Conference on Computer Vision and Pattern Recognition Workshops-AMFG, pp. 912−917.

Ohn-Bar, E., Trivedi, M.M., 2014. Beyond just keeping hands on the wheel: Towards visual interpretation of driver hand motion patterns. In: IEEE International Conference on Intelligent Transportation Systems, pp. 1245−1250.

Ohn-Bar, E., Martin, S., Tawari, A., Trivedi, M.M., 2014. Head, eye, and hand patterns for driver activity recognition. In: IEEE International Conference on Pattern Recognition, pp. 660−665.

Ohn-Bar, E., Tawari, A., Martin, S., Trivedi, M.M., 2015. On surveillance for safety critical events: In-vehicle video networks for predictive driver assistance systems. Comput. Vision Image Understanding 134, 130−140.

Ohn-Bar, E., Trivedi, M.M., 2016. Looking at humans in the age of self-driving and highly automated vehicles. IEEE Trans. Intell. Veh. 1 (1).

Oron-Gilad, T., Ronen, A., Shinar, D., 2008. Alertness maintaining tasks (AMTs) while driving. Acc. Anal. Prev. 40, 851−860.

Parasuraman, R., Riley, V., 1997. Humans and automation: use, misuse, disuse, abuse. Hum. Factors 39, 230−253.

Parasuraman, R., Sheridan, T., Wickens, C., 2000. A model of types and levels of human interaction with automation. IEEE Trans. Syst. Man Cybern. A Syst. Hum. 6 (3), 286−297.

Raksincharoensak, P., October 2013. Drive recorder database for accident/incident study and its potential for active safety development. In: FOT-NET Workshop, Tokyo, Japan.

Rass, S., Fuchs, S., Kyamakya, K., 2008. A game-theoretic approach to co-operative context-aware driving with partially random behavior, Smart Sensing and Context. Lecture Notes in Computer Science Volume 5279, pp. 154−167.

Recarte, M.A., Nunes, L.M., 2003. Mental workload while driving: effects on visual search, discrimination, and decision making. J. Exp. Psychol. Appl. 9, 119−137.

Regan, M.A., Williamson, A.M., Grzebieta, R., Charlton, J.L., Lenné, M.G., Watson, B., et al., 2013. Australian 400-car naturalistic driving study: innovation in road safety research and policy. In: Proceedings of Australian Road Safety Research, Policing & Education Conference, pp. 1−13.

Research Institute of Human Engineering for Quality Life, Driving behavior database (in Japanese) (Online). Available: <http://www.hql.jp/database/drive/> (accessed January 2017).

Romera, E., Bergasa, L.M., Arroyo, R., 2016. Need data for driver behaviour analysis? Presenting the public UAH-DriveSet. In: IEEE 19th International Conference on Intelligent Transportation Systems (ITSC), pp. 387−392.

Romera, E., Bergasa, L.M., Arroyo, R., 2016. Can we unify monocular detectors for autonomous driving by using the pixel-wise semantic segmentation of CNNs? arXiv preprint arXiv:1607.00971.

Satzoda, R.K., Gunaratne, P., Trivedi, M.M., 2014. Drive analysis using lane semantics for data reduction in naturalistic driving studies. In: Procedings of IEEE Intelligent Vehicles Symposium, pp. 293−298.

Siordia, O.S., De Diego, I.M., Conde, C., Reyes, G., Cabello, E., 2010. Driving risk classification based on experts evaluation. In: Proceedings IEEE Intelligent Vehicles Symposium, pp. 1098−1103.

Siordia, O.S., De Diego, I.M., Conde, C., Cabello, E., 2012. Wireless in-vehicle compliant black box for accident analysis. IEEE Veh. Technol. Mag. 7 (3), 80−89.

Siordia, O.S., De Diego, I.M., Conde, C., Cabello, E., 2014. Subjective traffic safety experts' knowledge for driving-risk definition. IEEE Trans. Intell. Transp. Syst. 15 (4), 1823−1834.

Stoichkov, R., 2013. Android smartphone application for driving style recognition. Project Thesis, Technische Universität München, Munich.

Supancic, J.S., Rogez, G., Yang, Y., Shotton, J., Ramanan, D., 2015. Depth-based hand pose estimation: Data, methods, and challenges. In: IEEE International Conference on Computer Vision, pp. 1868−1876.

Syed, F.U., Filev, D., Ying, H., 2007. Fuzzy rule-based driver advisory system for fuel economy improvement in a hybrid electric vehicle. In: Annual Meeting of the North American, pp. 178−183.

Syed, F., Nallpa, S., Dobryden, A., Grand, C., McGee, R., Filev, D., 2010. Design and analysis of an adaptive real-time advisory system for improving real world fuel economy in a hybrid electric vehicle. SAE Technical Paper, No. 2010-01-0835. Available at: <http://papers.sae.org/2010-01-0835/>.

Takeda, K., Hansen, J.H., Boyraz, P., Malta, L., Miyajima, C., Abut, H., 2011. International large-scale vehicle corpora for research on driver behavior on the road. IEEE Trans. Intell. Transp. Syst. 12 (4), 1609−1623.

Taniguchi, T., Nagasaka, S., Hitomi, K., Chandrasiri, N.P., Bando, T., 2012. Semiotic prediction of driving behavior using unsupervised double articulation analyzer. In: 2012 Intelligent Vehicles Symposium, Alcalá de Henares, Spain, June 3−7, pp. 849−854.

Taubman-Ben-Ari, O., Mikulincer, M., Gillath, O., 2004. The multidimensional driving style inventory—scale construct and validation. Acc. Anal. Prev. 36 (3), 323−332.

Tawari, A., Trivedi, M.M., 2014a. Robust and continuous estimation of driver gaze zone by dynamic analysis of multiple face videos. In: IEEE Intelligent Vehicles Symposium, pp. 344−349.

Tawari, A., Mogelmose, A., Martin, S., Moeslund, T., Trivedi, M.M., 2014b. Attention estimation by simultaneous analysis of viewer and view. In: IEEE International Conference on Intelligent Transportation Systems, pp. 1381−1387.

Tawari, A., Sivaraman, S., Trivedi, M.M., Shannon, T., Tippelhofer, M. 2014c. Looking-in and looking-out vision for urban intelligent assistance: estimation of driver attentive state and dynamic surround for safe merging and braking. In: IEEE Intelligent Vehicles Symposium, pp. 115−120.

Tran, C., Doshi, A., Trivedi, M.M., 2012. Modeling and prediction of driver behavior by foot gesture analysis. Comput. Vision Image Understanding 116, 435−445.

Vaitkus, V., Lengvenis, P., Zylius, G., 2014. Driving style classification using long-term accelerometer information. In: 19th International Conference on Methods and Models in Automation and Robotics (MMAR), pp. 641−644.

Van Ly, M., Martin S., Trivedi, M.M., 2013. Driver classification and driving style recognition using inertial sensors. In: Intelligent Vehicles Symposium (IV), IEEE, pp. 1040−1045.

Verberne, F., Ham, J., Midden, C., 2012. Trust in smart systems: Sharing driving goals and giving information to increase trustworthiness and acceptability of smart cars. Human Factors 54, 799−810.

VIVA: Vision for intelligent vehicles and applications challenge. <http://cvrr.ucsd.edu/vivachallenge/> (accessed January 2017).

Wang, R., Lukic, S., 2011. Review of driving conditions prediction and driving style recognition based control algorithms for hybrid electric vehicles. Veh. Power Propulsion Convergence (VPPC)1−7.

Wang, W., Xi, J., Chen, H., 2014. Modeling and recognizing driver behavior based on driving data: a survey. Hindawi Publishing Corporation, Mathematical Problems in Engineering Volume 2014, Article ID 245641, 20 pages.

Won, J.-S., 2003. Intelligent energy management agent for a parallel hybrid vehicle. Doctor of Philosophy Graduate Studies of Texas A&M University. Available at: <http://oaktrust.library.tamu.edu/bitstream/handle/1969.1/271/etd-tamu-2003A-2003032522-1.pdf>.

Xu, L., Hu, J., Jiang, H., Meng, W., 2015. Establishing style-orientated driver models by imitating human driving behaviors. IEEE Trans. Intell. Transp. Syst. 16 (5), 2522−2530.

Yebes, J.J., Alcantarilla, P.F., Bergasa, L.M., González, A., 2010. Occupant monitoring system for traffic control in HOV lanes and parking lots. 13th International IEEE Conference on Intelligent Transportation Systems Workshop (ITSC-WS 2010), pp. 1−6.

Zadeh, L.A., 1965. Fuzzy sets. Inform. Control 8, 338−353.

Zador, P., Krawchuk, S., Voas, R., 2000. Final report—automotive collision avoidance system (ACAS) program. National Highway Traffic Safety Administration, Washington, DC, USA, Technical Report DOT HS 809 080, 2000.

SUBCHAPTER 9.2

User Interface

Alfonso Brazález, Olatz Iparraguirre and Joshua Puerta
Ceit-IK4, San Sebastián, Spain

9.2.1 INTRODUCTION: FEEDBACK CHANNELS

The human−machine interface (HMI) allows the driver to interact with the vehicle. Car manufacturers and designers are focused on the development of innovative control elements for an intuitive use of the different functionalities, offering integrated systems and complete interaction with the driver. The HMI should optimally process and present information, allowing, when it is necessary, the prioritization of the information. Road safety must be assured in the HMI design. The system should be designed to support the driver and should not give rise to potentially hazardous behavior for the driver or other road users.

The physical interaction between the driver and the car is achieved by the communication channels: visual, acoustic, haptic, and gestural. Acoustic and haptic channels can be used as inputs or outputs, depending on the nature of the interaction, but visual and gestural channels are employed just as inputs to the driver. Nowadays, three main parts of the vehicle are commonly used for the installation of this kind of interactive device: instrument cluster; steering wheel (including surrounding buttons) and central console; and additional visualization device (simple screen, touch screen, head–up display (HUD), etc.). Some manufacturers include additional devices for controlling menus in such screens. In general, the visual channel has to be considered as the priority feedback channel, followed by the acoustic and haptic channels. Gestural interaction is less used.

9.2.1.1 Visual Channel

The principal visual output devices are: instrument panel cluster display, central panel display, or HUD. Independently on the position of the display, it is important to achieve a good reading visibility, contrast, resolution, and field of view. In this way, the information provided to the driver is clear and legible. The properties for visual devices would be brightness/intensity, visual acuity/spatial, contrast sensitivity, color specification, perceptual organization, figure–ground organization, and grouping principles (proximity, similarity, continuity, closure, common elements with common motions tend to be grouped). The requirements and recommendations of the visual signals and design principles are described in a further section.

9.2.1.2 Acoustic Channel

The main advantage of using auditory and haptic feedback is to provide information and warnings avoiding visual distraction. Simple reaction time for auditory stimuli is shorter than for visual stimuli (Salvendy, 1997) and could have an effect on the overall response time. Moreover, taking advantage of the sound direction, acoustic feedback can attract a driver's attention to the location of the hazard even if the hazard is out of the driver's field of view.

Like the visual channel, the acoustic channel also has to respect some general requirements in terms of sound pressure level, frequency range, and spatial resolution. First of all, it is important that the sounds used are well distinguishable from the other vehicle sounds (both informing sounds, like arrow sounds, and alerting sounds, such as empty fuel level), secondly the sounds' loudness levels need to be calibrated in order to transmit to the driver the proper level of urgency without creating annoyance.

Usually, the acoustic feedback is associated with another output, when it is not sure that the feedback itself is unequivocally related to an event. Therefore, it is common that the acoustic stimuli are associated with a visual as well as with a haptic feedback.

Tonal sounds (earcons) can typically be used for prompting or reinforcing other modalities while auditory icons are normally applied when a rapid response is required. It should be noted that auditory icons may require a wider frequency range of the audio reproduction system than tonal sounds.

Some examples of signals that could be employed are: sweeping sounds for immediate acoustic feedback, patterns of segments with constant pitch for short-term acoustic feedback, two-times chimes, high—low nonrecurrent for long-term acoustic feedback, and preconfigured messages.

9.2.1.3 Speech Recognition

Speech recognition can be considered a specific use case of the acoustic channel. The car is a challenging environment to deploy speech recognition. A well-developed speech recognition system should cope with the noise coming from the car, the road, and the entertainment system, and include the following characteristics (Baeyens and Murakami, 2011).

- *The microphone should be pointed at the driver position.* This assures that the incoming signal of the speech is as high as possible.
- *Push to talk button.* The speech interaction is initiated by the driver by means of pushing a button. Nevertheless, other occupants or voices from the entertainment devices can initiate any other action, with the proper command.
- The entertainment system should be muted in order to avoid interaction with the speech recognition system.

The car is a noisy environment, even when applying these basic techniques. Different driving speeds, varying road conditions, wipers, and air-conditioning are examples of noise sources. For this reason, the speech recognition system should include a noise cancellation algorithm. It could be useful to model the typical and specific sounds inside the car to handle nonstationary noises like wipers, etc.

Nowadays, the voice interface is mainly limited to the interpretation of several specific commands. The current challenge is to go a step forward in the use of the natural language. So far some devices appeared, although they are not globally spread out yet.

9.2.1.4 Haptic Output

Haptic signals as output signals are mainly used for alerting the driver (e.g., vibration on the steering wheel in case of lane departure) or for displaying a specific range of operation (e.g., force on accelerator pedal indicating exceedance of the current speed limit). In some cases, haptic signals are also used as a display for the driver to monitor the actions of a Driver Assistance System (e.g., steering wheel is turning when lane-keeping system is active).

These signals can appear in very different forms: tactile like vibrations, torque, or forces, and kinesthetic, such as accelerations, braking pulse, and turning rates.

In addition, haptic signals can be applied on different devices including the steering wheel, brakes, and accelerator pedals, as well as on the seat or on the seatbelt. Haptic signals are most often displayed on the device where an action of the driver is required.

Furthermore, acceleration or brake pulses that the driver perceives as body motion (kinesthetic perception) can also be classified as haptic output signals.

The parameters of the haptic feedback depend on the shape of the stimuli. A wide range of possible signal profile is shown in Fig. 9.2.1.

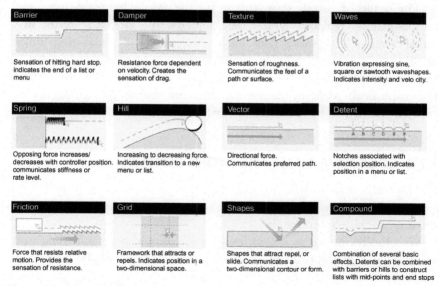

Figure 9.2.1 Haptic effects.

However, since the effects of haptic signals strongly depend on the specific hardware characteristics, general requirements could not always be derived from literature. Therefore, the application needs to be parameterized for a special adaptation of each hardware device.

9.2.1.5 Multimodality

In summary, the visual channel has to be considered as the priority feedback channel, followed by the acoustic and haptic channels. In some cases the three channels could work separately (or only one output channel is available), but it is better to think about them working together in order to give a more appropriate feedback to the driver: this case is called multimodality.

In multimodality mode, considering the different level of information/warning to be provided to the driver, a sort of prioritization of the output channels has to be planned.

It is recommended to use one channel for presenting information when the time to reaction it is not critical, and in this case the visual output is preferred. Usually two simultaneous channels are associated to the combination of visual with haptic or acoustic one, when redundancy is crucial for improving the reception of the message. In this case, visual output shall only be supportive and not the primary source.

The results of a study show that in the use cases where auditory information can be replaced with haptic information, this may result in a higher level of acceptance (Brockman et al., 2012). Concerning multimodality, it is recommended to arbitrate the output channels that should be used for performing the desired action by the driver or the system. This issue deals with the design and implementation of a prioritization strategy, which will be addressed in further sections of this chapter.

9.2.2 COGNITIVE LOAD AND WORK LOAD

The mental workload reduction of the driver is the guiding principle for reducing distractions while driving in order to achieve an optimal human—machine interface (HMI).

Specifically, mental workload involves various processes including neurophysiologic, perceptual, and cognitive ones. These can be defined as the proportion of the information processing capability used to perform a task. The workload is affected by individual capabilities and characteristics (e.g., age, driving experience), motivation to perform the task, strategies applied on task performance, as well as physical and emotional state.

There is a direct relationship between driver mental workload and the risk of having an accident (Kantowitz and Simsek, 2001). The most common types of collisions involve, as a rule, lack or loss of attention (Kantowitz, 2000). The connection between driving task demands, performance, and human capacity is well known and has been largely studied. For example, human attention has a limited capacity, and studies suggest that talking on the phone causes a kind of "inattention blindness" to the driving scene.

The workload of information processing can bring risks when unexpected driving hazards arise (Horrey and Wickens, 2006). Under most driving conditions, drivers are performing well-practiced automatic driving tasks. For example, without thinking about it much, drivers slow down when they see yellow or red lights, and activate turn signals when intending to make a turn or lane change. Experienced drivers automate some tasks: staying within a lane, noting the speed limit and navigation signs, and checking rear- and side-view mirrors. People can do these driving tasks safely with an average cognitive workload.

Driving a vehicle is a complex task that could include large deviations in mental workload (Baldwin and Coyne, 2003; Vervey, 2000). The drivers must process huge information coming from outside the vehicle: traffic intensity, traffic signs, etc. People connectivity, entertainment devices, and navigation systems lead to an increase of incoming information inside the car. This increasing information leads to the consequent demand of visual and auditory resources, which directly affects the mental workload.

The permanent adaptive control approach is an important basis for understanding the relationship between mental workload, driving performance, and task demand (as perceived by the driver) (e.g., Waard, 2002). In this context, there are moments on driving performance where the simultaneous rise in number of tasks increases demand on the driver, resulting in an increase of mental effort invested, and in a decrease of performance efficiency.

A correlation could be derived between task demand and mental workload, i.e., when there is a high task demand, there is also a high level of mental workload. However, drivers tend to develop a set of strategies that enable them to manage mental workload and regulate their performance. Thus, it is not possible to establish a correlation between task demand and mental workload, as this association is always dependent on the strategies and internal factors of each operator.

There is another process defined as dissociation, which represents the moment when task demand increases and driver workload decreases (Parasuraman, 2001). Considering the perspective of an integrated system in which task demand, adaptive strategies, workload and performance are closely related or interdependent; a change in each of these factors consequently affects all systems.

User-centered design is the key for developing user-friendly HMI systems with the aim of reducing driver workload. Human factors engineering addresses the physical and cognitive workload of the driver while designing interactivity that is intuitive for a broad range of users. Engineering activities in several fields, such as human factors, systems, software and hardware development, industrial design, and market research, must be coordinated to provide a human-centered design, in order to improve the HMI for the driver. Moreover, the use of a driver workload evaluation module would be optimal for a safe HMI development.

9.2.3 INFORMATION CLASSIFICATION AND PRIORITIZATION

The main strategy for reducing driver workload is the prioritization of the information. The International Organization for Standardization (ISO) and the Society of Automotive Engineers (SAE) have already established some standards about prioritization. SAE J2395 (SAE J2395, 2002) employs a method based on three criterions: safety relevance, operational relevance, and time frame. Depending on the combination of those three parameters linked to the information, SAE J2395 stablishes a Priority Order Index (POI) from a predefined table.

There is also a standard described in ISO/TS 16951:2004 (16951 ISO/TS, 2004), where the priority index of each message is defined based on the values from Criticality and Urgency ratings. The term Criticality is associated with the severity of the effect if the message is not received or is ignored by the driver; while Urgency is related to the time within which driver action or decision has to be taken if the benefit intended by the system is to be derived from the message.

Additionally, various scheduling algorithms (Zhang and Nwagboso, 2001; Sohn et al., 2008) have been developed to prioritize Intelligent Transport System (ITS) messages on board the vehicle. Sohn et al. (2008) managed the display of IVIS messages (that range from collision warnings and navigation instructions to tire pressure and e-mail alerts) based on their importance, duration, and preferred display time. They consider the dynamic value of the message based on the display time as a variable that changes dynamically. Zhang and Nwagboso (2001) focused on ITS messages prioritization over CAN by using a fuzzy neural network. They established that the priority of the message is related to the incident that occurs onboard or off-board the vehicle, so they included information about vehicle state or conditions in the prioritization process.

Furthermore, the information can be prioritized by the importance of the information in relation with the context of driving and the urgency of the information. The context can include aspects of external environment of the driving situation and the state of the driver itself, measuring parameters as fatigue or estimating the workload. Nevertheless, the most urgent information should have priority over any other input to the driver. The priority can be ranked taking into account different issues, for example the information related to a behavioral change. In this field, the information about lane changes or speed adjustment has more priority than not directly related driving behavior. On the other hand, the information about route planning and driving strategies has less priority than information related to maneuvers and control.

A complete prioritization criteria could include a dynamic estimation of the prioritization of the message based on the situation awareness. The importance of the message can be modified depending on the situation. For instance, there is not the same relevance to receiving a message of a possible traffic jam when there is no traffic or when there is a high density of the traffic. The modification of the initial value assigned to each message can increase the safety, the reliability, and the efficiency of the designed HMI.

9.2.4 IMPLEMENTATION ISSUES

The implementation of the HMI could be classified into some common modules. Firstly, it is usual to have an input module for the classification of entering variables which may need some filtering. This module could also provide a common interface with the assessment module. The assessment module is responsible for implementing the complete logic of the HMI module. This is the main module, where the prioritization criteria is calculated and evaluated, giving as an output the messages to be shown. Finally, the representation module decides how to present physically the information. The complete schema is shown below (Fig. 9.2.2).

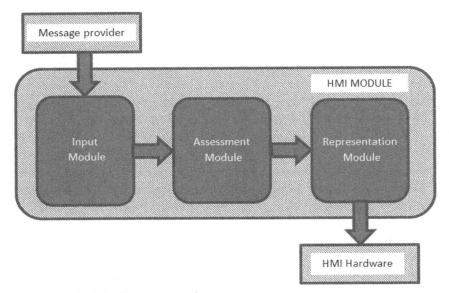

Figure 9.2.2 HMI implementation schema.

This architecture can be considered as a global framework for the implementation of an HMI module. The "Message provider" is the set of systems that provides inputs to the HMI module. For instance, each subsystem of the vehicle such as ABS, navigation, ESP, ASR, etc. is a message provider, and therefore, sends the information to the HMI processing unit. Once the message is received, it is filtered and treated in the "Input Module." Additionally, this module can provide a standard interface to the "Assessment Module." In highly advanced HMI developments, the assessment module can take into account the situation of awareness and evaluate the workload of the driver.

Finally, the results from the HMI logic, calculated in the Assessment Module, are provided to the Representation Module, where the knowledge is located of the different types of physical hardware of the HMI. This module decides how to present the message to the driver.

9.2.5 GUIDELINES AND STANDARDS

For the development of Human–centered design of the HMI there are some generic guidelines that can be used. The European Commission provided in 2008 the recommendations on safe and efficient in-vehicle information and communication systems. These principles promote the introduction of well-designed systems into the market, and take into account both the potential benefits and associated risks (Commission of the European Communities, 2008).

These principles also take into account the capabilities and constraints of all stakeholders in their efforts to design, install, and use in-vehicle information and communication systems. They are applicable to the development process, addressing issues like complexity, product costs, and time to market, and in particular take into account small system manufacturers. Since the driver ultimately decides whether to buy and use, for example, an integrated navigation system, a nomadic device, or a paper map, the intention is to promote good HMI design rather than prohibit the inclusion of some functionalities by simplistic pass/fail criteria.

The principles are not a substitute for any current regulations and standards, which should always be taken into consideration. These principles, which may be reinforced by national legislation or by individual companies, constitute the minimum set of requirements to be applied.

The United States Department of Transportation (DOT), through its Intelligent Vehicle Highway Systems (IVHS) program, promoted the definition of human factors guidelines for the design of in-vehicle information system (Green et al., 1995). These guidelines were the result of the detailed examination of two systems (traffic information and car phones) and the analysis of another three additional systems (route guidance, road hazard warning, and vehicle monitoring). Each of the five selected systems were examined separately in a sequence of experiments.

In Japan, the Japan Automobile Manufacturers Association (JAMA) provided in 2004 guidelines for in-vehicle display systems (Japan Automobile Manufacturers, 2004). These guidelines are based on the effect of the visual displays on the driving safety, including avoiding distractions and not obstructing the field of view. It can be assumed that it was derived from the massive inclusion of nomadic devices in passengers' cars at the beginning of the 21st century.

A complete review of guidelines and standards was performed during the Aide project. Some of the guidelines presented before are also included in this report. It covers also the standards for design, process, and performance, but it is not specific for vehicles. The review includes the standards for auditory signals, ergonomic requirements, etc. The report gives also the regulation and directives concerning type approval in Europe (Schindhelm et al. 2004).

REFERENCES

16951 ISO/TS, 2004. Road vehicles — Ergonomic aspects of transport information and control systems (TICS) — Procedures for determining priority of on-board messages presented to drivers. International Organization for Standardization, 2004.

Kantowitz, B.H. 2000. Attention and mental workload. In: Proceedings of the Human Factors and Ergonomics Society Annual Meeting, Human Factors and Ergonomics Society, pp.456−460.

Baeyens, B., Murakami, H., 2011. Speech recognition in the car:. Challenges Success Factors The Ford SYNC Case 2011 (7), 7−9.

Baldwin, C.L., Coyne, J.T., 2003. Mental workload as a function of traffic density: comparison of physiological, behavioral, and subjective indices. In: Driving Assessment 2003: The Second International Driving Symposium on Human Factors in Driver Assessment, Training and Vehicle Design, pp.19−24.

Brockman, M. et al., 2012. Deliverable D3. 1 | Results from IWI Evaluation Executive Summary. interactIVe.

Commission of the European Communities, 2008. ESoP - European Statement of Principles on human-machine interface. Official J. Eur. Union.

Green, P. et al., 1995. Preliminary human factors design guidelines for driver information systems. Nasa, (December). Available at: <http://scholar.google.com/scholar?hl = en&btnG = Search&q = intitle:Preliminary + Human + Factors + Design + Guidelines + for + Driver + Information + Systems#0>.

Horrey, W.J., Wickens, C.D., 2006. Examining the impact of cell phone conversations on driving using meta-analytic techniques. Human Factors: The Journal of the Human Factors and Ergonomics Society 48 (1), 196−205. Available at: http://hfs.sagepub.com/cgi/doi/10.1518/001872006776412135.

Japan Automobile Manufacturers, A., 2004. Guideline for In-vehicle Display Systems — Version 3.0 1 (2), 1−15.

Kantowitz, B.H., Simsek, O., 2001. Secondary-task measures of driver workload. In: Hancock, P.A., Desmond, P.A. (Eds.), Stress, Workload, and Fatigue. Lawrence Erlbaum Associates, Inc, pp. 395−407.

Parasuraman, R., Hancock, P.A., 2001. Adaptive control of mental workload. In: Kantowitz, B.H. (Ed.), Stress, Workload, and Fatigue. Lawrence Erlbaum Associates, Inc, pp. 305−320.

SAE J2395, 2002. ITS in-vehicle message priority. The Engineering Society For Advancing Mobility Land Sea Air and Space, SAE INTERNATIONAL.

Salvendy, G., 1997. Handbook of human factors and ergonomics. John Wiley.

Schindhelm, R. et al., 2004. Report on the review of available guidelines and standards (AIDE deliverable 4.3. 1).

Sohn, H., et al., 2008. A dynamic programming algorithm for scheduling in-vehicle messages. IEEE Trans. Intell. Transp. Syst. 9 (2), 226−234.

Waard, D.De, 2002. Mental workload. In: Fuller, R., Santos, J.A. (Eds.), Human Factors for Highway Engineers. Pergamon Press, pp. 161−175.

Vervey, W.B., 2000. On-line driver workload estimation. Effects of road situation and age on secondary task measures. Egonomics 43 (2), 187−209.

Zhang, A., Nwagboso, C., 2001. Dynamic message prioritisation for ITS using fuzzy neural network technique. SAE Technical Paper., 2001-1−0.

CHAPTER 10

Simulation Tools

Contents

Intelligent Vehicles
DOI: http://dx.doi.org/10.1016/B978-0-12-812800-8.00010-2

SUBCHAPTER 10.1

Driving Simulators

Alfonso Brazález[1], Luis Matey[1], Borja Núñez[1] and Ana Paúl[2]
[1]Ceit-IK4, San Sebastián, Spain
[2]CTAG - Centro Tecnológico de Automoción de Galicia, Porriño, Spain

10.1.1 INTRODUCTION

The continuous improvement of hardware performance is a well-known fact that is allowing the development of more complex driving simulators. The immersion in the simulation scene is increased by high fidelity feedback to the driver. A high quality visual system in terms of, high image refresh rate, realistic environment, and good 3D modeling increases the immersion in the simulation session. The behavior given by the mathematical models affects the scope of the simulation and the fidelity of the simulation itself.

In this framework, the application of new methods and the need to provide more realism have generated new requirements for simulator performances. The immersive character in the simulators is obtained by the stimulation of the sensorial organs of the driver, so sensations experienced by the human being are most similar to those that the simulator user feels driving the actual vehicle. Various senses are stimulated in the simulator: the visual sense, the hearing sense, the tactile sense and the vestibular system. The motion system usually consists of a six degrees of freedom (dof) motion platform that reproduces the sensations of linear accelerations and angular velocities that the driver feels in the actual vehicle (Liu, 1983). There are simulators with more complex motion systems, even without motion systems, or with simpler motion platforms of three degrees of freedom.

Some simulators provide haptic feedback to establish the interactions between the driver and some controls of the vehicle in driving simulators. Steering wheel torque feedback, configurable joysticks, and the actuators in the accelerator, clutch, and brake pedals enhance the degree of realism in the simulation and they allow the driver to feel realistic forces and torques on his arms and legs.

Through increasing the immersion in the simulation, driving simulators will get a bigger transfer rate of results from the virtual environment to real life in several fields. Automotive simulators are an important research tool in design, development, and validation stages. Naturalistic driver studies can achieve certain objectives, but experiments are a more appropriate approach to test hypotheses (Gelau et al., 2004). Some information such as subjective, physiological, or other performance data are not collected in naturalistic studies. Although the real road and FOTs (Field Operational Tests) give more realistic surroundings, simulators are necessary in order to reduce experimental costs and risks. On real roads, experiments suffer extra noise (Carsten and Brookhuis, 2005), while in a simulator with a fully controlled scene it is possible to produce the exact desired situation. Thus, getting a realistic environment is the key requirement of automotive simulators.

Apart from the investigation field, the main application of automotive simulators is training purposes. They can be used to instruct both novice and experienced drivers. According to the GADGET project, driving training is divided into four levels: maneuvers, traffic situations, context goals, and skills (Peräaho et al., 2003). In practice, however, it was concluded that most simulators only covered the first two levels, whereas the other two levels typically are omitted from the driver training because of the limitation of simulators (Lang et al., 2007). Increasing the configurability of the HMI enables new possibilities in training simulators especially in these two highest levels.

Focusing on the field of research, automotive simulators provide a wide range of possibilities in different disciplines (Slob, 2008). On the one hand, it is possible to analyze and investigate human factors (driver behavior and HMI); on the other hand, it is possible to use them to design and validate environmental issues, like tunnels, positioning road signs, or road planning. Finally, simulators are useful tools to develop, validate, and evaluate technical and technological innovations.

Simulators present two main benefits on R&D performance (Thomke, 1998). First, their use reduces costs and time, facilitating design iterations; second, they help to achieve a more effective learning in the R&D process, because they can increase the depth and the quality of the experimental analysis. From the point of view of experiments, a simulator adds other advantages: it is possible to reproduce hazard situations without any risk for the driver and a simulator enables to keep all the vehicle and environment parameters under control.

In the automotive area, the approach to technical innovations is very wide; it covers different tasks, such as the integration of new subsystems and applications, communication related issues, or ADAS/IVIS development and testing. Simulator capabilities have a special importance in all the cases. The more configurable a simulator is, the faster and easier will be the creation of new scenarios and their implementation for experiments.

Alongside the technical success, new systems implementations must reflect the study of human factors to guarantee their safety and usability. In this field, there is no standard and general methodology to measure the validity of a certain simulator (Eskandarian et al., 2008); this concept is strongly linked to the task to be performed. While the motion platform and a high resolution of the visuals can help to enhance the realism (Kaptein et al., 1996), the main conclusion is that the validity must be evaluated depending on the driving task being studied.

Although a normalized classification does not exist yet, there are different perspectives to categorize the existing simulator types: depending on whether they have motion or not; determined by the type of vehicle (car, bus, or truck); depending on the visual type; or subject to the simulation scope.

10.1.2 ARCHITECTURE OF DRIVING SIMULATORS

There are many configurations and scopes of use of driving simulators, but usually the main subsystems are: visual system, simulator control station, mathematical models, vehicle cabin, audio system and the motion system if the simulator provides motion feedback. Moreover, when the simulator is coordinated, it can be used as a host system for real-time coordination of every subsystem (Fig. 10.1.1).

The Simulator Control Station is the simulator interface. From this station different exercises can be selected. Usually, it provides an edition mode for the creation of new exercises; based on the configurability

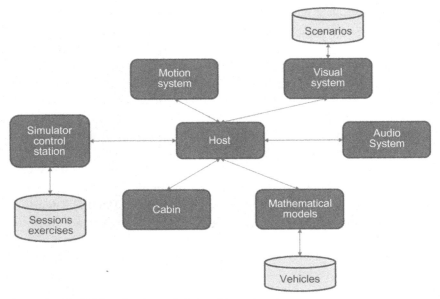

Figure 10.1.1 Driving simulator global architecture.

of the simulator modules, it can change vehicle parameters or weather conditions and set new incidences. In the Control Station, the simulated sessions can be stored for further evaluation and analysis.

The visual system provides visual feedback to the driver. Nowadays, it is possible to simulate different lighting conditions, weather conditions, and moving elements in the scene with a high degree of fidelity. From the hardware point of view, visualization systems are more efficient with a high resolution and visualization performance. It is common to have a 360 degrees of field of view. It is important to have a rear view though the rear mirrors and for front view, it is necessary for at least 180 degrees, covering the side views. The gaming sector has improved the development of dedicated hardware for real-time image rendering, with relatively low cost systems. Moreover, from the software point of view it is possible to program directly in the Graphic Processing Unit (GPU) of the image system. It allows for a better performance for image rendering in real time reducing the traffic data on the network.

Scene elements are critical for having a better feeling of immersion. The surrounding traffic provides additional elements for interaction. Many incidences can be simulated from other vehicles: a vehicle stopped in the shoulder, a vehicle at low speed, a vehicle passing through a red light, etc. A realistic microscopic traffic simulation increases the driving simulator capabilities. Also the pedestrian simulation leads to higher capabilities of the simulator, because it allows for the interaction with other elements of the scene. Various incidences related to pedestrians can be simulated, for example people crossing streets or passengers in a bus simulator.

The mathematical models are a key module for the performance of the simulator. It affects the simulation scope. Depending on the needs of the simulator, a high accuracy mathematical model could be required. Moreover, if the simulator should be used for the validation of new subsystems, where some hardware must be integrated as Hardware In the Loop (HIL), a well-defined interface should be provided as it was simulated. Complete vehicle models should reproduce the effect of the tires with a detailed mathematical model of the contact between road and tire, using different approaches of the Pacejka Magic Formula. For the vehicle dynamics, a multibody simulation must be performed, including the effect of the suspension with its geometry, parameters, and nonlinearities. The complete powertrain should be modeled; engine and transmission elements need to be simulated and any other system, such as e.g., ASR. Additional subsystems such as ABS or ESP can be simulated for a high fidelity simulator.

Not all driving simulators are equipped with a motion system. This tries to provide some motion feedback to the driver. The strategy of this system is to translate vehicle linear accelerations and angular velocities into driver moving sensations. Typically washout filter algorithms inherited from flight simulators were used, but there are some limitations due to large longitudinal and lateral accelerations in the case of road vehicles. It is necessary to develop appropriated algorithms in order to ensure the driver feels realistic motion feedback (Ares et al., 2001), and some additional degrees of freedom have been added to typical 6 dof motion platforms. Many driving simulators mount the 6 dof motion platform over a large excursion translation system in lateral and longitudinal. Additionally the motion platform can be rotated along his the vertical axis, reaching a 9 dof motion system.

It is important to have a real cabin, or at least real controls, for a better immersion of the driver in the simulation scene. This could avoid the initial gap of the driver for trusting in the simulator. A period of adaptation for the driver with several driving sessions is always needed, prior to obtaining any valid results of the simulation. A real vehicle, with a complete vehicle control set reduces the training sessions required for the adaptation. Once the driver feels he or she is actually driving, the results of the simulator can be transferred to real situations. In the cabin real vehicle communications can be installed for the integration of different subsystems. Usually CAN networks must be used for the connection of different ECUs or the integration of the instrument cluster or any other interface.

The audio system is not a minor feedback for the driver. If this system is switched off, the situation is like a deaf person is driving. Many inputs are received from hearing the engine, and sometimes the perception of gear changing is performed from the engine sound. Additionally, some malfunctions are identified from the specific sound that it generates.

10.1.3 APPLICATIONS

The use of driving simulators has evolved from the first applications related to driver training or analysis of the effect of different substances such as alcohol or drugs, to more complex research studies focused on human factors and driver behavior during primary and secondary driving tasks. Nowadays, driving simulators are used not only in research studies, but also in the different stages of the design, development, and validation of in-vehicle systems, as well as in the design of infrastructure elements (Paul et al., 2009).

Regarding human factors, the studies in the driving simulators have allowed examination of the human—machine interface (HMI), i.e., the communication between the user and the vehicle or its technologies covering the following aspects:

- Workload analysis.
- Usability.
- Distraction due to secondary tasks.
- Reaction times (e.g., to avoid a collision when offering a warning).
- The effects of driver information on driving performance.
- The location of HMI elements on the dashboard.
- Selection of communication channel (acoustic, visual, haptic).

The evaluation of In-Vehicle Systems (IVIS) and Advanced Driver Assistance Systems (ADAS) have been broadly applied in several research and development projects, showing that driving simulator studies are, together with tests on test tracks and tests in real life, valid tools (Engen et al., 2009). In fact, dynamic driving simulators play an important role in the initial phases of development, since driver behavior and driving performance (lateral control, longitudinal control, interaction with other vehicles, etc.) provide early valuable results in a virtual environment.

Several European Projects have introduced experimentation with driving simulators to analyze new in-vehicle functionalities. For example, in the AIDE project, funded by the Sixth Framework Programme, the effect of the combination of warnings coming from different ADAS functions were tested (Paul et al., 2008). Thus, four ADAS were selected for this study, Frontal Collision Warning (FCW), Lane Departure Warning (LDW), Curve Speed Warning (CSW), and Blind Spot Detection (BSD), in order to analyze user reaction and possible conflicts when simultaneous ADAS warnings were presented (Fig. 10.1.2).

Figure 10.1.2 Example of warnings provided by different ADAS functionalities.

In this study the warning conditions were manipulated to create different levels of theoretical mental workload. Two strategies were considered (independent variable) in a conflict situation: simultaneous activation of warning signals for the driver and prioritization of warnings. Such critical situations can only be reproduced in high performance driving simulators, with the required levels of safety and repeatability.

Another example of the application of driving simulators to the design of new ADAS is the INTERACTIVE Project (2013), launched under the Seventh Framework Programme, where information, warning, and intervention strategies were developed, according to aspects such as: layer of driving task, level of assistance and automation, situation awareness, mental workload, sequence of interaction, etc. (Fig. 10.1.3).

Figure 10.1.3 Selection of ADAS validated in driving simulator (INTERACTIVE Project).

Apart from the evaluation of the HMI and warning strategies conducted in driving simulators, the vehicle HIL simulations have proven an added value in several phases of the development process of ADAS, such as sensor verification, rapid control prototyping, model validation, functional level validation, fine-tuning of control algorithms, production sign-off tests, and preparation of test drives (Gietelink et al., 2006). This is possible thanks to the representative environment obtained, where test scenarios can be varied very easily in accurate and reproducible conditions.

Moreover, in the development of ADAS, different driving simulation platforms may be used with the combination of HIL and driver-in-the-loop (DIL), in order to create special testing scenarios that include high speed traffic flow, low-frictional load, etc., with high controllability and repeatability (Jianqiang et al., 2010). This allows to speed up the development process and to reduce the associated development costs.

Finally, HIL simulation can be used to support the development, testing, and verification of many functions and algorithms related to autonomous driving, by extending conventional HIL simulation to vehicle interaction with other vehicles in traffic and with a simulated surrounding environment sensed by simulated sensors. (Deng et al., 2008).

In this sense, there is no doubt that driving simulation is an essential and powerful tool in the design, development, and validation of current and future in-vehicle technologies, especially in the field of connected and automated road transport, where new challenges regarding user behavior, vehicle operation, complex scenarios, and innovative interaction capabilities are rapidly arising.

REFERENCES

Ares, J., Brazalez, A., Busturia, J.M. 2001. Tuning and validation of the motion platform washout filter parameters for a driving simulator. In: Driving Simulation Conference, pp. 295–304.

Carsten, O.M.J., Brookhuis, K., 2005. Issues arising from the HASTE experiments. Transp. Res. Part F Traffic Psychol. Behav. 8 (2), 191–196.

Deng, W., Lee, Y.H., Zhao, A., 2008. Hardware-in-the-loop Simulation for Autonomous Driving. IEEE 2008.

Engen, T., Lervåg, L.-E., Moen, T., 2009. Evaluation of IVIS/ADAS using driving simulators comparing performance measures in different environments.

Eskandarian, A., Delaigue, P., Sayed, R., Mortazavi, A., 2008. Development and verification of a truck driving simulator for driver drowsiness studies. CISR, The George Washington University.

Gelau, C., Schindhelm, R., Bengler, K., Engelsberg, A., Portouli, V., Pagle, K., et al., 2004. AIDE Deliverable 4.3. 2: Recommendations for HMI Guidelines and Standards. Technical Report, AIDE, Adaptive Integrated Driver-vehicle InterfacE.

Gietelink, O., Ploeg, J., De Schutter, B., Verhaegen, M., 2006. Development of advanced driver assistance systems with vehicle hardware-in-the-loop simulations. Veh. Syst. Dyn. 44 (7), 569–590, July 2006.

INTERACTIVE Project, 2013. Available at <http://www.interactive-ip.eu/project.html>.

Jianqiang, W., Shengbo, L., Xiaoyu, H., Keqiang, L., 2010. Driving simulation platform applied to develop driving assistance systems. IET Intell. Transp. Syst. 4 (2), 121–127.

Kaptein, N., Theeuwes, J., Van Der Horst, R., 1996. Driving simulator validity: Some considerations. Transp. Res. Rec. J. Transp. Res. Board 1550, 30–36.

Lang, B., Parkes, A.M., Cotter, S., Robbins, R., Diels, C., HIT, M.P., et al., 2007. Benchmarking and classification of CBT tools for driver training. TRAIN-ALL.

Liu, Z.Q., 1983. A Study of Washout Filters for a Simulator Motion Base. University of Toronto.

Paul, A. Sanz, J.M., Gago, C., García, E., Díez, J.L., Blanco, R., et al., 2009. Risk analysis of road tunnel using an advanced driving simulator to assess the influence of structural parameters in tunnel safety. DSC Europe 2009, Montecarlo.

Paul, A., Baquero, R., Díez, J.L., Blanco, R., 2008. Analysis of integrated warning strategies for ADAS systems through high performance driving simulator. DSC Europe 2008, Monaco.

Peräaho, M., Keskinen, E., Hatakka, M., 2003. Driver competence in a hierarchical perspective; implications for driver education. Report to Swedish Road Administration.

Slob, J.J., 2008. State-of-the-Art driving simulators, a literature survey. DCT Report, 107.

Thomke, S.H., 1998. Simulation, learning and R&D performance: evidence from automotive development. Res. Policy 27 (1), 55–74.

SUBCHAPTER 10.2

Traffic Simulation

Javier J. Sánchez-Medina[1], Rafael Arnay[2], Antonio Artuñedo[3], Sergio Campos-Cordobés[4] and Jorge Villagra[3]
[1]Universidad Las Palmas de Gran Canaria, Las Palmas de Gran Canaria, Spain
[2]Universidad de La Laguna, Santa Cruz de Tenerife, Spain
[3]CSIC, Madrid, Spain
[4]TECNALIA, Bizkaia, Spain

10.2.1 WHAT IS TRAFFIC SIMULATION AND WHY IS IT NEEDED

Traffic simulation has been a very active field in Intelligent Transportation Systems since the 1980s. But what are we talking about when we say traffic simulation? It has two parts, traffic and simulation.

Traffic, in this context is a collective word referring to a collection of transportation means operating simultaneously in a defined bounded geographical area. That may be a city, a metropolis, a number of interconnected cities, a region, etc. In that geographical circunscription we have a number of possible mobility options. In the context of this book, we are primarily talking about terrestrial transportation means.

Also, each application may restrict the range of transportation means or modes. For example it may be considered composed of only cars, or of all kinds of private vehicles, even man-powered like bicycles, or it can also be expanded to pedestrians, cablecars, etc. The multimodality level of a particular traffic simulation heavily depends on the application.

Obviously, the more modes that are included in a particular traffic simulation, the more complex it may be in terms both of mathematical model definition and algorithmic implementation.

The other part of it is the word simulation. A simulation is the recreation of a real-world phenomenon through the instantiation of a previously defined model of it. In other words, we create a mathematical model abstracting a natural phenomenon, then we carefully initialize all the variables involved, and we "run it" with the purpose of getting a reasonably accurate idea of evolution, generally across time, of that phenomenon, paying attention to some aspect of it, like maybe occupation, density, traffic volume, average speed, or many others.

Each simulation tries to answer the question "What would happen if" in the real world. That is very important and useful for many reasons. For example, thanks to the power of modern computers, it allows the evaluation of a number of alternatives and their outputs, before risking the implementation of any of them in the real world. Also, it may permit not just numerous setups, but also extreme case ones. For example, a researcher may simulate an emergency response in case of a dramatic event, that hopefully will never happen in real world.

Having said all of that, before simulations, we need models and that is the hardest part of it. There has to be a previous model definition, including all the necessary elements and concepts. That model may have probably passed through some validation process where it needs to be evidenciated that it is accurate enough for its planned application.

That famous George Box quote says: "All models are wrong, but some are useful." A model is a formal representation of a natural phenomenon. In principle, every phenomenon will always be too complex to be completely accurately represented. In other words, as soon as we create a representation, we are stepping down from absolute accuracy. Furthermore, we do not want absolute accuracy. There will always be a balance between accuracy and performance. Sometimes a model needs to be primarily fast, because it will be used in real-time applications. Some other times it may be very slow, even meant to be run only once, as much as it is very accurate.

All of that can be quickly extended to traffic. Any vehicular traffic situation is such a complex and random process, especially because of the human intervention in most of them. We definitely need to equip ourselves with models to describe and analyze that process and also to make future forecasts.

If we look at how traffic modeling and simulation has evolved through history, in the very first stages it was about creating mathematical models, generally coming from physics and fluids dynamics, but after a few years, especially after the exponential growth of computing power, more ambitious and detailed simulation paradigms were explored.

That physics-based approach is the so-called macroscopic traffic modeling. Traffic was understood as a continuum. That kind of model was quite efficient for several tasks and applications, but very soon a wildly different approach was presented, where vehicles were modeled individually with more or less the same level of detail, and more importantly, the interaction between them was also modeled. That discrete way of modeling traffic was called microscopic traffic modeling.

Even from the 1960s the very first traffic simulators where designed following one of the two philosophical approaches to traffic modeling. At some point, some hybrid models where developed incorporating the strengths and leaving out the weaknesses of both of them. That was when Mesoscopic models were defined and applied.

Back in 1956, one can find in Wilkinson (1956) one of the first publications that could be considered to be on traffic simulation modeling, where trunk traffic was modeled as a random Gaussian process, mainly described by mean and variance, assuming some level of noise.

We have to wait until the 1960s to see the first works on computer-aided simulations of traffic, although these were still rudimentary. For example, we have Stark (1962), where a custom made very simplistic vehicle simulation of nine blocks was published. It is interesting to note that at that time the simulation visualization was done by taking pictures of an oscilloscope screen every quarter of a second.

Another interesting work is Shumate and Dirksen (1965) where a programming language is presented (SIMCAR), basically for designing highways and simulating vehicles and even different driving styles, again it is rudimentary and with very little scalability.

Within microscopic traffic simulation we have the so-called car-following models. In May and Harmut (1967) some car-following models are evaluated and compared. Car following models could be referred to as the ancestors of modern microsimulation, where the front to bumper distance is modeled, showing quite realistic effects like stop and go waves.

Later, in the 1970s, we got the first simulation frameworks like FREFLO (Payne, 1979; Mikhalkin, 1972) where freeway traffic is modeled mathematically in terms of variables like density, space-mean-speed, and flow rate. Another example of macroscopic simulation was SATURN-a (Hall and Willumsen, 1980), which has a relative that is still around.

Also in the 1970s, the first microscopic simulation frameworks such as TEXAS (Rioux and Lee, 1977) showed up. Traffic simulation was prevented from simulated bit zones because of the cost of the computing power they had. They are very slow in comparison to microscopic simulators.

It was in the 1980s and especially in the 1990s when there was a real revolution regarding traffic simulators, mainly because of the exponential reduction of computing hardware costs, making powerful computers more and more affordable. Also parallel computing advances and IBM's PC standardization helped a lot to a make computing more and more available for increasingly more complex traffic simulation frameworks.

Talking about the present, there are a few live challenges for traffic simulation these days. The first one is multimodality. Every traffic simulation framework needs to support multimodality, which is incorporating a wide palette of transportation means models, including pedestrians, bicycles, electric vehicles, and more. That is very important because mobility is gradually tending to become more and more diverse regarding transportation means, at least these days when some paradigm shift revolutions are in place, like transportation electrification or driverless automation. Depending on the application, every different transportation means may need to be modeled. An electric car has a quite different dynamic behavior than a gas powered one.

Another important challenge is about real-time managing of traffic. A usual demand from traffic managers is real-time monitoring of the current state of traffic. To do so, we need simulators to be computationally efficient and scalable and a part of the scientific community is devoted to that goal.

Optimization is a very important topic. With traffic being so complex, it can be hardly managed with analytical tools. Instead, it usually needs nondeterministic optimization techniques, i.e., Genetic Algorithms (Sanchez-Medina, 2008; Sanchez-Medina, 2010).

Finally, it is worthy mentioning two of the most exciting things that have happened regarding traffic simulation and modeling in recent years. The first one is the influence Open Data is having on the development of this area. In the last years more and more local governments in the World have decided to put their data in open access. This is extremely interesting, because it allows researchers all around the globe to access and use even real-time traffic data feeds for their research, scaling the number of publications and discoveries.

The second very exciting event is the open simulation shift that has been led by SUMO (Krajzewicz, 2002). In Section 10.2.4 we will talk longer about it, but it is worth advancing that this is truly game-changing in a so far quite closed traffic simulation framework business. The SUMO developers' community is a marvelous example of what are the benefits of free software: constant updates, collaboratively developed plugins, and open developers and practitioners fora.

10.2.2 CLASSIC TRAFFIC SIMULATION PARADIGMS

10.2.2.1 Macroscopic Simulation

Macroscopic simulation is the branch of traffic simulation that relies on the so-called macroscopic models. The "Macro" part in it tries to express that this kind of model views traffic from a distance, considering it as a continuum or fluid. Therefore, the objective of this kind of simulation is the spatiotemporal representation of mainly three real variables: volume $q(x,t)$, speed $u(x,t)$, and density $k(x,t)$. Volume is regarding the number of vehicles passing through a specific point in space. Speed has to do with the space traversed by a particular vehicle in a fixed period of time. Finally, density has to do with the number of vehicles occupying a fixed area (a lane, a multiple lane street, etc.). Formally speaking, the basic formula in macroscopic modeling announced by Gerlough and Matthew (1975) is the conservation or continuity equation:

$$\frac{\partial q}{\partial x} + \frac{\partial k}{\partial t} = 0 \qquad (10.2.1)$$

This kind of formulation is inherited from hydrodynamics. It basically means that, if there are no inputs or outputs, the number of vehicles must remain the same across the pipe (highway, street, etc.).

That very simple equation is considering an equilibrium situation without effects like congestion formation, stop and go waves, etc.

Therefore, it was modified by Payne (1979) in the early 1970s, resulting in the following:

$$\frac{\partial k}{\partial t} + u\frac{\partial q}{\partial x} = \frac{1}{T}[u_e(k) - u] - \frac{\nu}{k}\frac{\partial k}{\partial x} \qquad (10.2.2)$$

In a summarized fashion, u represents the speed–density relationship that can be as announced by May and Harmut (1967) as follows:

$$u = u_t\left[1 - \left(\frac{k}{k_{jam}}\right)^{\alpha}\right]^{\beta} \qquad (10.2.3)$$

In Eq. (10.2.2) it is intended to reflex two very important elements on the nonequilibrium traffic flow effect, namely acceleration and inertia. The right side of Eq. (10.2.2) has two parts. The first one reflects the action of the driver adjusting speed aiming at the equilibrium speed. In that part, T is the so-called relaxation time, and ν means the anticipation parameter.

The second half of the right-hand side of Eq. (10.2.2) reflects how drivers' reactions affect downstream traffic conditions.

According to Barceló (2010), Payne's model seems to present accuracy problems particularly in dense traffic in on-ramps or lane drops.

Microscopic models and microsimulators were the first in the field and consequently, some of the more venerable traffic simulation frameworks are based on macroscopic models, at least in origin. For example, FREFLO (Payne, 1979; Mikhalkin, 1972) and SATURN-a (Hall and Willumsen, 1980) were some of the first microsimulators back in the 1970s, mainly for highway traffic simulation. SATURN is still around these days.

Some other macroscopic traffic simulation frameworks are TRANSYT-7F (Wallace, 1998) or METANET (Spiliopoulou, 2015).

Nowadays, macroscopic simulation is itself is becoming less and less common for a simple reason: microscopic simulation is more accurate and even when in general it is much heavier in terms of computing power, it is also true that computing power is becoming more and more available at reasonable costs.

However, for some applications, or in combination with other microscopic models (mesoscopic simulation), they are still in use and one can find relevant and interesting literature on them. For example, a new generation of macroscopic simulators are based on a new paradigm. They are the gas-kinetic (GKT) traffic flow models, for example, the one used by Delis (2015) for modeling traffic flow with adaptive cruise control.

10.2.2.2 Microscopic Simulation

Microscopic simulation models (multiagents) simulate the movement of individual vehicles based on vehicle traceability and on theories of change of lane. Typically, vehicles enter the transport network using a distribution probabilistic of arrivals (a stochastic process) and are followed during their passage through the network in small time intervals (e.g., one second or a fraction of a second). After entering each vehicle, each is normally assigned a destination, a type of vehicle, and a type of driver. These models are effective in evaluating traffic congestion, complex geometric configurations, or the impact of transport improvements that are beyond the limitations of other types of tools. However, these models have a high cost in time, money, and can be difficult to calibrate.

The definition and implementation of such simulators, involves knowledge of different scientific and engineering fields:

- *Microscopic Modeling of Traffic itself.* This approach constitutes the basis for traffic flow theory (e.g., Herman and Potts, 1900). As we will see later, there exist many advanced available software, both open and commercial, that are able to manage accurate and fast simulations of large geographical areas.

- *Computational Physics.* Experience of the adoption of simple and very fast models of physical processes, with lower simulation computing requirements. While physics models manage particles, here we manage people with a similar order of elements (e.g., regional and municipality microscopic simulation). To establish a compromise between model interaction detail, simulation/interaction speed, and computational requirements is mandatory.

- *Microscopic Behavioral Modeling of Demand/Agent-Based Modeling.* We can find as many definitions of "agent" as researchers, some of them are "a discrete entity with its own goals and behaviors, with capability to adapt and modify its behavior" (Macal and North, 2005) or "anything that can be viewed as perceiving its environment through sensors and acting upon that environment through actuators" (Russell and Norvig 2002). As we will discuss later, these models combine elements from game theory, complex systems, emergence, computational sociology, multi-agent systems, and evolutionary programming. Agent-based modeling uses simple rules that can result in different sorts of complex behavior. The key point is the autonomously, emergence, and complexity. Samples of these models are cellular automata models (Nagel and Schreckenberg, 1992) or the gravity model (Wilson, 1971).

* *Complex Adaptive Systems/Coevolutionary Algorithms.* Traveling and moving by means of any transport mode, from public transport, car, or walking, involves fundamentally game-theoretic reasoning: individual decisions are evaluated in specific interaction scenarios (e.g., congestion, innovative shared transport, activity grouping) resulting from decision made by the collective rather than in isolation. Different gaming strategies as Nash-equilibrium approaches, where individual decision maximizes the gain of the other, and other such as dominant strategies or mixed strategies have been deployed in transport analysis have been developed in transport assignment from the mid of the last century. As we mentioned before, metaheuristics methods combined with equilibrium logic implements schedule coevolutionary search schemes.

Some of the most well-known microsimulators are HUTSIM (Kosonen, 1999, 1996), VISSIM (Park et al., 2003), CORSIM (Owen, 2000), SESIM (Flood, 2008), AIMSUN (AIMSUN, 2017), Transims (Rilett and Kyu-Ok, 2001), and Cube Dynasim (Citilabs, 2017). The MATSIM (Balmer, 2009) are developed under the multiagent paradigm, which is of great relevance if the emergence of traffic behavior under certain traffic demands is the subject of research. SUMO (Krajzewicz, 2002, 2006) is an open source traffic simulator which is based on the Gipps-model extension (Krauß, 1998) and, more recently, also on the IDM model.

10.2.2.3 Mesoscopic Simulation

These models combine the properties of macroscopic and microscopic models. As in microscopic models, the traffic unit is the individual vehicle. Its movement, however, follows the simplification performed by the macroscopic models, and is determined by the average speed of the route. Dynamic speeds or volume ratios are not considered. Therefore, mesoscopic models provide less fidelity than microscopic models, but are superior to typical analysis techniques.

Some of the relevant meso-simulators are DYNAMIT-P-X (Ben-Akiva, 2002), DYNASMART-P-X (Mahmassani and Jayakrishnan, 1991), and MesoTS (Meng, 2012). In any case, most of the previously mentioned microscopic simulators provide hybrid (micro—meso) integrated solutions.

10.2.3 SOME (TRADITIONAL) SIMULATION FRAMEWORKS

10.2.3.1 CORSIM

Corridor Simulator (CORSIM) (Halati, 1997) is a microscopic simulation framework developed by the Federal Highway Administration (FHWA) in the United States. Therefore, it is a sort of standard simulation framework for many research groups, especially in that country when they have to deal with the administration. Historically CORSIM is the evolution of two older models: FRESIM (FREeway SIMulation) and NETSIM (NETwork SIMulation). FRESIM (Halati, 1990) is a microscopic simulator for highways traffic. NETSIM (Rathi and Santiago, 1990) does the same but for urban traffic.

FRESIM's antecesor is INtegrated TRAffic Simulation (INTRAS) (Wicks and Andrews, 1980), a microscopic traffic simulator from the early 1980s.

They main characteristic of CORSIM is that it is fundamentally based on Car-Following models, also known as time-continuous models. Car-Following models are defined through differential equations to describe the position and speed of every vehicle. The aim of this kind of model is to approximate the bump to bump distance (s_α) between two consecutive cars like in Eq. 10.2.4:

$$s_\alpha = x_{\alpha-1} - x_\alpha - l_{\text{alpha}-1} \qquad (10.2.4)$$

$x_{\alpha-1}$ is the position of the vehicle in front (leader), x_α the position of the current vehicle, and $l_{\text{alpha}-1}$ the length of the leader vehicle.

CORSIM comes along with Traffic Software Integrated System (TSIS), a MS Windows–based application that takes care of the visualization layer of CORSIM. That is a very important addition because as both a scientist and a manager, it is extremely useful to get a visual of what is happening in a simulation to truly understand the possible effects of every setup on real-world traffic. TSIS includes a rich set of useful tools from the traffic network design and simulation to its analysis.

There are hundreds of works based on CORSIM in literature for both freeway and urban traffic. A key element with CORSIM is its calibration. The CORSIM framework requires the calibration of a big number of variables, before its exploitation at a particular case. There are parameters regarding the drivers, the vehicles, the roads, etc. Therefore, there are literarily hundreds of works on CORSIM calibration methods, using Artificial Intelligence techniques like in Cobos

(2016), Statistical Techniques (Paz, 2015), Bayesian methods (Bayarri, 2004), and more.

Currently, CORSIM is being maintained by the University of Florida's McTrans Center. Here are some of the listed capabilities of CORSIM in McTrans website:

- Public presentation and demonstration.
- Freeway and surface street interchanges.
- Signal timing and signal coordination.
- Diverging diamond interchanges (DDI).
- Land use traffic impact studies and access management studies.
- Emergency vehicles and signal preemption.
- Freeway weaving sections, lane adds, and lane drops.
- Bus stations, bus routes, carpools, and taxis.
- Ramp metering and High Occupancy Vehicles (HOV) lanes.
- High occupancy toll (HOT) lanes.
- Unsignalized intersections and signal warrants.
- Two-lane highways with passing and no-passing zones.
- Incident detection and management.
- Queuing studies involving turn pockets and queue blockage.
- Toll plazas and truck weigh stations.
- Origin–destination traffic flow patterns.
- Traffic assignment for surface streets.
- Statistical output postprocessing.
- Adaptive cruise control.
- Importing and exporting to TRANSYT-7FTM (Wallace, 1998) (TRAffic Network StudY Tool, version 7 F), the descendant of TRANSYT, developed by the Transport Research Laboratory in the U.K. in 1969.

10.2.3.2 MATSIM

The conventional and widely extended trip-based model of travel demand forecasting has been the reference model in urban mobility planning for the last decades. Nevertheless, this model was conceived for evaluating the impact of infrastructure investment options at the strategic planning stage. In fact, this model is not able to deal with real day-to-day issues such as time-dependent and spatial neighbourhood effects or collective decisions.

On the other hand, from a social and behavioral explanation of mobility, we have the following principles:

- The travel demand is conducted by the specific needs and wish of the individual.
- Social relationships influence on displacements and mobility habits and patterns.
- There exists some relevant constraints around travel: spatial, temporal, collective, facilities, and transportation accessibility barriers, among many others.
- While other models do not imply any kind of sequence or dependence on travel, it is needed to reflect the sequencing of activities in time and space.

The activity-based model (Rasouli, 2016; Castiglione, 2015) gives response to these questions and adds to the classic four stages model questions (mode, route, location, and timing), the following decisions:

- *Activity type choice*: Which activity should I do?
- *Activity chain choice*: In which order should I do my activities?
- *Activity starting time choice*: When should I start the activity?
- *Activity duration choice*: How long should I do the activity?
- *Group composition choice*: Who should I take along in the activity?

Summarizing, this approach or model is aimed at identifying and predicting for how long and with whom an activity is conducted, additionally to the classic parameters.

The "Multiagent transport simulation toolkit" (MATSim) simulation (Horni, 2016), is based on the agent concept and adopts the just mentioned activity-based approach. Each traveler, here the concept of driver is extended to any person traveling by any transport mode, of the target population is modeled as an individual agent able to take independent decisions.

The simulation consists of two sides mutually coupled:

- On the demand side, agents predefine a preliminary and independent plan that specifies its intentions during the time period under analysis. This plan is the output of an activity-based model that comprises route choice among other stages as seen, depending on expected network, public transport, or road, conditions.
- On the supply side, a mobility (including traffic flow) simulation or real operation takes place, executing all the plans of the predefined agents.

A learning mechanism for the agents is implemented by the iterative coupling of demand, defined as the agent generation and supply, obtained by traffic flow simulation. Basically, it takes the candidate agent's plan, evaluates their performance and adopts the best options (including metaheuristics methods, to evolve the set of solutions, avoiding local optimums).

The control flow or process of MATSim is composed by the following iterative activities:

- *Initial Demand*. MATSim requires a synthetic population of agents, each with individual transport-related attributes and daily activity plans, being representative of the population. These parameters are managed for each activity instance.
- *Simulation*. The traffic flow simulation runs the expected plans, emulating the interactions between agents and transport system according to its characteristics and constraints.
- *Scoring*. MATSim uses a simple utility-based approach to calculate a plan score, with positive values for time dedicated to perform "productive" activities and negative for traveling and delays on displacement activity locations.
- *Replanning*. Sequence of configurable algorithms that iterate on the population plans, Usually, adopting population-based heuristics, the algorithm considers sets of individuals (population) where each one represents a solution to the problem and evolves the set leading to improvements for average plan scores and travel times.

MATSim requires complete data containers or inputs to perform any transit simulation:

- A simulation configuration (e.g., parameters, modules to use for each step, iterations);
- A multimodal network and transit schedules (e.g., public transport agencies, lines, services, and vehicle/rolling stock characteristics and in the other side, the navigable road network);
- Time-dependent network attributes (manages parameters per segment, such as free speed, lanes, and capacity, that can change during the day, due to incidences or dynamic traffic adaptive solutions);
- Mobility plans (defines subpopulations, person attributes, their mobility plans, and transport demand of the analysis population);
- Some facilities that specify where the agents realizes the different activities; and
- Counts, taken from real operation that allows to compare and calibrate the simulation and specific scenarios.

MATSIM adopts a modular concept, in a broad sense of the word, referring to components at different levels, from functions, components or third party extension tools and frameworks. In any case, we can substitute a module by a specific functionality. Some relevant extensions currently available cover: freight management, car sharing, joint trips, parking, electric vehicles, pricing, emission calculation, travel time calculation, advanced analysis, multimodal transport, traffic signaling, among many other.

MATSim is written in the Java programming language and distributed under the GNU Public License (GPL), being available for download, use, and extension. Extensive documentation is accessible for developers, including specification of key-aspects of MATSim, configuration and underlying theory, guidelines, details for most of the extension packages, and data/samples for testing.

10.2.3.3 AIMSUM2

Urban congestion has a high impact on pollutant and energy consumption KPIs. Cities and their citizens have a fundamental role to play, since they concentrate the largest number of vehicles and the greatest problems of congestion, generating in these urban centers much of the total emissions of the planet. There is ample room for improvement, through a more intelligent use of means of transport and the integration of advanced technology as support for the improvement of mobility services.

By Active Transportation and Demand Management (ATDM) (US DoT, 2017), we integrated different strategies to provide solutions for congestion by combining public policy and private sector innovation to encourage people to change their transport habits, increase the share of sustainable mobility, prevent breakdown conditions, improve safety, and maximize transport efficiency and performance in general.

By definition ATDM implementation strategies fall under three major categories:

- *Active Demand Management (ADM)*, by using information and technology to dynamically manage demand, including redistributing travel, or reducing vehicle trips by influencing mode choice, adoption of more sustainable transport modes;
- *Active Traffic Management (ATM)*, that tries to dynamically manage recurrent and nonrecurrent congestion based on current and predicted traffic ; and finally

- *Active Parking Management (APM)*, parking facilities management to optimize performance and utilization of those facilities while influencing travel behavior.

All these strategies are based on estimations of traveler behavior; external factors, effects, and effectiveness of the actions themselves are subject to a high uncertainty.

AIMSUN (AIMSUN, 2017) is a widely used commercial transport modeling software, developed and marketed by Transport Simulation Systems (TSS). It integrates microscopic and mesoscopic components allowing dynamic simulations. AIMSUN provides the tools to carry out traffic assessment, in terms of environmental impact, capacity, or safety analysis, of some of the main actions to implement the previously mentioned categories:

- Feasibility studies for High Occupancy Vehicle (HOV) and High Occupancy Toll (HOT) lanes (ADM).
- Impact analysis of infrastructure design such as highway corridors (ADM) (Silva, 2015).
- Toll and road pricing (ADM).
- Evaluation of Variable Speed policies and other Intelligent Transportation Systems (ITS) (ADM).
- Bus Rapid Transit (BRT) schemes (ADM).
- Workzone management (ATM).
- Signal control plan optimization and adaptive control evaluation (ATM).
- Assessment and optimization of Transit Signal Priority (TSP) (ATM).
- Proactive Traffic Management, evaluating in real time the effect of decisions.

The simulator is highly configurable and extensible with new features and capabilities. By default, it can manage different traffic networks, demand modeling as flows at sections or O/D matrices, etc. but also can be extended by programming, enabling the modification of the behavioral models and the addition of new functionalities to the application. A detailed description of such parameters and extension capabilities can be found in the tool's manual (AIMSUN, 2014).

The commercial references of its application to urban and interurban transport planning are large. Specifically, for C-ITS deployment studies (Aramrattana and Maytheewat, 2016), the execution in combination with network simulators is mandatory (e.g., OMNeT++ simulator), and we can find an exhaustive list of such experiences in Segata (2014).

More recently, AIMSUN is being used to evaluate deployments of new mobility solutions, such as electromobility or autonomous driving vehicles. In the case of EVs, an extension of AIMSUN was implemented in the context of FP7 EMERALD project (Boero, 2017) to evaluate and optimize the impact of recharging infrastructure design in urban and interurban traffic management. This project supported the development of new transport models and algorithms (specifically oriented to FEVs), evaluation of intelligent transport systems and cooperative systems (V2I/I2V), and control plan optimization. Other additions were the third dimension in the maps and several types of behavior driven and consumption equations for each FEV type. In the context of autonomous vehicles, the project FLOURISH managed by the UK government have adopted AIMSUN to support the assessment of different scenarios from motorway to urban use; in this case the focus is on the user, their demands, expectations, and challenges for specific collectives, such as elderly people.

10.2.4 OPEN TRAFFIC SIMULATION: SUMO

"Simulation of Urban Mobility" (SUMO) (Krajzewicz, 2012) is a microscopic, multimodal, space–continuous, and time-discrete road traffic simulator. It is open source software licensed under the GNU GPL (General Public License) that is mainly developed by the German Aerospace Center (DLR). The development of this simulation tool started in 2000 having in mind portability and extensibility as main design criteria. Moreover, the need for handling large road networks required the taking into account of the execution speed and memory footprint as further guidelines.

The simulation scenario for SUMO has to be defined through a road network and a traffic demand. The road network can be defined either manually by generating XML files describing the network or by importing the network from other formats such as: OpenStreetMap (OSM), PTV VISUM and VISSIM, OpenDRIVE, MATsim, ArcView, etc. Besides that, SUMO is able to generate random networks under some rules (random-networks, spider-networks, and grid-networks).

The traffic demand can be defined in some different ways depending in the available input data: trip definitions, flow definitions, randomization, OD-matrices (VISUM/VISION/VISSIM formats), etc. Also for traffic demand XML files are used. SUMO supports different vehicle types such as motorcycles, trucks, buses, bicycles, or railways. Pedestrians are also supported. Moreover, some useful tools are provided for traffic demand modeling. For example, ActivityGen allows generating traffic demand from a description of the population data in the net through some parameters such as population's age brackets, school locations, bus lines, or work hours. Within the simulation, the vehicle movements are based on the longitudinal and lateral models. Both models can be chosen for each vehicle type among some that are already implemented in SUMO.

The outputs that SUMO can generate in each simulation include a number of different data: vehicle-based information over time (vehicle positions, pollutant emission values based on HBEFA database), lane/edge-based network performance information (vehicular noise emission based on HARMONOISE model), simulated detectors, and traffic lights information, among others. A useful graphical user interface (GUI) is also included in SUMO package. Besides making easier the basic use of the simulator, this GUI is very useful for monitoring the evolution of the simulation through a 2D representation as well as diverse aspects of the simulation (e.g., lane and vehicle coloring based on current occupancy or, $CO_2/CO/NOx/PMx/HC$ emissions, noise emission, average speed, etc.) at runtime (Fig. 10.2.1).

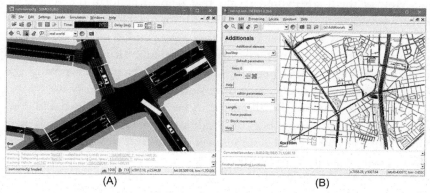

(A)   (B)

Figure 10.2.1 (A) SUMO-GUI; (B) NetEdit tool.

Due to its flexibility, SUMO is currently one of the most used simulation frameworks in the research, academic, and industry sectors. Among its applications, it is worth mentioning traffic forecasting, traffic management evaluation, route choice and re-routing evaluation, logistics, and traffic surveillance methods.

The possibility of adding extensions makes SUMO extend to new applications. One of the main extensions is "Traffic Control Interface" (TraCI). This interface enables the online interaction between SUMO and external applications. It allows to manipulate the behavior of simulated objects and to retrieve values at runtime. TraCI is available in different programming languages: C++, Java, Python, and Matlab. This interface is commonly used for providing the simulation platform with new functionalities such as external 3D visualization, traffic lights control simulation, or couplings with communication networks simulators (ns3, JiST/SWANS, or OMNeT++). For example, TraCI was extensively used in iTETRIS project (iTETRIS, 2017) whose goal is to couple SUMO with a communication network simulator using a middleware (iCS) for V2X applications. This work has been extended within the COLOMBO project (COLOMBO, 2017). In addition, SUMO has been used in further European projects: AMITRAN (AMITRAN, 2017), DRIVE C2X (DRIVE, 2017), among others; which shows the continuously growing use of this simulation framework for different mobility-related purposes.

Besides TraCI, SUMO package includes a great number of tools for different purposes: traffic assignment, dealing with real life induction loop data, traffic analysis, importing data, traffic light systems, trip generation, graphical evaluation of SUMO-outputs, working with sumo output files, etc. Most of them are Python scripts that help to perform different tasks. Some remarkable tools are: osmWebWizard (for quickly creating simulation scenario from a web browser by selecting a geographic region on a OSM map and specifying random traffic demand), sumolib (it is a set of modules for working with SUMO networks), tools for making easier parsing and visualizing simulation results, and conversion tools for data analysis.

The extensive use of this simulator makes it well maintained and constantly developing, fixing bugs and adding new features in each new update. In the last versions some new remarkable features have been added. Some examples are the availability of the NetEdit tool (since version 0.25.0) for graphical network creation and edition, and the inclusion of a mesoscopic model (available since version 0.26.0).

10.2.5 FUTURE TRENDS AND HOPES

The purpose of this section is not to divine the future development of traffic modeling and simulation, but to comment on some of the trends that seem to be present in the years to come regarding this topic.

First, we must say that simulation will likely be more and more online in the next decades. The ongoing revolution in the fields of Big Data and Data Stream Mining will possibility deeply affect traffic simulation, in particular when applied to the traffic managing, decision making, emergency managing, and advanced travelers information systems (ATIS).

The Internet of Things, sensor networks, and pervasive computing, including personal smart devices, are all expanding areas; they are likely to be useful sources of information on the current traffic situation for every transport network. That amazing torrent of information needs to be exploited. It is very urgent that data stream mining, which is the brand in Data Mining and Big Data thought to cope with live data, serves updated models and predictions on traffic states.

With the same purpose, simulation frameworks will need to be configured to feed from online information and to run in real time. Some simulation platforms, like SUMO, will need to improve its performance to guarantee the real-time restrictions, maybe through parallelization and/ or rewriting its libraries in faster programming languages (Romero-Santana, 2017).

Also, another hot topic where traffic simulation is playing a mayor role is on the Connected Vehicle move. It seems quite clear that driverless mobility is transiting from the autonomous car paradigm, where the intelligent vehicle is equipped with enough sensors and computation to accurately perceive its environment, calculate trajectories, predict surrounding vehicles' intentions, etc., to a connected car paradigm, where perception, 3D reconstructions, etc. can be shared with the infrastructure and all the other connected vehicles, exponentially extending the "safety bubble" of each one of them. In this new connected setup, at some point future forecasts will be required to predict traffic state, to dynamically propose alternative routes, to warn of possible future hazards, etc. Fast online traffic simulation will definitely play an important role there.

Multimodality is also a greatly challenging topic for traffic modeling and simulation. New kinds of vehicles, with new dynamic behavior, such as GPL, electric, or hybrid powered platforms, will need to be considered in order to accurately simulate their behavior. Also, driverless driving is challenging since it has been proven that driverless cars will behave quite differently (usually more conservatively) than human drivers (Kaber and Endsley, 2004). Finally, many cities in the world are working hard to foster greener transportation modes into the system, with the very interesting benefits for both traffic management and quality of life of citizens it may bring along (Rietveld, 2000).

One final piece to be incorporated in a near future in simulation frameworks are Unmanned Aerial Vehicles (UAV). There are already studies and proof of interest coming from both academia and industry on incorporating UAVs for surveillance, fast delivery, and other applications (Shim, 2005; Coifman, 2004).

REFERENCES

AIMSUN, 2014. Aimsun 8 Users' Manual, TSS-Transport Simulation Systems, 2014.
AIMSUN, 2017. AIMSUN — the Integrated Transport Modelling Software, <http://www.aimsun.com/site/>.
AMITRAN, 2017. AMITRAN Project. <http://www.amitran.eu>.
Aramrattana, M., 2016. Modelling and simulation for evaluation of cooperative intelligent transport system functions.
Balmer, M., Rieser, M., Meister, K., Charypar, D., Lefebvre, N., Nagel, K., 2009. MATSim-T: architecture and simulation times.
Barceló, J., 2010. Models, traffic models, simulation, and traffic simulation. Fundamentals of Traffic Simulation. Springer, New York, pp. 1—62.
Bayarri, M.J., James, O.B., Molina, German, Rouphail, Nagui M., Sacks, Jerome, 2004. Assessing uncertainties in traffic simulation: a key component in model calibration and validation. Transp. Res. Rec.32—40.
Ben-Akiva, M., Bierlaire, M., Koutsopoulos, H.N., Mishalani, R., 2002. Real time simulation of traffic demand-supply interactions within DynaMIT. Transportation and Network Analysis: Current Trends. Springer, pp. 19—36.
Boero, M., et al., EMERALD- Energy ManagEment and RechArging for Efficient eLectric Car Driving Project, vol. 2017. <http://www.fp7-emerald.eu/> .
Castiglione, J., Bradley, M., Gliebe, J., 2015. Activity-based travel demand models: a primer.
Citilabs, 2017. Software for the modeling of transportation systems, <http://www.citilabs.com/>.
Cobos, C., Erazo, C., Luna, J., et al., 2016. Multi-objective memetic algorithm based on NSGA-II and simulated annealing for calibrating CORSIM micro-simulation models of vehicular traffic flow.
Coifman, B., McCord, M., Mishalani, M., Redmill, K., 2004. Surface transportation surveillance from unmanned aerial vehicles. In: Paper Presented at Proceedings of the 83rd Annual Meeting of the Transportation Research Board.
COLOMBO, 2017. COLOMBO Project. <http://colombo-fp7.eu>.
Delis, A.I., Nikolos, I.K., Papageorgiou, M., 2015. Macroscopic traffic flow modeling with adaptive cruise control: development and numerical solution. Comput. Math. Appl. 70, 1921—1947.
DRIVE, 2017. DRIVE C2X Project. < http://www.drive-c2x.eu > .
Flood, L., 2008. SESIM: a Swedish micro-simulation model. Simulating an Ageing Population: A Microsimulation Approach Applied to Sweden. Emerald Group Publishing Limited, pp. 55—83. Chapter 3.
Gerlough, D., Huber, M., 1975. Transportation research board special report 165: traffic flow theory: a monograph. Transp. Res. Board.
Halati, A., Torres, J., Mikhalkin, B., 1990. Freeway simulation model enhancement and integration—FRESIM Technical Report. Federal Highway Administration, Report No.DTFH61-85-C-00094.
Halati, A., Lieu, H., Walker, S., 1997. CORSIM- Corridor traffic simulation model.

Hall, M., Willumsen, L.G., 1980. SATURN-a simulation-assignment model for the evaluation of traffic management schemes. Traffic Eng. Control21.

Herman, R., Potts, R.B., 1900. Single lane traffic theory and experiment.

Horni, Andreas, Nagel, Kai, Axhausen, Kay W., 2016. The multi-agent transport simulation MATSim. Ubiquity, London9.

iTETRIS, 2017. ITETRIS Platform. <http://www.ict-itetris.eu>.

Kaber, David B., Endsley, Mica R., 2004. The effects of level of automation and adaptive automation on human performance, situation awareness and workload in a dynamic control task. Theor. Issues Ergon. Sci. 5, 113−153.

Kosonen, I., 1999. HUTSIM-urban traffic simulation and control model: principles and applications, vol. 100.

Kosonen, I., 1996. HUTSIM: simulation tool for traffic signal control planning.

Krajzewicz, D., Bonert, M., Wagner, P., 2006. The open source traffic simulation package SUMO. RoboCup.

Krajzewicz, D., Hertkorn, G., Rössel, C., Wagner, P., 2002. SUMO (Simulation of Urban MObility)-an open-source traffic simulation. In: Paper presented at Proceedings of the 4th middle East Symposium on Simulation and Modelling (MESM20002).

Krajzewicz, D., Erdmann, J., Behrisch, M., Bieker, L., 2012. Recent development and applications of SUMO-simulation of urban mobility. Int. J. Adv. Syst. Measure.5.

Krauß, S., 1998. Microscopic modeling of traffic flow: investigation of collision free vehicle dynamics. D L R - Forschungsberichte.

Macal, C.M., North, M.J., 2005. Tutorial on agent-based modeling and simulation.

Mahmassani, H.S., Jayakrishnan, R., 1991. System performance and user response under real-time information in a congested traffic corridor. Transp. Res. Part A-Policy Pract. 25, 293−307.

May Jr, A.D., Harmut, E.M., 1967. Non-integer car-following models. Highway Res. Rec.

Meng, M., Shao, C., Zeng, J., Dong, C., 2012. A simulation-based dynamic traffic assignment model with combined modes. Promet - Traffic - Traffico 26, 65−73.

Mikhalkin, B., Payne, H.J., Isaksen, L., 1972. Estimation of speed from presence detectors.

Nagel, K., Schreckenberg, M., 1992. A cellular automaton model for freeway traffic. J. de Phys. I 2, 2221−2229.

Owen, L.E., Zhang, Y., Rao, L., McHale, G. 2000. Traffic flow simulation using CORSIM.

Park, B., Schneeberger, J.D., 2003. Microscopic simulation model calibration and validation: case study of vissim simulation model for a coordinated actuated signal system. Transp. Res. Rec.185−192.

Payne, H.J., 1979. FREFLO: a macroscopic simulation model of freeway traffic. Transp. Res. Rec.

Paz, A., Molano, V., Sanchez-Medina, J., 2015. Holistic calibration of microscopic traffic flow models: methodology and real world application studies.

Rasouli, S., 2016. Uncertainty in modeling activity-travel demand in complex urban systems. TRAIL Research School.

Rathi, A.K., Alberto, J.S., 1990. Urban network traffic simulations TRAF-NETSIM program. J. Transp. Eng. 116, 734−743.

Rietveld, P., 2000. Non-motorised modes in transport systems: a multimodal chain perspective for the Netherlands. Transp. Res. Part D Transp. Environ. 5, 31−36.

Rilett, L.R., Kim, Kyu-Ok, 2001. Comparison of TRANSIMS and CORSIM traffic signal simulation modules. Transp. Res. Rec.18−25.

Rioux, T.W., Lee, C.E., 1977. Microscopic traffic simulation package for isolated inter-sections. Transp. Res. Rec.

Romero-Santana, S., Sanchez-Medina, J.J., Alonso-Gonzalez, I., Sanchez-Rodriguez, D., 2017. SUMO performance comparative analysis: C Vs. Python. In: Paper Presented at In International Conference on Computer Aided Systems Theory, EUROCAST2017 (in press). Springer Berlin Heidelberg. Las Palmas de Gran Canaria. Spain.

Russell, S.J., Norvig, P., 2002. Artificial intelligence: a modern approach (International Edition).

Sanchez-Medina, J.J., Galan-Moreno, M.J., Rubio-Royo, E., 2008. Applying a traffic lights evolutionary optimization technique to a real case: "Las Ramblas" area in Santa Cruz De Tenerife. IEEE Trans. Evol. Comput. 12, 25−40.

Sanchez-Medina, J.J., Galan-Moreno, M.J., Rubio-Royo, E., 2010. Traffic signal optimi-zation in "La Almozara" district in saragossa under congestion conditions, using genetic algorithms, traffic microsimulation, and cluster computing. IEEE Trans. Intell. Transp. Syst. 11, 132−141.

Segata, M., Joerer, S., Bloessl, B., Sommer, C., Dressler, F., Cigno, R.L., 2014. Plexe: a platooning extension for veins. In: Paper Presented at Vehicular Networking Conference (VNC), 2014 IEEE.

Shim, D., Chung, H., Kim, H.J., Sastry, S., 2005. Autonomous exploration in unknown urban environments for unmanned aerial vehicles. In: Paper Presented at AIAA Guidance, Navigation, and Control Conference and Exhibit.

Shumate, R.P., Dirksen, J.R., 1965. A simulation system for study of traffic flow behavior.

Silva, A.B., Mariano,, P., Silva, J.P., 2015. Performance assessment of turbo-roundabouts in corridors. Transp. Res. Proc. 10, 124−133.

Spiliopoulou, A., Papamichail, I., Papageorgiou, M., Tyrinopoulos, I., Chrysoulakis, J., 2015. Macroscopic traffic flow model calibration using different optimization algo-rithms,. Oper. Res. 17.

Stark, M.C., 1962. Computer simulation of traffic on nine blocks of a city street. Highway Res. Board Bull.

U.S. DoT, 2017. Active Transportation and Demand Management. <http://www.its.dot.gov/research_archives/atdm/index.htm>.

Wallace, C.E., Courage, K.G., Hadi, M.A., Gan, A.C., 1998. TRANSYT-7F user's Guide. Transportation Research Center. University of Florida, Gainesville, Florida.

Wicks, D.A., Andrews, B.J., 1980. Development and testing of INTRAS, a microscopic freeway simulation model. Volume 2: User's Manual. Final Report.

Wilkinson, R.I., 1956. Theories for toll traffic engineering in the U. S. A. Bell Syst. Technol. J. 35, 421−514.

Wilson, A.G., 1971. A family of spatial interaction models, and associated developments. Environ. Plan. A 3, 1−32.

SUBCHAPTER 10.3

Data for Training Models, Domain Adaptation

Antonio M. López, David Vázquez and Gabriel Villalonga
CVC-UAB, Barcelona, Spain

10.3.1 TRAINING DATA AND GROUND TRUTH

Nowadays it is rather clear that sensor-based perception and action must be based on data-driven algorithms. In other words, we must use machine learning techniques for developing the algorithms that automatically perform the required tasks, from perception to action. Obviously, resorting to machine learning implies relying on data. For the sake of simplicity, in the following we assume we are discussing visual data, i.e., images, but the considerations we are going to introduce can be also extrapolated for other types of data, such as LIDAR, RADAR, etc. However, working with images is specially challenging, not surprisingly, since the sense of sight and how to understand the world through it is incredible difficult for machines.

Applying machine learning for developing a visual task, i.e., to learn a visual model, implies having three different datasets of images: (1) training; (2) validation; (3) testing. Equivalently, we can consider that a single dataset can be randomly split into those three to have several combinations. The training images are used to learn the desired visual model, i.e., its parameters. Such models usually have hyper-parameters that are set by trial and error (a very primitive and costly form of learning). Given a trial of hyperparameters the model parameters are learned and tested on the validation dataset. Then, it is selected the trial of hyperparameters that shows the best results on this dataset. Finally, by applying the learned model to the testing dataset, we obtain a proxy of its expected accuracy in real-world conditions.

In terms of relative sizes, usually the training set is much larger than the others. The validation set used is the smallest one. A typical split of all available data can be 60% for training, 10% for validation, and 30% for testing. The reason is that the machine learning algorithms in general produce better models when more data is available provided it has been collected randomly, i.e., without any undesirable bias (obviously less data but better selected may produce better models than lots of redundant data). Thus, it is desired to use most of the data for training. Note that training, validation, and testing datasets cannot overlap for ensuring that the measured accuracy of the learned model makes sense in terms of generalization, i.e., in terms of how the model will behave under previously unseen data. Altogether, this implies that most of the data should be used for training. Moreover, when the absolute amount of data is low, then, as mentioned before, the train−validation−test splitting is performed several times to come up with different models, which brings a more realistic assessment of both the machine learning method in use and the usefulness of the learned models in terms of their accuracy. In the following, we will use the term "training" to refer to both "training" and "validation" as defined here, because these stages are part of the process of developing a model, while testing is used for assessing its performance. In this way we simplify the terminology without losing generality.

Once we have introduced the critical need for training data we have to add what, in fact, is the most challenging point. It is not only that we need images for training, but also the ground truth associated to them. Fig. 10.3.1 draws the idea for two specific vision-based tasks: object detection and semantic segmentation. In the former case, the ground truth consists of the bounding boxes (BBs) framing the objects (cars in the example). In the latter case, the ground truth consists of the silhouettes of all the semantic classes in consideration (road surface, sky, vegetation, building, vehicles, pedestrians, etc.); in other words, a class must be assigned to each pixel of each image used for training. What is the problem? These ground truths are provided manually, which is a tiresome procedure prone to errors. Obviously, both for validation and testing ground truth is also required, but the most time-consuming part is due to training since, as we have mentioned, this is the stage that requires most of the data. It is worth mentioning that in the machine learning literature there are proposals that try to train models without the use of ground truth; however, the models that are really accurate do need such ground truth. These are the so-called supervised machine learning methods, in contrast to the unsupervised ones (no ground truth used). Well-known examples of supervised machine learning methods are support vector machines (SVM), logistic regression, Adaptive Boosting (AdaBoost), Random Forest, and Convolutional Neural Networks (CNN).

In the fields of advanced driver assistance systems (ADAS) and autonomous driving (AD), we can find several examples of datasets with ground truth publicly available. In the ADAS community a popular pioneering

Figure 10.3.1 Manual annotation (labeling) of: (top) bounding boxes (BBs) that frame objects (cars in this case); (bottom) silhouettes of all the semantic classes, i.e., pixel-wise assignment of semantic classes (road surface, vehicles, etc.).

example was the Daimler Pedestrian dataset of Enzweiler and Gavrila (2009), which includes 3915 BB-annotated pedestrians and 6744 pedestrian-free images (i.e., image-level annotations) for training, and 21,790 images with 56,492 BB-annotated pedestrians for testing. Another pioneering example corresponds to the pixel-wise class ground truth provided in Brostow et al. (2009) for urban scenarios; giving rise to the popular CamVid dataset which takes into account 32 semantic classes (although only 11 are usually considered) and includes 701 annotated images, 300 normally used for training and 401 for testing. A few years after, the KITTI Vision Benchmark Suite of Geiger et al. (2016) was an enormous contribution for the research focused on ADAS/AD given the high variability of the provided synchronized data (stereo images, LIDAR, GPS) and ground truth (object bounding boxes, tracks, pixel-wise class, odometry). More recently, Daimler has led the release of the so-called

Cityscapes dataset of Cordts et al. (2016), which tries to go beyond KITTI in several aspects. For instance, it includes 5000 pixel-wise annotated (stereo) images covering 30 classes and per-instance distinction, with GPS, odometry, and ambient temperature as metadata. In addition, it includes 20,000 more images but where the annotations are coarser regarding the delineation of the instance/class contours. This kind of dataset is difficult to collect since driving through 50 cities covering several months and weather conditions was required. In order to appreciate how difficult it is to provide such ground truth, we can mention the fact that annotating one of those images pixel-wise may take from 30 to 90 minutes of human labor depending on the image content. Thus, assuming an average of 60 minutes, annotating the 5000 images mentioned before, requires 5000 working hours for a person. Fig. 10.3.2 shows examples of Cityscapes images: each color represents a different urban semantic class (e.g., light pink means sidewalk, dark pink road, red pedestrian, etc.). In the top there is an example of a finely annotated image, in the bottom we see a coarsely annotated one. Note how the silhouettes of the classes are not accurately traced in the coarse case.

In order to shorten the annotation time and be more robust to erroneous annotations, we can think about crowdsourcing this task. For instance, this was the approach followed in the computer vision community where tools such as Amazon's Mechanical Turk (AMT) and LabelMe of Russell et al. (2008) were used to annotate popular publicly available datasets such as ImageNet (see Deng et al., 2009), and PASCAL VOC (see Everingham et al., 2010). However, crowdsourcing usually seeks low cost and, therefore, is not based on professional annotators. As a consequence, methods to automatically assess the quality of the ground truth are still required. In fact, since ADAS and AD face mobility safety, companies must rely on a more professional pipeline with many qualified annotators involved in the annotation of the data. In addition, not all kinds of ground truth can be provided by relying on manual annotations. For instance, we may need to develop a dense (pixel-wise) depth estimation algorithm or an optical flow one. A person cannot manually provide the pixel-wise ground truth desired to train and/or test such algorithms.

The reader may appreciate already how difficult is ground truth collection. However, we can see that the situation is even worse by introducing a very relevant point not yet mentioned here. As we introduced in Section 9.1.4, deep learning and, in particular, CNN architectures are the core of the state-of-the-art of many computer vision tasks, including

Figure 10.3.2 Ground truth examples from Cityscapes dataset.

those related to ADAS/AD, such as object detection and semantic segmentation. The starting point of this breakthrough was the task of image classification (i.e., assigning a single label to a full image), for which the AlexNet of Krizhevsky et al. (2012) just smashed the previous state-of-the-art. This work already pointed out one of the reasons for the success of deep CNNs in general, namely the massive availability of data with ground truth. In particular, AlexNet was trained on the ILSVRC dataset, with about 1000 images of 1000 categories; overall, about 1.2 million images for training, 50,000 for validation, and 150,000 for testing, all of them with image-level class annotation.

The publicly available datasets of reference in ADAS/AD, i.e., KITTI and Cityscapes, are orders of magnitude away from ILSVRC. Moreover, image-level ground truth is too poor for ADAS/AD tasks where object-wise (BBs) ground truth is a minimum, but most of the times pixel-wise ground truth (Fig. 10.3.2) is required. Nowadays, even for ADAS/AD tasks, the fine-tuning approach is followed; i.e., taking a deep CNN such as AlexNet and somehow reusing it by exposition to the more scarce annotated data acquired for the new (ADAS/AD) tasks.

10.3.2 VIRTUAL WORLDS AND DOMAIN ADAPTATION

Due to all these considerations, a totally different way of addressing the ground truth acquisition problem has been assessed. It started timidly in 2010 but nowadays there is an explosion of works in this line, with even workshops devoted to it. We refer to the use of virtual worlds for generating realistic images (and potentially data from other simulated sensors) with automatically generated ground truth. Fig. 10.3.3 illustrates the idea with the pioneering work of Marin et al. (2010). In this case, a modification of the videogame Half-Life 2 was used for automatically generating pixel-wise ground truth for the pedestrians contained in virtual-world RGB images. These images are acquired on board a virtual car that drives along an urban scenario of the virtual city. In Marin et al. (2010) it was demonstrated that using the state-of-the-art pedestrian detector at that time (i.e., pyramid sliding window, HOG/Linear-SVM classifier, non-maximum suppression), the accuracy of the classifier trained on the virtual environment and the accuracy of an analogous classifier trained on real-world images (having access to the same number of samples) was statistically the same.

Further extensions of Marin et al. (2010) demonstrate that the results were not always directly as good as expected (see Vazquez et al., 2014). In particular, the accuracy obtained by different object detectors was lower when training with virtual-world data and testing in a given real-world dataset, than when training with the data of such real-world dataset. However, it was demonstrated by Vazquez et al. (2014) that the accuracy gap was not really specifically due to virtual-to-real differences. Virtual-to-real was shown to be a special case of a more generic problem, namely sensor-to-sensor differences. In other words, training an object detector with images of a given camera model and testing with images of another camera model, also ends up in worse results than if training and testing images come from the same camera. Note how important this problem is; if we annotate a large dataset of images for ADAS/AD and later we change the camera, we may need to annotate again another large dataset

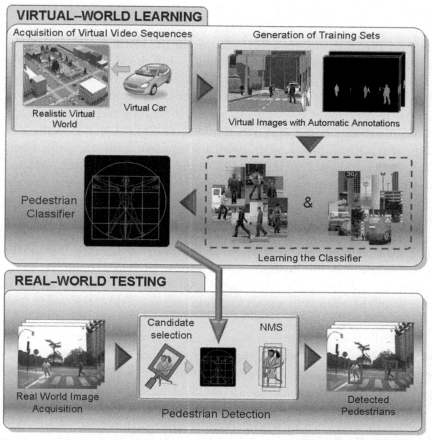

Figure 10.3.3 Example of pedestrian detector where the pedestrian classifier is trained in a virtual world, and then plugged into a detection pipeline to process real-world images.

to achieve the same accuracy. In fact, the problem is even more generic than sensor-to-sensor differences, the discrepancies in the statistics of the image content cause accuracy drops as well. As a matter of fact, the computer vision community started to realize this problem, which was just ignored for a long time (see Saenko et al., 2010, 2011). In particular, the so-called domain adaptation (DA) and transfer learning (TL) techniques started to gather relevance among the computer vision community since they pursued reusing previous knowledge (in the form of model or annotated data) for performing accurately in new domains and tasks, but using either many data without annotations (unsupervised DA/TL) or few data with annotations (supervised DA/TL). For instance, in Vazquez et al. (2014) supervised DA based on active learning was used to adapt virtual and real domains and, therefore, recovering the above mentioned accuracy gap of the pedestrian detector developed in the virtual world. In Xu et al. (2014) a more sophisticated technique was used to adapt deformable part-based models (DPM) from virtual to real domains, in this case and contrarily to Vazquez et al. (2014), without revisiting the source (virtual-world) data. Fig. 10.3.4 illustrates the idea: an initial DPM is learned by using virtual-world data with automatic ground truth for pedestrians. The model is applied to real images. Some pedestrians are not detected and background regions are classified as pedestrians due to the domain gap (virtual-to-real). To solve this gap the initial DPM is refined by either actively collecting errors with a human oracle in the loop, or a procedure is able to automatically collect annotations without human intervention. The DPM refinement can be done progressively by iterating this procedure. As guiding information, in Vazquez et al. (2014) and Xu et al. (2014) the proposed DA techniques saved 90% of the annotation effort that would be needed to obtain the same accuracy in the real-world (target) domain.

Overall, these experiments showed that appearance models trained in virtual worlds act as strong priors with the potential of saving a large amount of human annotation effort. Interestingly, the winner of the first pedestrian detection challenge in the KITTI dataset was based on a

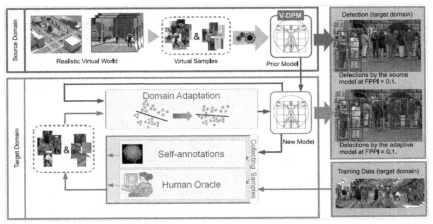

Figure 10.3.4 Domain adaptation when the source domain is a virtual world and the target domain is the real world.

virtual-to-real-world domain-adapted classifier (Xu et al. (2016)). Note also that automatically collected ground truth is more precise than that collected by humans. It is also worth mentioning that even deep CNNs require DA/TL (see Tommasi et al., 2015).

Driven by the success of the use of virtual-world data and domain adaptation for training the appearance pedestrian models, in the last two years new works have been presented going far beyond pedestrian detection. For instance, in Ros et al. (2016) a very large city was created, named SYNTHIA, which allowed generating hundreds of thousands of RGB images (random and arranged as sequences) with all kinds of interesting ground truth automatically generated: pixel-wise class ID, instance ID, and depth; vehicle odometry; and 360 degrees views. In order to force variability, the city includes many pedestrian models, vehicles, city styles, highways, vegetation, lighting conditions, and four seasons. Fig. 10.3.5 shows different snapshots of the city content and ground truth. Using data from SYNTHIA and basic domain adaptation techniques, Ros et al. (2016) show that it is possible to boost the accuracy of deep CNNs designed for semantic segmentation. Season- and lighting-dependent images together with vehicle odometry can be used to train place recognition methods that may be part of vehicle localization in maps (a key component nowadays of prototypes of self-driving cars).

Analogously, Gaidon et al. (2016) presented a virtual environment that mimics KITTI, termed as Virtual KITTI, showing its usefulness for

Figure 10.3.5 SYNTHIA: RGB image, ground truth for Class ID and depth; images acquired from the same camera location at different seasons.

designing object trackers (e.g., for tracking cars). Following the same line of work, Richter et al. (2016) show semantic segmentation results with deep CNNs using the GTA-V videogame world. Interestingly, other lower level visual tasks such as depth estimation and optical flow estimation are being currently addressed by the use of deep CNNs trained on virtual data, see Mayer et al. (2016). Note that ground truth for such tasks cannot be collected by humans.

In this setting one of the arising questions is how the degree of photorealism of the virtual images affects training visual models. By comparing SYNTHIA images and GTA-V ones (Fig. 10.3.6), in Lopez et al. (2017)

Figure 10.3.6 Comparing photorealism. GTA-V (top) is more photorealistic than SYNTHIA (bottom).

it is shown that even for the most realistic video-games, the virtual-to-real domain gap is still an issue. This is not surprising, since we have mentioned before that there may be sensor-to-sensor domain gaps even for real-world sensors.

We would like to highlight that virtual environments are gaining attention not only for understanding the sensor raw data, but also for learning to act (see Dosovitskiy and Koltun, 2016); in other words, given an image, a deep learning architecture directly outputs the control commands for self-driving (e.g., steering angle, brake/accelerate, etc.), without explicitly creating an intermediate 3D understanding of the driving scenario.

Finally, we would like to note that virtual worlds are not only useful for training models, in fact, they can be a very convenient tool for exhaustive simulations that allow the setting of hyperparameters, debugging the behavior of algorithms, experimenting on corner cases, etc.; in other words, the more traditional functionalities assigned to simulators of any kind. Obviously this is a more standard use of virtual environments, and the revolution has been to see that they can be also used for training models, especially visual deep models.

REFERENCES

Brostow, G.J., Fauqueur, J., Cipolla, R., 2009. Semantic object classes in video: A high-definition ground truth database. Pattern Recogn. Lett. 30 (20), 88—89.

Cordts, M., Omran, M., Ramos, S., Rehfeld, T., Enzweiler, M., Benenson, R., et al., 2016. The Cityscapes dataset for semantic urban scene understanding. In: IEEE Conference on Computer Vision and Pattern Recognition.

Deng, J., Dong, W., Socher, R., Li, L.-J., Li, K., Fei-Fei, L., 2009. Imagenet: a large-scale hierarchical image database. In: IEEE Conference on Computer Vision and Pattern Recognition.

Dosovitskiy, A., Koltun, V., 2016. Learning to Act by Predicting the Future. arXiv 1611, 01779.

Enzweiler, M., Gavrila, D.M., 2009. Monocular pedestrian detection: Survey and experiments. Trans. Pattern Recogn. Mach. Anal. 31 (12), 2179—2195.

Everingham, M., Van Gool, L., Williams, C.K.I., Winn, J., Zisserman, A., 2010. The PASCAL visual object classes (VOC) challenge. Int. J. Comput. Vis. 88 (2), 303—338.

Gaidon, A., Wang, Q., Cabon, Y., Vig., E., 2016. Virtual worlds as proxy for multi-object tracking analysis. In: IEEE Conference on Computer Vision and Pattern Recognition.

Geiger, A., Lenz, P., Stiller, C., Urtasun, R., 2016. Vision meets robotics: The KITTI dataset. Int. J. Robot. Res. 32 (11), 1231—1237.

Krizhevsky, A., Sutskever, I., Hinton, G., 2012. ImageNet classification with deep convolutional neural networks. In: Annual Conference on Neural Information Processing Systems.

Lopez, A.M., Xu, J., Gomez, J.L., Vazquez, D., Ros, G., 2017. From virtual to real-world visual perception using domain adaptation — the DPM as example. Domain adaptation in computer vision applications, Springer Series: Advances in Computer Vision and Pattern Recognition, Edited by Gabriela Csurka.

Marin, J., Vazquez, D., Geronimo, D., Lopez, A.M., 2010. Learning appearance in virtual scenarios for pedestrian detection. In: IEEE Conference on Computer Vision and Pattern Recognition.

Mayer, N., Ilg, E., Hausser, P., Fischer, P., Cremers, D., Dosovitskiy, A., et al., 2016. A large dataset to train convolutional networks for disparity, optical flow, and scene flow estimation. In: IEEE Conference on Computer Vision and Pattern Recognition.

Richter, S.R., Vineet, V., Roth, S., Vladlen, K., 2016. Playing for data: ground truth from computer games. In: European Conference on Computer Vision.

Ros, G., Sellart, L., Materzyska, J., Vazquez, D., Lopez, A.M., 2016. The SYNTHIA dataset: a large collection of synthetic images for semantic segmentation of urban scenes. In: IEEE Conference on Computer Vision and Pattern Recognition.

Russell, B.C., Torralba, A., Murphy, K.P., Freeman, W.T., 2008. LabelMe: a database and web-based tootl for image annotation. Int. J. Comput. Vis. 77 (1—3), 157—173.

Saenko, K., Kulis, B., Fritz, M., Darrell, T., 2010. Adapting visual category models to new domains. In: European Conference on Computer Vision.

Tommasi, T., Patricia, N., Caputo, B., Tuytelaars, T., 2015. A deeper look at dataset bias. In: German Conference on Pattern Recognition.

Vazquez, D., Lopez, A.M., Marin, J., Ponsa, D., Geronimo, D., 2014. Virtual and real world adaptation for pedestrian detection. Trans. Pattern Recogn. Mach. Anal. 36 (4), 797—809.

Xu, J., Ramos, S., Vazquez, D., Lopez, A.M., 2014. Domain adaptation of deformable part-based models. Trans. Pattern Recogn. Mach. Anal. 36 (12), 2367—2380.

Xu, J., Ramos, S., Vazquez, D., Lopez, A.M., 2016. Hierarchical adaptive structural SVM for domain adaptation. Int. J. Comput. Vision 119 (2), 159—178.

CHAPTER 11

The Socioeconomic Impact of the Intelligent Vehicles: Implementation Strategies

Maria E. López-Lambas
Universidad Politécnica de Madrid, Madrid, Spain

Contents

11.1 INTRODUCTION

According to the definition provided in Chapter 1, Introduction, the iVehicle (iV) is one that is able to capture information about its conditions and/or its surroundings (more or less near), to process the information, and then to make decisions, either providing information or taking actions. A distinction between a connected vehicle and an automated vehicle can also be made: whilst the former includes vehicle-to-vehicle (V2V) and vehicle-to-infrastructure (V2I) communications, the latter refers to the capability of a vehicle to operate and maneuver independently in real traffic situations, using onboard sensors, cameras, associated software, and maps in order to detect its surroundings (Declaration of Amsterdam, 2016).

Although new technologies can change the transport market radically, providing a range of benefits in terms of safety, traffic congestion, etc., a number of questions remain unsolved; questions related to the legal framework, liability, insurance, security, etc.; even software hacking/misuse are reported as a major concern (Kyriakidis et al., 2015). Nevertheless, the same authors report that among a sample of 5000 respondents from 109 countries, 69% would estimate that fully automated driving would reach a 50% market share between 2015 and 2050. This means that, beyond the uncertainties related to the new technology, the market is willing to accept it.

Besides a brief review of the connected vehicle, this Chapter will deal mainly with the iV understood as autonomous vehicle, since it does not only make decisions, but incorporates the highest number of driving assistance systems, turning it into the highest expression and the most refined version of these kind of vehicles.

This Chapter does not intend to be an exhaustive list of barriers, weaknesses, or even threats that prevent the successful implementation of the intelligent vehicle, but deals with the main ones that actually hamper the successful implementation of the intelligent vehicles and their potential benefits, pointing out the main impacts on the society and economy of such a dramatically new and disruptive transport system.

11.2 FROM CONNECTED TO AUTONOMOUS VEHICLE

Big data, Internet of Things (IoT), and a society increasingly connected are the driving forces of a revolution in our lifestyle, affecting jobs and communications. In this hyper-connected society the introduction of the connected vehicle is a further step towards a change of paradigm in mobility, business, economy, and habits, and also represents the first step on the way to a greener, more efficient, and safer transport, whether it be driverless or simply connected. But, unlike autonomous cars, which are yet to be developed in terms of technology and regulatory framework, the connected vehicle is already a reality.

The level of automated driving is increasing step by step: from driver assistance systems—such as adaptive cruise control (ACC) or predictive emergency braking, on the market for more than ten years—to highly and fully automated, still under development. In 2015, 200 million connected vehicles were in use. Connected cars take advantage of smartphones and apps which, in turn, prevent their success since drivers use the smartphone to solve their connectivity needs instead paying for the extra costs associated with embedded connectivity (EVERIS, 2015). As with the autonomous vehicle (AV), a connected car increases safety, reduces energy consumption and associated emissions and costs, by providing the most efficient routes.

As for automated vehicles, it was in 1939, during the Universal Exposition in New York (Futurama) that General Motor's vision about a future of transport with driverless cars was presented. After years of research, in 1997, on the I-15 highway of San Diego, California, 20 vehicles including car, buses, and trucks, were shown. In the absence of a common set of standards and specifications, the fact is that the accelerated deployment and use of ITS, makes automated driving the most promising way of intelligent transport.

Currently, according to the Institute of Electrical and Electronics Engineers (IEEE), automated vehicles will account for 75% of cars on the road by 2040 (ASEPA, 2014). Furthermore, the journal Wired foresees that no driving licenses will be needed from 2040 on, and PricewaterhouseCoopers concludes that the current US fleet of 245 million cars will drop to 2.4 million with the introduction of fully autonomous driving (ASEPA, 2014). Market forecasts cannot be more optimistic.

As stated in Chapter 8, Automated Driving, regarding the different levels of automation, the United States Department of Transportation (DOT, 2016) draws a distinction between levels 0—2 and 3—5, depending on who has primary responsibility for monitoring the driving environment, i.e., the human operator or the automated system, respectively. Vehicles with automation levels above 3 must also incorporate connected vehicle technologies, and therefore many of the technologies overlap.

Nevertheless, technology must yet overcome some barriers to be fully legal, marketable, and reliable. The next chapter addresses the main benefits, as well as the concerns and challenges to face in the future regarding this topic.

11.3 SOCIAL ISSUES

11.3.1 Acceptance of the Innovations

Maybe the first question to deal with when facing such a disruptive technology as the driverless experience, is the fear of what is viewed as unknown. This perception could, in fact, limit the adoption of the autonomous vehicle, given the innate human suspicion about what is novel and innovative. At the other end would be the eventual misuses of the new systems by adopting high risk behavior as users consider that with better safety measures they may pay less attention.

Nevertheless, data suggest the contrary since, according to a world survey (CETELEM, 2016), three out of four drivers think that the autonomous vehicle will be soon a reality, with those in the developing countries being more optimistic. The less confident seem to be, paradoxically, the pioneers, such as the Japanese (63%), British and Americans (61%). Potential users with a mature car market (Europe, U.S.A. and Japan) are more careful as regards the full adoption of the 100% autonomous vehicle between 2020 and 2025.

11.3.2 Safety

It is expected that autonomous vehicles will reduce the number of injuries and deaths, avoiding human error, which is the cause that lies behind 90% of crashes (Forrest and Konca, 2007), or 94% according to the US Department of Transportation; not to mention the costs of both injuries and deaths in terms of insurance, rehabilitation, medical leave, etc. To set an example, when the brakes are applied a second earlier, one driver traveling at 50 km/h can reduce the crash energy by 50% (Forrest and Konca, 2007).

Highly Automated Vehicles-HAVs-(SAE levels 3—5) can take advantage of data and experience from other vehicles on the road, as well as sensor technologies such as V2V and V2I, which will contribute to reduce the number and severity of accidents (DOT, 2016).

In addition, safety is also good for reducing congestion: by reducing the number of car accidents, congestion also decreases, avoiding traffic delays and reducing fuel costs and the related emissions.

However, a possible rebound effect could arise due to misuse of the new systems, which may occur when highly risky behaviors are adopted in the belief that everything will get organized and fixed automatically.

11.3.3 Effects on Employment

The automated market will impact on jobs, with a potential reduction of employment in the traditional car activities. What about truck drivers, taxi drivers, public transport? How will this new model impact on these sectors? With no need for drivers, a balance should be made since many low-skilled or semiskilled employees, from taxi and delivery drivers to repair or maintaining sectors, will very likely lose their jobs, despite the obvious saving of personnel costs for the companies. Indeed, artificial intelligence (AI) will bring about changes that will disrupt the lives of millions of people around the world.

Recently, a report from the Executive Office of the President (White House, 2016), set out five main economic effects:

- Positive contributions to aggregate productivity growth.
- Changes in the skills demanded by the job market, including greater demand for higher-level technical skills.
- Uneven distribution of impact across sectors, wage levels, education levels, job types, and locations.
- Churning of the job market as some jobs disappear while others are created.
- The loss of jobs for some workers in the short-run, and possibly longer depending on policy responses.

But the automotive industry has the potential to create new services, new business models that could be set up, e.g., through the opportunity to buy shares in a vehicle and then request specific times to use it, as a sort of timeshare. A large corporation could buy into shares of cars all over the world so wherever an employee went they would have a vehicle available (Forrest and Konca, 2007). So, individual and legal entities will own the vehicles, while others will have multiple owners.

As for the public transport, it is certainly not the same to drive a rail-based system as a bus, since the role of the driver is different, the driver of the latter being more involved given the fact that they must interact with other vehicles on the streets. Anyhow, in both cases drivers will be no longer needed, resulting in many people losing their jobs.

Moreover, a driver's license will be no longer necessary for young people, putting an end to the driving schools. And the same goes for the Governments who will lose taxes associated with the manufacturing and use of vehicles.

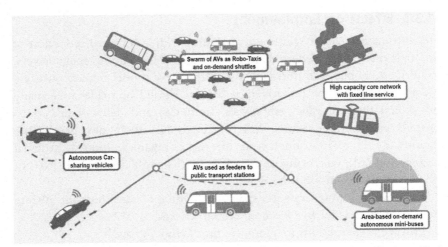

Figure 11.1 Possible applications of autonomous vehicles as part of diversified public transport system. *Adapted from UITP, 2017. Policy brief. Autonomous vehicles: a potential game changer for urban mobility. <http://www.uitp.org/autonomous-vehicles> (accessed 13.02.17).*

The main strategies to cope with these threats must be based on education and training for the new jobs and the empowerment of the workers to ensure shared growth (White House, 2016). The same report also highlights the need of continuous engagement between government, industry, technical, and policy experts and the public to carry on those policies.

According to UITP (2017), AVs could play a key role in urban mobility if employed as "robo-taxis" or as car-sharing schemes (Fig. 11.1).

In short, the arrival of the autonomous car will bring more efficiency to the transport system, but the society must be prepared to assume the cost in terms of employment. Ultimately, as stated by the report from the White House (White House, 2016), *"the challenge for policymakers will be to update, strengthen, and adapt policies to respond to the economic effects of Artificial Intelligence."*

11.4 LEGAL ISSUES

Legal issues are a chief concern for the implementation of the AV. In the case of a system failure, responsibilities must be clearly defined.

Nevertheless, the development of legislation usually is a long and slow process, whereas a clear and flexible enough legal framework is needed in order to provide legal certainty but be easily adaptable to a fast moving industry.

The main issues regarding the legal aspects of the highly automated and autonomous driving are summarized below.

11.4.1 Liability/Insurance

As stated by Schellekens (2015), liability answers the question of *whether the costs of accidents are borne by the victim or can be transferred to another actor, typically somebody who is in one way or another (co)responsible for the occurrence of the damage.*

Insurance systems differ largely from one country to another. Nevertheless, broadly speaking we can distinguish between *"product liability"* and *"traffic liability"*. In the former case, "the producer is liable for damage caused by a defect in his product," while in the latter the holder or driver the one liable for damages or injuries occurred in the operation of the vehicle (Schellekens, 2015). It depends on the kind of coverage: first-party or third-party insurance. In the first case, if obligatory, the manufacturer is broadly covered, but this would make autonomous vehicles even more expensive.

Actually, hands-free driving systems do exist and if an accident occurs, the liability lies on the driver. The key and real change will come when the cars will be not only autonomous but connected, without the driver monitoring. Then, the car will communicate with the rest of the cars around and there will be no human being to avoid an imminent accident. Then and only then could the liability transfer to the manufacturer and/or infrastructure provider.

Nevertheless, perhaps the main risk will arise when autonomous cars start to share the space with the traditional ones. In fact, some believe that AVs should not share infrastructure with pedestrians, bicycles, and even conventional cars, while others think that for the system to function properly all the vehicles have to be autonomous.

At any rate, in the future all the vehicles should incorporate a device similar to a black box that is able to identify who is in charge at all times, human or autopilot, pointing at who is liable therefore.

11.4.2 Test and Validation

A collateral activity linked to liability regards the question of the validation procedures and testing requirements, a question which exceeds the technical field. Again, highly automated driving will not be allowed without legal changes. Actually, one of the main existing limitations is the United Nations Vienna Convention on road traffic (1968), that sets out that drivers must keep control of the car at all times. Certainly, this is not the case in Spain, which did not sign the aforementioned Convention; a situation that puts the other member States of the European Union at a disadvantage since they cannot develop a technology that will be implemented in a few years.

Nevertheless, a set of nonhomogeneous rules and legislation to test automated driving already exists in Europe (ERTRAC, 2015), and while this is a step forward, it still means that there is another domain where harmonization is needed. To avoid this problem, and thinking of the near future, in March 2016 the United Nations Economic Commission for Europe (UNECE) introduced some amendments to the Convention in order to allow autopilot driving as long as the driver can rectify.

In the USA, the Department of Transportation has presented some recommendations, not mandatory, about the legal changes needed to allow these vehicles to move freely on the road. Those recommendations include best practices for the safe predeployment design, development, and testing of highly automated vehicles, prior to sale or operation in public roads (DOT, 2016). Moreover, the deployment strategy identifies potential new tools, authorities, and regulatory structures for the introduction of new technologies. The Guidance is applied to both test and deployment.

Car manufacturers and even third parties are advised to pass safety tests that combine simulations, tested in closed–circuits and roads.

11.5 PRIVACY AND HACKING

11.5.1 Privacy

A different question regards privacy which, to put it in other words, means that to enable the information society to work, a proper balance between privacy and data flow is needed (Gil, 2016). In short: who owns data? Individuals, who the information refers to or the companies that collect, manage, and create those data? The best solution seems to be that provided by the World Economic Forum, according to which everybody should have duties and rights over the information since, although referring to an individual, data are created from the interaction between different parties (Gil, 2016).

This points straightforwardly to individual empowerment: a new business model in which an individual, as any other economic agent, manages their own information for their own ends, and shares a part with the companies, to communicate what they want, how and when, and obtain joint profits (Gil, 2016).

Obviously, since a lot of private information is involved, cyberattacks are a real threat which must be counteracted by deploying top level security measures. This point leads to the next one: hacking.

11.6 HACKING

Data protection may be threatened when managing large amounts of data from specific users. With millions of connected cars in the near future, improvements in security must be made to avoid hackers' attacks. Instead, while engineering departments in the automotive industry pay a great deal of attention to safety, less awareness is devoted to cybersecurity.

The main threats may include not only theft, but ransomware, traffic disruptions, etc. Nevertheless, the most common cyberattacks deal with gathering information through unauthorized access rather than sabotage (ENO, 2013). Indeed, as reported by Schellekens (2016), in 2015 American hackers seized the control of different car models (Chrysler and Tesla) to remotely control some features such as ventilation, engine, and brakes.

However, according to an on-line survey by the University of Michigan Sustainable Worldwide Transportation Department—with questions about vehicle security and data privacy—, respondents were more concerned about hacking to gain control of vehicles than hacking to get access to personal information (UMTRI, 2014).

Anyway, both technical protection and regulation can counteract these attacks. In the first case, different security measures should be deployed to prevent cyberattacks through the improvement in V2V and V2I protocols. Regarding regulation, there are command-and-control measures—rules stating what to do and by what means—and procedural standards. Also economic instruments, such as liability rules in case of accidents due to hackering, are instruments of regulation too (Schellekens, 2016). To put it another way: who is to blame in the case of an accident caused by a hacker? The car manufacturer or the hacker? (Schellekens, 2016). Note that in this case safety and security are intertwined, since a leak in the second affects the first. Besides, self-regulation and communication to the users about relevant cybersecurity information (in a sort of "consumer empowerment") may lead in the long run to avoidance of cyberthreats.

Very recently, Waymo (Google's AV division) launched its driverless vehicles programme, according to which the cars manufactured by the company only get connected with the cloud to send and receive information, such as traffic data, before closing the connection. The system does not depend on an external network, which make the cars more reliable.

11.7 ECONOMIC ASPECTS

Regarding iV, a cost—benefit analysis would be very complex to be carried out, since it is difficult to disaggregate the causes of some effects, as well as to identify which part is directly or indirectly involved in those effects. To give an example, improving the active safety, will result in fewer traffic accidents, but it is difficult to know which part of this cause (active safety) is directly attributable to the effect (accident reduction), since there are other possible reasons, such as the economic crisis, fuel prices, point system on driving licenses, unemployment, etc., all circumstances which contribute to traffic reductions. In the same way, alcohol, speeding, inexperience, weather conditions, etc. can result in traffic accidents, all of them factors that are not mutually exclusive (Eno, 2013).

Certainly, there are predictive models that can calculate these effects, but such an analysis falls beyond the scope of this document. Hence, this section will only deal with the rough main impacts on congestion, fuel and infrastructure, vehicles and insurance costs.

Nevertheless, this section deals with some of the main costs items considered in the automotive industry and mobility.

11.7.1 Congestion

According to TomTom, urban traffic congestion can increase the travel time by about 60% in many cities, and even 100% in the night peak hour (CETELEM, 2016). Regarding congestion costs, the autonomous vehicles will avoid traffic congestion through an intelligent vehicles network and infrastructure that will inform drivers on the traffic and road conditions, helping them to select the less congested route.

Indeed, according to the value of time theory, the higher the value of time, the greater the potential benefits from time savings, although this depends, obviously, on the value that we place on time: time of day, trip purpose, etc. are factors affecting this value. According to the Conference Board of Canada (CBC, 2015) we can assume a conservative one of 6$/h. Following the example provided by the CBC, assuming cars traveling at free-flow, i.e., 90 km/h on highways and 40 km/h on local and arterial roads, with an average of 65 km/h, 4.97 billion driver hours per year could be saved. In monetary terms, according to Fagnal and Kockelman (2015), a mere 10% market penetration, could result in $27 billion for the US economy annually.

To set an example, in the Spanish case, assuming a value of time of 9€/hour (Valdés, 2012; Salas et al., 2009), with an average congestion of 18 hours/driver/year (DGT, 2016), and a total fleet of 22 million cars (DGT, 2014), about 3,564 M€ per year could be saved. Note that most predictions from car manufacturers announce that the AV will be ready sometime between 2020 and 2025 (Schellekens, 2015) and, according to ERTRAC (2015), the global market for automated vehicles is estimated at 44 million by 2020.

11.7.2 Fuel

Regarding fuel savings, these will result from different sources, such as less time looking for parking space, less congestion, optimization of both journey and travel time, fewer traffic crashes due to the enhanced safety, etc. Furthermore, if the focus is on a shared autonomous vehicle (SAV), results show that each SAV can replace around 11 conventional vehicles. Nevertheless, these results can be counteracted by the induced demand adding up to 10% more travel distance than comparable non-SAV trips, resulting in largely positive emissions impacts (Fagnal, Kockelman, 2014). Moreover, some argue that AV will allow people to live even further from the city.

On the other hand, an AV could enable drivers to make other nondriving activities while in their vehicles, which means using their time for more worthwhile and even pleasant purposes than driving (Schellekens, 2015). Travel time will not be a negative externality; instead, this means leisure or productive time, or both, not to mention the reduction (or disappearance) of stress, with the related consequences on health.

Actually, the combination of an autonomous driving system with multimedia platforms will make the travel time more pleasant (Jungwoo,

2015). Indeed, the preferred situations for using an automated car are highways and congestion, when the driving conditions are either monotonous or stressful (Payre et al., 2014): reading the newspaper, surfing on the internet, or chatting with other passengers are some of the activities that will allow the driver a certain relief.

11.7.3 Infrastructure Costs

With the driver as a mere supervisor, by 2025 AV will be on the roadways, which must be adapted accordingly by then. It is in this scenario where investments on the road network are required, in order to detect the danger and warn the driver to take back control functions for 10 s; otherwise, the car will be parked on the hard shoulder of the road.

Certainly the high investments required in terms of infrastructures makes AV technology extremely expensive and slow in terms of both implementation and deployment. As for Spain, according to the experts, to get AV fully functional will demand a huge amount of money, about 3000 M€ to be more specific, and this only for the horizontal signing, i.e., those that the AV must read to operate. So the question is: if you do not have an AV why pay for it? It seems to be more rational that those having an AV pay higher registration and annual taxes for example than those who do not have.

Similarly, another question remains unsolved: given the need for cooperation among stakeholders (i.e., responsible for infrastructures, system integrators, car or components manufacturers, etc.), how costs and incomes will be shared? Who will pay for what?

11.7.4 Vehicle Cost

The main reason for the high cost of the AV is the cost of the LIDAR technology, which makes the acquisition of one of these cars nearly impossible to the average consumer, and it is likely to continue to do so in the next decade. But, despite the high investment needed to implement the autonomous driving, it is expected that, as with electric vehicles, technological advances and large-scale production will likely contribute to reduce prices (Fagnal and Kockelman, 2015) following the rule of the economies of scale. Actually, according to Eno (2013), EV costs have already dropped by 6%—8%.

Indeed, Volvo has confirmed the introduction of autonomous system in their premium cars as an option at an extra cost of 9000€

(Diariomotor, 2016). According to the same source, Mercedes has already made available a piloted driver assistance package at 2,837€.

11.7.5 Insurance Costs

In the current scenario, what is the driver and what is the vehicle remains unclear, which affects liability in the event of an accident. Software manufacturers, telecommunication companies, maps providers, etc., all come into play, which also has consequences on a fragmentation of the risk.

Most likely, a part of the liability will keep on being laid on the owner, but a driver's liability is less clear. It will depend on whether someone is driving or is autopilot driving. In this latter case all the circumstances should be assessed in order to determine who is responsible: the car manufacturer, the GPS companies, etc.

However, very recently, after the National Highway Traffic Safety Administration has reported a decrease in the number of road accidents by 40% since the launch of the first Tesla's AV—the car company is planning to run its own business model, offering a single price covering purchase, maintenance, and insurance (El Confidencial, 2017). According to Tesla Motors, if the autonomous vehicle has proven to be safer, it is fair to reduce the insurance costs consequently. Therefore, if the insurance companies do not lower the price of the policies proportionally to the risk, Tesla will do it. Furthermore, the company is preparing the AV's second generation, which is expected to reduce even more the percentage of accidents, from the current 40 down to 90%.

11.8 LIVEABILITY

In the boom of the so-called shared economy, autonomous cars will also change the traditional model of car ownership. With cars parked 90—95% of the time, a self-driving car will be ordered only when needed, or so they say. It should be recalled that, according to Eurostat, around three quarters of population of the EU (72.4%, to be precise) lives in cities and, as stated in the EC's Green Paper (EC, 2007), *"urban areas represent the backbone of economic wealth creation. They are the places where business is done and investments are made,"* and where 85% of the EU's gross domestic product is created. After all they are the so-called *"drivers of the European economy."* Hence, with a lot of free space on the streets, and the possibility to requalify former parking spaces, cities will increase their value, and there will be more urban space turned into green areas.

According to UITP (2017), every citizen could get to their destination with 80% fewer cars at least; but this can only happen if all forms of shared mobility are actively promoted: subsidies, incentives (priority parking, for example), promotional campaigns, etc., and any other measure to help reduce single car occupancy and avoid empty autonomous cars.

Besides, impacts on equity and inclusiveness are also of importance, since autonomous technology should allow the access to remote areas or regions where today public transport cannot, not only enabling their development but also ensuring mobility and accessibility for all (ERTRAC, 2015), providing affordable mobility options that enable anyone entering the city when living in the suburbs.

11.9 CONCLUSIONS

Despite the unquestionable advantages, AV can also produce undesired effects, such as induced demand that will bring about an increase in the number of vehicle miles traveled (VMT). This could be the case if mobility for those too young to drive, the elderly, the disabled, etc., is provided, which presumably will turn into an automobile-oriented development, lack of space, obesity, etc. (ENO, 2013). As stated by Fagnal and Kockelman (2015), VMT per AV "*is assumed to be 20% higher than that of non-AV vehicles at the 10% market penetration rate, and 10% higher at the 90% market penetration rate.*"

Furthermore, some questions around the impact of AV on sprawl are yet to be answered: will people live even further or will they concentrate around high density areas, since highly-mobile autonomous zones are expected to develop there?

On the other hand, presumably the reduction in walking due to self-park and door-to-door services in general, will significantly affect health, inducing cardiovascular diseases, overweight, obesity, and other disorders associated with physical inactivity. This could be counteracted by the expected reduction in the number of people killed or injured in car accidents, and for the repurposing of some infrastructures, such as on-street parking areas turned into bike lanes, which may encourage people to cycle, or reconversion of parking lot space into pedestrian zones, which will promote walking due to enhanced safety, walking space, and less pollution.

Regarding liability, to avoid lawsuits arising from the different treatment between "human" drivers and "computer" drivers, the solution could be to consider the computer driver liable only if a human driver under the same circumstance would be liable (Greenblatt, 2016).

Some questions remain uncertain, such as that known as the "ethical algorithm," i.e., how the vehicle should act in case of an accident? Avoiding killing 10 pedestrians crossing the street or lacerating the "*driver*"? Should the car make that kind of decision? In fact, according to different surveys, many declare their willingness to drive an AV only if they could activate/deactivate the *autopilot*.

Besides, the eventual rebound effects remain a threat, since the availability of a car for impaired people, for the elder, the youngest, etc., may turn into an increased number of kilometers traveled, a perfect example of a rebound effect—not to mention the negative externalities in terms of environmental impacts if the fuel does not come from renewable sources. To avoid this threat, a shared fleet scheme with AVs integrated—instead of competing—with traditional public transport services, will contribute to meet a sustainable and better mobility as well as equity (UITP, 2017).

At the European level, the European Road Transport Research Advisory Council (ERTRAC, 2015) has set a number of key challenges and objectives, in the framework of a roadmap to be fully developed by 2025, based on research and innovation, but also on the societal challenges. The ERTRAC Strategic Research Agenda faces the future in an integrated manner, with the public and private sector working together towards a common objective: the full deployment of automated driving. Safe and robust technology, a legal and regulatory framework, acceptance, common validation procedures and testing requirements, infrastructure requirements, and manufacturing remain at the core the ERTRAC approach to improve the efficiency of road transport.

All the authorities involved need to be more proactive in preparing the authorization of driverless operation, in order to allow the full operation of this new mode of transport. In short, this means that to extend research funding, to set appropriate and harmonized standards for liability, security, and data privacy, and to develop guidelines for certification would seem to be the way forward in the right direction for a successful automation technologies deployment. Put another way, a common legal context is the first question to be addressed in order to establish a sound and reliable framework of common rules for the technical approval procedures of automated vehicles and infrastructures, and for the liability of the operation. Additionally, the future is already upon us.

REFERENCES

ASEPA, 2014. Sistemas de asistencia al conductor y de gestión inteligente del tráfico. Monografías 6. Asociación Española de Profesionales de Automoción, 2014.

CETELEM, 2016. Observatorio. El coche autónomo. Los conductores, dispuestos a ceder la conducción a la tecnología. Available at <www.elobservatoriocetelem.es> (accessed 23.11.16).

Conference Board of Canada, 2015. Automated Vehicles. The coming of the next disruptive technology. The Van Horne Institute, January 2015.

Declaration of Amsterdam, 2016. Cooperating in the field of connected and automated driving. Available at <https://english.eu2016.nl/documents/publications/2016/04/14/declaration-of-amsterdam> (accessed 1.02.17).

Diario Motor, 2016. <http://www.diariomotor.com/imagenes/2015/12/volvo-s90-2016-75.jpg> (accessed 13.02.17).

Dirección General de Tráfico (DGT), 2014. Series históricas- Parque de Vehículos. <http://www.dgt.es/es/seguridad-vial/estadisticas-e-indicadores/parque-vehiculos/series-historicas/> (accessed 23.11.16).

Dirección General de Tráfico (DGT), 2016. Revista de Tráfico y Seguridad Vial. <http://revista.dgt.es/es/noticias/nacional/2016/10OCTUBRE/1011las-ciudades-con-mayores-atascos.shtml#>. WEAHurLhCUk (accessed 23.11.16).

El Confidencial, 2017. <http://www.elconfidencial.com/tecnologia/2017-02-28/tesla-coche-autonomo-seguro-mantenimiento_1339883/> (accessed 03.03.17).

Greenblatt, N.A., 2016. Self-driving cars and the law. <www.spectrum.ieee.org/whitepapers> (accessed 30.12.16).

ENO Center for Transportation, 2013. Preparing a Nation for Autonomous Vehicles. Opportunities, barriers and policy recommendations. www.enotrans.org.

ERTRAC, 2015. Automated driving roadmap. ERTRAC Taskforce Connectivity and automated driving 21/01/2015.

European Commission, 2007. Green Paper "Towards a new culture for urban mobility" (COM 2007 -551 final).

EVERIS, 2015. Everis Connected Car Report. A brief insight on the connected car market, showing possibilities and challenges for third-party service providers by means of an application case study.

Fagnal, Daniel J., Kockelman, Kara, 2014. The travel and environmental implications of shared autonomous vehicles, using agent-based model scenarios. Transp. Res. Part C 40 (2014), 1—13.

Fagnal, D.J., Kockelman, K., 2015. Preparing a nation for autonomous vehicles: opportunities, barriers and policy recommendations. Transp. Res. Part A 77, 167—181.

Gil González, E., 2016. Big data, privacidad y protección de datos. Imprenta nacional de la agencia estatal. Boletín oficial del estado.

Forrest, A., Konca, M., 2007. Autonomous car and society. Worcester Polytechnic Institute.

ITS International, 2017. <http://www.itsinternational.com/> (Last accessed 13.02.17).

Kyriakidis, M., Happee, R., de Winter, J.C.F., 2015. Public opinion on automated driving: results of an international questionnaire among 5000 respondents. Transp. Res. F 32 (2015), 127—140.

Payre, W., Cestac, J., Delhomme, P., 2014. Intention to use a fully automated car: attitudes and a priori acceptability. Transp. Res. Part F 27 (2014), 252—263.

Salas, M., Robusté, F., Saurí, S., 2009. Impact of a pricing scheme on social welfare for congested metropolitan networks. Transp. Res. Rec. J. Transp. Res. Board 2115, 102—109.

Schellekens, M., 2015. Self-driving cars and the chilling effect of liability law. Comput. Laws Security Rev. 31 (2015), 506–517.

Schellekens, M., 2016. Car hacking: navigating the regulatory landscape. Comput. Laws Security Rev. 32, 307–315.

UITP, 2017. Policy brief. Autonomous vehicles: a potential game changer for urban mobility. <http://www.uitp.org/autonomous-vehicles> (accessed 13.02.17).

University of Michigan Sustainable Worldwide Transportation (UMTRI, 2014). A survey of public opinion about autonomous and self-driving vehicles in the U.S., the U.K., and Australia. Available at <https://deepblue.lib.umich.edu/bitstream/handle/2027.42/108384/103024.pdf> (accessed 20.02.17).

U.S. Department of Transportation (DOT), 2016. Accelerating the next revolution in roadway safety. Available at <https://es.scribd.com/document/324670391/AV-Policy-Guidance-PDF> (accessed 17.01.17).

Valdés, C., 2012. Optimization of urban mobility measures to achieve win–win strategies. Doctoral Thesis. Universidad Politécnica de Madrid, September 2012. Available at: <http://oa.upm.es/14220/1/Cristina_Valdes_Serrano.pdf> (accessed 30.11.16).

White House, 2016. Artificial intelligence, automation and the economy. Executive Office of the President. December, 2016. Available at: <https://www.whitehouse.gov/sites/whitehouse.gov/files/documents/Artificial-Intelligence-Automation-Economy.PDF> (accessed 31.12.16).

FURTHER READING

EUROSTAT. Statistics on European cities. <http://ec.europa.eu/eurostat/statistics-explained/index.php/Statistics_on_European_cities#Demography> (accessed 30.12.16).

INRIX, 2014. The future economic and environmental costs of gridlock in 2030. An assessment of the direct and indirect economic and environmental costs of idling in road traffic congestion to households in the UK, France, Germany and the USA. <http://inrix.com/wp-content/uploads/2015/08/Whitepaper_Cebr-Cost-of-Congestion.pdf> (accessed 08.01.2017).

KPMG and Center for Automotive Research (CAR), 2012. Self-driving cars: the next revolution. kpmg.com / cargroup.org (accessed 08.01.17).

RACC (Real Automóvil Club de Cataluña, 2016).

Shin, J., et al., 2015. Consumer preferences and willingness to pay for advanced vehicle technology options and fuel types. Transp. Res. Board Part C 60, 511–524.

Vienna Convention on Road Traffic. Vienna, November 8th 1969, entry into force on 21 May 1977.

CHAPTER 12

Future Perspectives and Research Areas

Felipe Jiménez
Universidad Politécnica de Madrid, Madrid, Spain

Contents

12.1 INTRODUCTION

The intelligent vehicle is already a reality, but it is necessary to know what will be its evolution in the following years. Of course, it can offer significant benefits, but in the same way of any new technology, it can carry associated risks (Gill et al., 2015). In addition, its global deployment not only passes through the vehicle itself but also extends to the infrastructure, so the public and private sectors should plan for the arrival of autonomous and cooperative vehicles, in their most developed state in a time horizon not excessively far away.

However, talking about evolution in the future is highly risky, especially when dealing with long-term predictions since there are many variables that can become highly relevant over the years (or not) that significantly change the evolution of these systems. For example, the evolution of mobile telephony and communications allows the deployment of previously unthinkable services, so that the explosion of other technologies can revolutionize the panorama with respect to the forecasts that have been made. On the other hand, the forecast of effects of certain advances is, in some cases, complicated. As an example, new communication technologies could prevent a high number of trips, both in the

Intelligent Vehicles
DOI: http://dx.doi.org/10.1016/B978-0-12-812800-8.00012-6

personal and professional spheres. Thus, in the workplace, telework or "virtual" contacts could reduce the requirements of traveling during work. However, this has not been the case since the possibility of access to more information and more contacts has fostered broader relations.

What is clear is that the future of road transport goes through profound changes, although presumably, gradual, in which new technologies have a preponderant role. This chapter presents, firstly, the current trends in the evolution of road transport, as they provide a vision of the path that is being adopted. In addition, future developments will depend on how the barriers identified will be solved, and some of the key issues will be highlighted. The focus of the chapter will be oriented, unlike the previous chapter, on technical issues, although they cannot be independent of other issues such as economic, social, political, etc.

12.2 CURRENT TRENDS

Several differences can be found technologically speaking between current vehicles and those of barely 20 or even 10 years ago. The number of sensors, controller units, and actuators involved in new systems oriented to safety and comfort has grown exponentially. In this way, systems that, at first, equipped only high performance vehicles such as antilock braking or stability control, are now present in all new passenger cars. Other systems such as navigators already have a very high market penetration (Pérez and Moreno, 2009). The new vehicles will come equipped with the emergency call system and these are just a few examples of the endless list of driving assistance systems. In this way, on the roads, we already find vehicles equipped with:

- Assistance systems such as exterior mirrors with blind spot detection systems, tire pressure control, lane maintenance system, active precollision system with millimeter radar, or facial recognition system.
- Vehicle control systems such as slope start assistance and precollision braking assistance, in addition to the already well-known antilock system or stability control or electronic traction control or adaptive cruise control.
- Information systems as an indicator of ecodriving, night vision systems in addition to navigators that handle more and more information and are able to recalculate routes in real time depending on the conditions of the road.

In the same way, the roads have received technological improvements that, although not as striking as in the case of the vehicles, are highly effective. Such is the case for variable information panels (although future trends are aimed at providing that personalized information on the vehicle itself through wireless communications), speed control radars, pavement sensors, weather stations, and surveillance cameras for information collection, wiring with optic fiber for the transmission of information along the main roads, etc.

These improvements do not respond to isolated impulses from administrations or vehicle and component manufacturers, but rather respond to long-term trends towards more sustainable transport in the broadest sense of the term. These evolutionary models affect not only transport but society as a whole. Several studies have revealed the general guidelines for these trends.

Thus, in the year 2002, 20 major transitions in the world of transport were presented in an article in the United States, many of them made possible by ITS (Sussman, 2005). The following are highlighted:

- Change from information management and actions in wide temporal windows to work in real time. It should be noted that this information management is still a challenge, e.g., in achieving high levels of reliability in that information.

- Change from using information sporadically to make investment plans to direct use of continuous data for planning and exploitation, which is a result of a strong reduction in the costs of information technologies.

- Change from zonal management of traffic to management of wider regions that can produce greater benefits, although with greater complexity, with an integrated treatment that allows dealing with problems, such as the environmental ones, which are not possible in a more local form.

- The emphasis is changed from encouraging mobility to promoting accessibility, according to Ming Zhang, stated as follows: "Mobility refers to the ability to move from one place to another while accessibility refers to reaching the opportunities of destiny under time conditioning and cost." That is, expectations about transportation grow.

- The transition from economic development to sustainable development, which should take into account some effects today in the background and the way of certain technical advances are considered.

- Vehicles and infrastructure are no longer independent but cooperate, or require technological innovations but also strong institutional support and cooperation between the public and private sectors.
- Change from the priority concept of reducing the consequences of accidents to reducing the number of accidents, that is, promoting primary safety measures.

The degree of development of these transitions is different. High levels of transformation have been reached in some of them, such as broad traffic management, information management in a scenario closer to "real time," or the inexorable introduction of primary safety systems in vehicles. In other areas, the evolution has started later. This is the case, for example, with the integration of vehicles and infrastructure. However, the current trend points towards this inexorable path. It remains to be answered in some cases, how the change materializes in practice.

On the other hand, in 2006, from a work financed by the European Commission, the guidelines were set for achieving sustainable mobility (McDonald et al., 2006). Among them, the following ones can be highlighted:

- *Reducing road transport impacts*: ITS must provide tools for more efficient use of infrastructure, and managing traffic to reduce congestion and pollution. In addition, they must support other forms of mobility, such as car-pooling or car-sharing. Finally, assistance systems should increase the safety of vehicles.
- *Improving the use of existing infrastructures*: taking advantage of ITS capabilities, it is possible to increase the capacity of a road network without the need for additional civil works, at a much lower cost than conventional measures and with greater flexibility to adapt to changing conditions.
- *Intermodality*: the transport system must provide travelers with the possibility of taking the most convenient mode of transport, thus achieving a more balanced distribution of traffic. To this end, efforts are focused on encouraging that the different modes are complementary and, the provision of the information that facilitates this modal exchange is achieved from the ITS. This line is related to the one that promotes the use of public transport (Comisión de Transportes, 2003).

12.3 CURRENT RESEARCH AREAS

The challenges identified today must mark the way forward. That is to say, the resolution of the current problems will encourage the implementation of systems and services in the future, although they will certainly reveal new difficulties that need to be addressed. The responsible administrations are clearly aware of this fact and proof of this are, for example, R&D transport plans. Thus, in the Horizon 2020 work program in Europe, smart, green, and integrated transport are placed as major pillars of transport in the future. Among the program lines for the period 2014—15 (European Commission, 2014), the following can be emphasized:

- Advanced bus concepts to improve efficiency, mixing "classic" technologies with the latest in bus design.
- Traffic safety analysis and an integrated approach oriented towards safety of vulnerable users, since it continues to be a group that has not received the improvements that others have had.
- Cooperative ITS for safe, congested, and sustainable mobility where communications play a major role, especially by pursuing field tests that demonstrate the limitations of technical developments under previous programs.
- Safe and connected automation in road transport, where specific requirements are imposed on simple automation, limiting its foreseeable scope in the short term.
- Management of urban congestion, as one of the earliest challenges and which continues without a complete and satisfactory solution given the growth of cities and the requirements of mobility within them.
- Connectivity and shared information for intelligent mobility, aimed at bringing together all sources of information to provide higher quality services.

In addition, the introduction of green vehicles, mainly electric vehicles, opens a new field of work, such as the management of recharging points, the organization of fleets of these types of vehicles, taking into account their autonomy, etc.

On the other hand, when the technological development of some technologies is already considered mature, its validation in extensive field tests is essential. Many current efforts are being focused on this new vision rather than concentrating on actual development itself.

12.4 MAIN EXPECTED TECHNOLOGICAL LEAPS

Experience in this and other fields has shown that, with few exceptions, many of the changes that have been introduced have been made gradually, although there is always a milestone that is often identified as the initial trigger. In the field of transport and ITS, where so many actors intervene and the implications, both positive and negative, can be remarkable, the introduction of technological leaps is usually prudent. Therefore, the changes that will take place in a more or less near future should already be glimpsed. Below are some of the most representative ones.

Perhaps the most striking technological leap is focused on autonomous vehicles and, a step further, autonomous cooperative vehicles. However, some issues need to be pointed out. Firstly, the operation must be completely reliable and safe. This specification is a highly demanding requirement, which usually complies with limiting the application of these vehicles to controlled scenarios or simple applications or reduces the speed of operation. Thus, it is understood that autonomous driving is viable in dedicated lanes or platoons in situations of congestion, or parking maneuvers, which are already a reality in many cases. Collision avoidance safety systems are more complex, especially those that take control of the vehicle in emergency evasive maneuvers. On the other hand, the system that controls the vehicle must be economically viable, so that the cost of the sensors must comply with what the market can assume. Finally, legal and social barriers must be considered.

That is, according to the above, it can be said that fully automated driving is not a technological utopia. But its practical implementation on the roads should be carefully analyzed. Researchers and manufacturers advocate for automation in successive stages, firstly approaching longitudinal control, then simple side controls, to reach complete automation in scenarios of increasing complexity. To sum up, the change from conventional to assisted driving is now a reality and now the leap is being made towards automated driving, in its different stages: partial, high, or fully automated.

Another relevant technological leap is the greater exchange of information between vehicles and with the infrastructure or with other users, that is to say, the widespread implantation of the vehicle connected to its surroundings. This point already has important developments in the research and test fields. Undoubtedly, the use of these systems for the provision of information will be a reality in a short period of time. It will also be relevant to support the previously mentioned technological leap of autonomous driving, presenting the concept of cooperative autonomous driving under certain premises. However, it will take more time to use these communications to support safety applications. Mainly, latency and positioning accuracy requirements must be guaranteed at all times, which will require a strong technological effort given the current state-of-the-art, assuming that the final costs of the systems, as in the previous case, must be in the order of magnitude the market could assume.

In summary, cooperative systems could be seen as catalysts of autonomous driving and can provide solutions to some environments in which autonomous vehicles based only on in-vehicle sensors have several limitations. Information exchange between vehicles and infrastructure enables autonomous driving in more complex scenarios. The integration of these two technological leaps will shape the vehicles of the future and their interaction with the infrastructure as shown in Fig. 12.1.

Partly as technologies that catalyze previous advances, efficient, cheap, and reliable methods of collecting and processing information in real

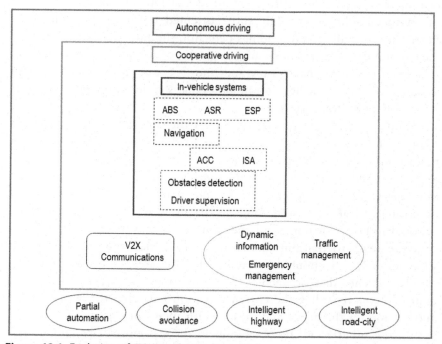

Figure 12.1 Evolution of ITS introduction in vehicles.

time, even when the volumes are large, continue to be necessary. The decrease in the price of increasingly accurate and reliable sensors is a must, both in vehicles and in infrastructure. New sensors in the infrastructure, the use of floating vehicles with access to more and more internal variables that can be shared with other vehicles or elements of the infrastructure, etc., have increased the volume of information. Thus, the integration of several sources and the fusion of information is critical, although this aspect requires data compatibility, not contemplated originally, and they are still, at present, a source of inefficiencies in data transmission. In addition, the high number of potential information emitters and receivers causes the communications scheme to be revised in order to support this simultaneous flow in real time, while discriminating to what extent certain information is relevant to be transmitted to some points and with what latency.

On the other hand, the dependence of many systems on the positioning of the vehicle will imply an evolution in these technologies, as well as in the accuracy and detail of the digital maps. Thus, positioning problems in complex environments must be solved and, for many applications, a centimetric precision must be guaranteed that will allow the deployment of advanced solutions, mainly those oriented toward improving safety. Furthermore, vehicle surroundings detection should be enhanced by new sensors and data fusion, taking advantage also of wireless communications, in order to have a reliable view and find free areas to move safely. This surveillance also affects decision modules that should reproduce a compatible behavior with human users.

Another relevant technological leap is the one that concerns specific solutions for conventional roads because of the limited work focused on them until now and the poor conditions that are sometimes found there (e.g., degraded painted lines). With the new improvements, the distinction between high-capacity roads and conventional roads becomes increasingly clear. Certainly, some of the measures will be extrapolated, although others will not be in the short term due to the lack of infrastructure (e.g., to transmit information). It is therefore considered that affordable solutions should be introduced in this area, while not providing all the benefits that are achieved on the main roads, but that offer improvements regarding the current situation regarding the availability of information for the driver and measures for increasing safety.

On the other hand, the evolution in the transport of passengers and goods is to continue with an increasing use of information in real time to improve the service and the exploitation of the fleets. From the point of view of technical requirements, these can be considered less relevant than those of other systems previously treated, although the economic and organizational barriers can influence their rapid implementation. Finally, the priority action of intermodality was already mentioned, for both travelers and goods. The promotion of this intermodality passes through greater and new flows of information.

12.5 OTHER EXPECTED AND/OR NECESSARY CHANGES

It should also be noted that technological leaps must be in line with social evolution. In this sense, it is relevant to analyze prospective mobility studies that advocate a significant transfer towards shared mobility and a change in the organization of cities (Fundación OPTI, 2009). Among the conclusions related to ITS, these studies highlight the information systems for real-time route planning and it is assumed that the contribution of ITS systems to the improvement of mobility will be decisive in terms of efficiency, safety, and environmental sustainability, including real-time management of infrastructures, communication with vehicles, monitoring of vehicles according to their characteristics, or control of freight traffic. In addition, there is a role for ITS in promoting other forms of mobility, such as electric vehicles for local and global energy management, models for car-sharing use, or other forms of operation of passenger and freight transport companies. Undoubtedly, depending on the aspects that are further strengthened, some forms of mobility will be conditioned by others, apart from, at least in part, technical solutions, which makes the prospective exercise even more complex.

Thus, in a mutual interaction, technological advances can significantly influence the choice of the transport mode. For example, autonomous vehicles combine the advantages of public transport and traditional private vehicles such as comfort, flexibility, and the possibility of other activities at the same time. In addition, the public transport market of certain groups of people (senior citizens, handicapped people, etc.) can be opened to the private vehicle and the use of the private vehicle can become safer and more convenient, since a more personalized service can be offered and more adjusted to the demand with smaller vehicles. In this way, public transport should explore other ways to be competitive against an individual (not necessarily private) use of the vehicle. It is estimated that price may be the main tool to influence behavior. In any case, autonomous vehicles can represent an opportunity for public transport and models such as car-sharing.

What seems to be a clear trend is the potential redefinition of the concept of vehicle ownership, seeking greater efficiency with vehicle sharing for the reduction of costs and additional infrastructure (e.g., car parks), which results in changes in the urban design. This change also enables the introduction of more technology, complex sensors, and actuators in vehicles because the final price of the vehicle is not a key issue for customers because it is a shared cost.

12.6 CONCLUSIONS

The overview of how road transport should be in the future is conceptually clear and was raised many years ago, even when many of the technologies that now allow the deployment of many of these measures had not been developed or deployed.

It is not possible to evolve in transport only with conventional measures where the safety of vehicles passes exclusively by improving their structural behavior and retention systems, such as safety belts, or road capacity problems by building new civil infrastructure only. That is, the role of measures such as the introduction of the safety belt is undeniable and constitutes one of the great milestones in the history of the automobile, having saved countless lives. However, evolution should not stop there and many of the solutions go through technologies that fall within the intelligent systems. Likewise, the improvement of infrastructures is essential, but not only at the level of civil works, but it is necessary to incorporate new technological equipment. It has already been shown that increasing road infrastructure to increase total capacity is not sustainable and there are trends that advocate changing the model of cities, seeking a "city for citizens" rather than a "city for vehicles". This undoubtedly goes through social structural changes that are slow and difficult to predict.

In spite of the great advances, the vision of the general lines has been maintained over the years, although the technologies that have been introduced, as well as the tests of deployment, have modified some details, but not the final objectives. Therefore, it is estimated that the guidelines already designed more than 20 years ago remain valid for the following years. Concerning their practical implementation, there are still doubts on how to solve some of the barriers present. The search for solutions to these barriers, as well as the possible introduction of new technologies, can qualify this deployment and these are difficult aspects to foresee now and it is risky to make estimates.

Furthermore, making estimates of time horizons is very risky and any forecast can lead to relevant errors as has happened in the case of previous predictions, e.g., the introduction of some systems in vehicles such as intelligent speed control systems. This difficulty lies in the multitude of parameters that can influence and that are difficult to estimate in the long term. There are three critical points that can drastically influence the implementation of ITS:

- Influence of new technologies not previously known or not applied to the transport sector, which will improve the performance of existing technologies in areas still to be fully solved such as accurate positioning (with an accuracy of more than an order of magnitude higher than current systems), communications, or environmental sensing, to name a few. It is identified asessential that the cost should be competitive for mass introduction in vehicles or infrastructure, so that they can evolve further than "laboratory prototypes."
- Society must accept and assimilate transitions. This aspect is applicable to any area, with transport being one of them. Some technological solutions may involve profound changes in the behavior of individuals (e.g., acceptance of automated driving) and society (e.g., if mobility patterns are modified). To achieve this acceptance, the introduction of changes must be progressive. The introduction of ITS in vehicles is strongly conditioned by the market. Clearly, regulatory aspects may condition the implementation of some measures, fostering their introduction (e.g., increased use of seat belts by making their use compulsory in many countries, or average speed reductions when road radars number increases; Elvik and Vaa, 2004). However, this solution is not applicable to any service.
- Finally, political and economic scenarios set an action framework. Thus, an economic crisis with decreases in private investments (in the purchase of vehicles) and public investments (relative to the improvement of the road network) can significantly delay the technological developments and the final implementation of systems and services.

What seems clear is that changes in road transport will not occur abruptly and drastically. Bearing in mind that safe, sustainable, efficient, and clean mobility is sought, this objective is measured by many perspectives that undoubtedly involve the promotion of intelligent transport systems. However, its introduction will take years. In this sense, a critical technical aspect is the need of a minimum number of equipped users and the obligation to consider the interaction between vehicles equipped with intelligent systems with others that do not have them, taking into account the difficulties of market penetration of systems whose effectiveness is affected by a low penetration rate.

Of course, this future success will undoubtedly go through a rational and gradual introduction. A clear example emerges with the issues of autonomous driving, where this automation should be partial in the first instance, and work in controlled scenarios and simple applications such as maneuvers at low speed or vehicle platooning, before moving to a complete autonomous driving, which, although technically demonstrated in some prototypes, still requires a more complete approach within the entire road transport system.

In any case, it seems clear that advances in smart vehicles must be in line with adequate infrastructure planning to support these technologies and exploit their full potential.

REFERENCES

Comisión de Transportes del Colegio de Ingenieros de Caminos, Canales y Puertos, 2003. Libro Verde de los sistemas inteligentes de transporte terrestre. Colegio de Ingenieros de Caminos, Canales y Puertos, Madrid (in Spanish).

Elvik, R., Vaa, T., 2004. The Handbook of Road Safety Measures. Elsevier.

European Commission. 2014. Horizon 2020. Work Programme 2014–2015. Smart, green and integrated transport.

Fundación O.P.T.I. 2009. Movilidad en las grandes ciudades. Estudio de prospectiva. Madrid (in Spanish).

Gill, V., Kirk, B., Godsmark, P., Flemming, B., 2015. Automated Vehicles: The Coming of the Next Disruptive Technology. The Conference Board of Canada, Ottawa.

McDonald, M., Keller, H., Klijnhout, J., Mauro, V., Hall, R., Spence, A., et al., 2006. Intelligent transport systems in Europe. Opportunities for future research. World Scientific.

Pérez, J.I., Moreno, A., 2009. La contribución de las TIC a la sostenibilidad del transporte en España. Real Academia de la Ingeniería, Madrid (in Spanish).

Sussman, J.M., 2005. Perspectives on Intelligent Transportation Systems. Springer.

INDEX

Note: Page numbers followed by "*f*" and "*t*" refer to figures and tables, respectively.